AVICULTURE
A HISTORY

Edited by

Ingvar Svanberg
&
Daniel Möller

ISBN-13: 978-0-88839-013-4 [Trade Hardcover]
ISBN-10: 0-88839-013-0

Library and Archives Canada Cataloguing in Publication

Aviculture : a history / Ingvar Svanberg, Daniel Möller, editors.

ISBN 978-0-88839-013-4 (hardcover)

1. Aviculture. 2. Aviculture--History. I. Svanberg, Ingvar, 1953-, editor II. Möller, Daniel, 1974-, editor

SF461.A85 2017 636.6'8 C2017-905926-2

Printed in the United States

Production: D. Williams
Cover Design: M. Lamont
Technical Edit: Josef Lindholm, III

Cover: Silver Pheasant, Scarlet Macaw, Red-breasted Parakeet, Guinea Fowl and monkey. Oil painting by Charles Collins, 1740

The authors would like to acknowledge the financial support of the Stiftelsen Olle Engkvist Byggmästare and Traditionsvetenskapliga nämnden at the The Society of Swedish Literature in Finland.

Published simultaneously in Canada and the United States by

hancock house

HANCOCK HOUSE PUBLISHERS LTD.
19313 Zero Avenue, Surrey, B.C. Canada V3Z 9R9
(604) 538-1114 Fax (604) 538-2262

HANCOCK HOUSE PUBLISHERS
#104- 4550 Birch Bay-Lynden Rd, Blaine, WA, U.S.A. 98230-9436
(800) 938-1114 Fax (800) 983-2262
www.hancockhouse.com | sales@hancockhouse.com

Blue-and-yellow Macaw (*Ara ararauna*), illustration from 1754.

Painted bunting (*Passerina ciris*) and Indigo bunting (*Passerina cyanea*), two finches which earlier were common in the bird trade for centuries. Illustration from 1750.

Contents

Figure 1: Romantic couple with a song bird engraving from 1779.

Preface

This book discusses the history of aviculture across the cultural spectrum of keeping and raising birds in captivity: For pleasure; companionship; as ornaments; religious purposes; or various economic or practical reasons. Since the dawn of mankind, humans have kept birds in captivity.Several species are truly domesticated, such as the Red Junglefowl (Poultry), Turkey, Mallard (Eurasian Domestic Duck), Muscovy Duck, Greylag (European Domestic) Goose, Swan ("Chinese") Goose, Helmeted Guineafowl, Blue Peafowl, Japanese Quail, Rock (Domestic)Pigeon, Barbary Dove, Canary, Bengalese (Society) Finch, Java Sparrow, and more recently, the Zebra Finch, Budgerigar, and Cockatiel. Others have been bred in captivity for many generations, for instance Estrildid finches, Fringillid finches, doves, quail, pheasants, and parrots.Today bird-keeping for pleasure appears to be declining in the West, mostly due to bird protection laws and growing awareness about conservation issues. The importation of wild-caught birds into the European Union has been very restricted since 2005, but in many other parts of the world, capturing and trading birds is still a thriving business. Although aviculture has had and still has a deep impact on human beings, it remains a neglected field within humanities and social sciences. Relatively little is published about the cultural and historical aspects of aviculture. This anthology is intended for a general audience of readers and it shows various aspects of keeping birds in captivity for pleasure, ornamental reasons or practical purposes around the world. It also deals with the great variety and complexity of the practice of keeping birds, and the specific cultures which have developed around it.

The first chapter gives a brief introduction to the questions we focus on in the book, together with a historic overview from prehistory to the early twenty-first century, including pet birds among natives in South America, Southeast Asia and Africa, sailors and their parrots, birds in religious rituals, primitive domesticates in various peasant societies, etc. The other chapters offer descriptive case studies in pre-modern and early modern ways of bird-keeping in various historical contexts. Also modern aviculture in zoological gardens is discussed, and specific bird categories within twentieth-century aviculture are described in some chapters. We encounter sophisticated bird-keeping in pre-Columbian societies, Norse trade with falcons, the European craze for songbirds, practices with captive birds used in human habitations to keep vermin under control, and we see how avicultural expertise is used for trying to save vanishing species by breeding them in captivity. Together these topics illustrate the great variety and complexities of bird-keeping practice.

The authors are specialists in aviculture and most of them hail from the countries about which they write. This book bridges the disciplines of cultural anthropology, ethnobiology, history, natural history and ornithology and is intended to benchmark the development of the subject for a broader audience, which until now has had few possibilities to become acquainted with it. The completion of this book was made possible by a grant from Stiftelsen Olle Engkvist Byggmästare and Traditionsvetenskapliga nämnden at the The Society of Swedish Literature in Finland.

Spring 2018

Ingvar Svanberg and Daniel Möller

Figure 2: Budgerigars (*Melopsittacus undulatus*), also known as pet parakeets, originally imported from Australia, are globally among the most popular birds in captivity.

Chapter 1. History of Aviculture

Ingvar Svanberg and Daniel Möller

> There is a distinctive charm in aviculture. It can be carried on anywhere: In a rude slum in St. Giles, in a Belgravian mansion, in a country Vicarage, in a suburban villa, and you will always enjoy yourself. What is more important, this enjoyment will not embitter your present, nor endanger your future, by getting you into a bad temper, whether you are successful or not; provided also you are a true bird lover. This is not the case with other sports.
> *Reverend C. D. Farrar,* Birdroom and Aviary *(London 1910), p. 9.*

For thousands of years, Human beings keep birds for economic, ornamental and entertainment reasons since several thousands of years. We do not know when the first humans began to hold birds in captivity and we can only speculate about the reasons behind this. Possibly Palaeolithic hunter-gatherers tamed birds and we can assume that some kind of symbiosis developed between them and various bird species during the common hunt for food. This phenomenon has been described in more recent cultures, e.g. with the indigenous Khoekhoe in southern Africa and the Greater Honeyguide (*Indicator indicator*). The cooperation probably developed at an early stage of evolution during the joint search for wild beehives, due to the birds guiding ability observed by humans. While the Khoekhoe gatherers harvested honey from the hives, the bird preferred the larvae.[1]

Scavengers like ravens and vultures had their own reasons for following groups of hunters and gatherers, when they roamed hunting territories. The possibilities for the birds to become at least semi-tame in these relationships were not scarce. Old World Vultures are very easy to tame and so are ravens, and both adapt easily to human habitats. Keeping captured birds as live decoys may also have occurred among early hunters. The use of live owls to attract other birds was known until quite recently.

The First Agricultural Revolution, which began around 10,000 BCE, provided new opportunities for keeping birds in captivity. Most certainly, Neolithic peasants, who lived a rather sedentary life, kept birds in their settlements. Similar to peasants in Europe, Asia, and South America until only recently, these Neolithic farmers probably brought nestlings of the family Anatidae, other water fowl, auks, and various gallinaceous birds to

Figure 3: Fishing with trained Cormorants in the Great Canal, Suzhou, China in 1901

their dwellings, fed them with available food, and fattened the birds before slaughtering them. The consumption of seafowl fledglings, intensively fattened at the end at the end of their nesting period, was also practiced much later in many coastal areas and island societies.[2] Most probably, the Neolithic farmers also brought into their homes young birds of various species, in the same way as some indigenous jungle people still do, for amusement and "toys' for their children. Children in different cultures still do this. Many birds are easily tamed if taken as nestlings, and they will be affectionate toward their masters or human households. Parrots, toucans and other species are today kept among aboriginal tribes in the Amazon region.3 Collecting eggs was of immense significance for coastal dwellers and islanders in many parts of the world. By creating various kinds of artificial nests, for centuries humans have been able to gather eggs for food, using a kind of intensive egg-harvesting. Sea birds become used to this arrangement and continue to produce new eggs for the owners of the nesting-boxes. The Common Goldeneye (*Bucephala clangula*), which often inhabited these artificial nesting-boxes along the Swedish Baltic coast, was called "the poor man's chicken". A nineteenth century Norwegian author, Jens Andreas Friis, when writing about Lapland (Sápmi), notes that the nesting box "is the bee-hive" of the fishing Saami, "which in due time will be robbed". Such egg-collecting boxes are known from Scandinavia, Sápmi, Russian Karelia, northern Russia and Estonia.[4]

Humans and aviculture

The purpose of this book is to place the tradition of taming and keeping birds in captivity into an historical perspective and social context. The history of aviculture is little known and many aspects are seldom discussed. Aviculture is here defined as the practice of keeping birds in captivity, where humans are using some kind of controlled conditions, for various reasons, and including a variety of ways related to keeping birds in human dwellings. Some are kept for pure economic causes, providing meat, fat, feathers and eggs. This is especially true about the early domesticated birds such as poultry, ducks and geese. Others are kept for amusement and to entertain their owners; for instance fortune telling parrots or pigeons are an old diversion during fairs and in pleasure gardens. Some are trained to perform tricks: "I have recently seen a troop of trained cockatoos that were absolutely marvellous", wrote the famous American author Alpheus Hyatt Verrill in a charming book from 1915 on keeping animals as pets. He continues: "One of these birds played 'Home Sweet Home' by pulling strings attached to bells; another danced to music, and one rang any number which was called to it on a bell". Competition between owners through fighting birds or exhibitions are other important reasons for keeping birds.[5]

Using birds for hunting (falconry) is an ancient practice, known as a royal sport already among the Assyrians. Central Asian nomads, especially Kazakh and Kyrgyz, still hunt with Golden Eagles (*Aquila chrysaethus*). This is an old tradition, often described by travelersr and ethnographers. In the eighteenth century , Bashkirs in the Ural region in Russia fed eagle nestlings until they were grown, and then sold them to Russian and Tatar merchants, who transported them down to the Kazakh border markets. The Kazakh elite was said to prefer these eagles for hunting and training, and they paid large sums for such birds.Also, commoners kept eagles and other raptors for hunting foxes, marmots and even young wolves. An already trained eagle was worth the equivalent of several horses. In his chapter on Icelandic trade with Gyrfalcons *(Falco rusticolus)*, Sigurður

7Ægisson analyses the globalization of the trade with raptors, extending from the far north to the Mediterranean region and the Middle East in the south.[6]

The use of captive birds that help fishermen to catch fish is another ancient activity, described first in Chinese sources from the Sui Dynasty (589–618 CE). Local fisherman among the Bai people in Unan use trained Great Cormorants (*Phalacrocorax carbo*), as do fisherman on the Li river near Guilin City in Southern China. On the Nagara River in Japan, fishermen use the native species (*Phalacrocorax capillatus*). Tamed Indian Snakebirds (*Anhinga melanogaster*) were employed for the same purpose by communities in the northeastern Indian state of Assam. In the eighteenth and nineteenth centuries, peasants in various parts of Europe used fish-eating ducks like the Common Merganser (*Mergus merganser*) to drive fish shoals into nets.[7]

The motives for keeping and taming birds are manifold, and they also include using them as "watch birds", controlling venomous snakes in the vicinity, or for religious rituals. The use of birds to get rid of slugs, fleas and other pests in yards and dwellings is described in a chapter by Sabira Ståhlberg and Ingvar Svanberg. Ostriches

Figure 4: Swans upping on the Thames in 1875.

(*Struthio camelus*) and ostrich races exist in some parts of the world, including the United States. Carrier pigeons are used widely for communication, a practice which goes back at least to Antiquity. The Pima people of southern Arizona kept captive Golden Eagles and Red-tailed Hawks (*Buteo jamaicensis*) for their feathers. The raptors were released after a couple of years. Canaries were used in coal-mines as a kind of early warning system. When the bird showed signs of distress, conditions were also unsafe for the miners.[8]

Well-trained or rare exotic birds provide a higher status to their owners and therefore rare and unusual birds are much sought after. Whereas "parrots are so common as to attract no interest or curiosity in the beholder, yet the toucans always attract attention and interest visitors", wrote Verrill in 1915. Exotic birds are often kept tame in ponds or yards, sometimes wing-clipped, while others are confined to cages and aviaries ("flying-cages"). Some are used for laboratory research and others, such as the Zebra Finch (*Taeniopygia guttata*), for genetic studies. During our modern and post-modern times an increasing number of taxa are becoming domesticated, but today mostly for display and companions.

Some recently domesticated species, such as various ratites (ostrich, emu, and rhea), have provided new economic species which are farmed for egg, oil, meat and leather.[9]

Early domestic birds

Poultry belong probably to the oldest domesticated birds. Their ancestor, Red Junglefowl (*Gallus gallus*), still occurs in South and Southeast Asia from Pakistan to Indonesia. This is the sole ancestor of domestic fowl and its domestication probably took place in the Indus Valley around 3000 BCE. Due to seaborne trade, fowl spread east- and westwards and are known from Mesopotamia and ancient Egypt in the fifteenth and fourteenth centuries BCE. During the first millennium BCE, fowl-keeping spread rapidly and it also reached the South American continent long before the arrival of Europeans. Many chicken breeds are now kept as hobby animals for exhibitions and there are miniaturized breeds (Bantams) especially suitable for hobby aviculture. The popular Silkies originate from Asia, but have long been appreciated as pet birds in Europe as well. In Japan the extremely long-tailed and magnificent Onagadori has been bred to perfection among hobby enthusiasts. The old Araucanian breed, kept by aboriginal peoples in South America for centuries, is another interesting chicken; it is famous for its green or light blue eggs. Pure-bred Red Jungle Fowl are easy to raise in captivity and therefore popular among aviculturists.[10]

With increasing industrialization of poultry and egg production, many former economic breeds have become rare, being replaced by more highly productive breeds. They are therefore of interest to hobbyists. Many care especially for the original local breeds that existed among the peasantry before industrialization affected animal production in the countryside. These rather uncultivated old breeds are of special interest not only as a kind of "historical document", but also as a resource for the future, when alternative genetic material is sought after. Local breeds are therefore a treasure for times to come. Some of these old and natural breeds are also popular outside their traditional regions.[11]

The Guineafowl is the African contribution to domestic birds used for economic purposes. The domesticated populations originate from various subspecies of Helmeted Guineafowl (*Numida meleagris*), found in many parts of Sub-Saharan Africa. The northwest African subspecies *N. m. sabyi* was probably domesticated by the Egyptians and reached the Mediterranean region during Antiquity. Romans regarded it as holy and it was used for religious sacrifices. The Guineafowl was popular as a household pet among the Romans, but a few centuries later it had disappeared in Europe, only to be reintroduced at the end of the Middle Ages. The West African subspecies *N. m. galeata* had been long domesticated in Africa, when the Portuguese brought it to Europe in 1455. It spread fast in southwestern Europe and was observed in England in 1550.[12]

The third gallinaceous species with a long history is the Turkey (*Meleagris gallopavo*). The wild species lives today along the eastern coast of North America down to Mexico. It was domesticated by native inhabitants in Central America and bones been found in the Tehuacán Valley and dated to 200 BCE; we know also that the Aztec kept turkeys. When Conquistador Hernan Cortés arrived in Mexico, turkeys could be found in every town, and the Aztecs offered them to hm as presents. This shows that the bird probably was domesticated much earlier. The Incas in Peru also kept turkeys, long before the arrival of the Spaniards. The turkey was eventually brought to Spain in the early sixteenth century and spread quickly throughout Europe, even as far north as Denmark. Interestingly enough, domesticated turkeys were brought back to North America with settlers from Europe. As with other domesticated birds, it is not only kept as a production animal – in the United States the turkey has a high cultural significance for the Thanksgiving celebration – but also as an ornamental bird. In Europse, various color mutations were already known in Europe already a century after its first appearance on that continent. Several breeds have been developed in North America, e.g. the old breed Kentucky Red, now known throughout Central and Western Europe as the Bourbon Red.[13]

Another early domesticated bird is the goose. The domestic goose of Europe descends from the Greylag Goose (*Anser anser*). Domestication took place in Ancient Egypt and probably also in Europe at least 2000 BCE. Some oure ancient local breeds survive. The *Ölandsgås* from the Baltic Island Öland is known to have lived there for several hundred years. The *Twentse landgans,* from the Dutch Twente region, goes back to at least the eighteenth century, while the *Føroysk gás* from the Faroe Islands is said to have descended from the birds brought to the islands by Vikings a thousand years ago. It is now kept both for meat production and by hobbyists. These breeds are all critically endangered, as is the Sebastopol Goose with its long curling feathers, and the small Shetland Goose.[14] In eastern Asia, the Swan Goose (*Anser cygnoides*) is a common domesticated production animal. It is appreciated for its feathers, meat, fat and eggs. It reached Europe with Portuguese sailors and the East India trade in the seventeenth and eighteenth centuries; the "Chinese Goose" became popular beyond the countries to which it had been thus imported. The species has now spread to farms for goose production in many parts of the world and it is also kept in zoological gardens and park collections for ornamental reasons.[15]

Figure 5: Barbary dove (*Streptopelia risoria*), 1700s

The Egyptian Goose (*Alopochen aegyptiacus*), native to the Nile River, is little known as a domesticated bird. It was featured in art already during the Old Kingdom of Egypt (third millennium BCE). This bird is probably, after the pigeon, the oldest domesticated bird in Africa and was kept for religious reasons and food by the Egyptians. Domesticated Egyptian Goose was brought to the island of Crete, but to what extent they were kept as ornamental birds by Greeks and Romans is not clear, although a mosaic from Pompeii possibly shows this goose. It was for some reason not spread as a domestic animal to other parts of Africa and remained forgotten for many centuries, being introduced again as an ornamental bird to Central Europe in the seventeenth and eighteenth centuries. The goose was bred in great numbers in Paris by 1839. Feral populations exist today in several countries and it is actually regarded as a pest in Great Britain.[16]

As with the domestic goose, we do not know when and where the Mallard (*Anas platyrhynchos*) was domesticated. Domesticated ducks were and are still common in Asia and continental Europe. Lots of local breeds exist and many old breeds survive thanks to hobby enthusiasts. Indian Runner Duck (*Anas platyrhynchos domesticus*) is a strange breed probably from Southeast Asia. Unlike other breeds this duck stands up like a penguin. Ducks are kept for meat, fat, eggs and feathers, but domesticated ducks can also be kept as pets, in gardens or backyards and some even inside houses. Attractive duck breeds have great potential as ornamental birds and are therefore often kept in ponds and canals in estates and city parks. Bantam ducks are available for aviculturists who possess limited space. Old breeds like Ancona Duck, Aylesbury Duck, Cayuga Duck, Shetland Duck, Welsh Harlequin and Swedish Yellow Duck (*svensk gul anka*) are now critically endangered and therefore of increasing interest for small zoological gardens, 4-H-farms, educational centers, public parks and hobby enthusiasts.[17]

Muscovy Duck (*Cairina moschata*) is an American contribution to the domesticated duck species. Its native distribution as a wild species stretches from Mexico in the north to eastern Peru and northern Paraguay in the south. When and where the indigenous peoples domesticated it is not known. One theory is that the domestication took place in coastal Ecuador. It was brought to Spain in the 1530s and soon spread throughout southern Europe and northern and western Africa. According to Pierre Belon's *L'Histoire de la nature des oyseaux* from 1555, it had "recently" reached France. White and pied mutations are featured in early seventeenth century book illustrations. This bird was known as Indian duck or Turkish duck in Germany in 1603, where it was commonly kept in manor parks; in Sweden it could be found in a royal park in 1654 and Carl Linnaeus kept Muscovy Ducks in his garden menagerie in the 1750s

Figure 6: Golden pheasant (*Chrysolophus pictus)*, 1746

in Uppsala. It is today a common economic bird in certain parts of Africa, East Asia, North America and Europe, but elsewhere kept for ornamental purposes and because of its reputation for eating slugs.[18]

A very early domesticated bird is the Pigeon (*Columba livia*), known already in Antiquity. Roman authors describe how to breed and keep pigeons. Pigeons have spread widely and many various breeds exist; some are kept for food, but most as hobby animals. Messenger pigeons possess a long tradition, but have no practical purpose any more. Old dove cots in various forms and shapes are still part of the local architecture in many parts of the world. No other domestic animals, with the exception of the dog, exist in so many different variations and breeds as the pigeon. Natural historian Charles Darwin as well as champion-boxer Mike Tyson have kept pigeons. "I have kept every breed which I could purchase or obtain", wrote Charles Darwin in Chapter 1 of his famous *On the Origin of Species* (1859). It is an ancient hobby which still thrives in different environments, from villages in the Iraqi countryside to skyscrapers of downtown New York, and it is still gaining in popularity in many parts of the world.[19]

Further species are today under domestication for economic reasons and especially ratites seem to have some potential. The Ostrich *(Struthio camelus)* was kept and bred in captivity already by ancient Egyptians, and Romans kept them for fighting in the Coliseum. During the Byzantine period the bird was farmed in Egypt. Pierre Belon reported in 1553 that both Libyans and Numidians were raising ostriches in order to sell their feathers. Somali nomads have kept semitamed herds of ostriches, which they tended in the same way as cattle and dromedaries.

More than a century ago there was a renewed

interest in farming ostriches for feather production, starting in Algeria about 1860, developing in South Africa and spreading to widely different places like southern California and central Sweden in the early twentieth century. Carl Hagenbeck, a famous German animal dealer, opened an ostrich farm in 1909 at Stellingen near Hamburg, in order to improve the domestication of the species, but the market collapsed in the wake of World War I. Ostrich farms are again rapidly spreading in many parts of the world. Meat and eggs are the most important products.

Emu (*Dromaius novaehollandiae*) farming is also becoming popular for similar reasons and also for oil. Emu farming started only in 1976 and the species is in addition used as ornamental birds, especially in zoological gardens.[20]

Ornamental birds in an expanding world

One important reason for keeping birds is their ornamental value, which can be combined with culinary interests. Rulers have kept menageries with exotic birds for status; often different animals were given to them as gifts. As Vernon N. Kisling shows in his chapter on aviculture in Ancient Old World societies, such animal and bird collections are known already from Mesopotamia, the Old Kingdom of Egypt, Ancient India, Shang Dynasty China, Ancient Persia as well as the Graeco-Roman world. The tradition of royal menageries prevailed also in medieval Europe and Asia. Marco Polo reported in the late thirteenth century that Kublai Khan kept gyrfalcons and hawks for pleasure in his royal park at Shangdu.

Royalty in South Asia have kept Indian Peafowl (*Pavo cristatus*) for probably more than 3,000 years. King Salomon is said to have had a fleet of ships that once in three years brought back "gold, and silver, ivory, and apes, and peacocks" (1 Kings 10:22). The species reached Europe as a domestic animal already during Antiquity and Peafowl were widely kept as ornamental birds by the Romans, also being mentioned as a garden bird in Charlemagne's *Capitulare de villis vel curtis imperii* (802 CE) and in Scandinavia during late Medieval times. Peafowl is one of the most popular ornamental birds globally in zoological gardens, private collections, city parks and farms. A white mutation has been known for centuries. The Peafowl is also often used for symbolic purposes, for instance as logo for Sri Lankan Airlines and as coat of arms for the German city of Neuwied.[21]

Another ornamental bird with a long tradition is the Barbary or Ringneck Dove (*Streptopelia risoria*), which is a fully domesticated bird. Its wild ancestor is most likely the African Collared Dove (*Streptopelia roseogrisea*), at home in arid African and Arabian regions. The domestication seems to have occurred some 2,500 years ago, probably somewhere in Nubia or Abyssinia, and the bird reached Europe through Egypt. White Barbary Doves are popular among magicians. During the last century the number of color variations has increased; a recent hand-book mentions over forty recognized color mutations.[22]

Early menageries in England, France and the Netherlands showed African Ostriches, Pelicans, Crowned Cranes (*Balearica pavonina*), Demoiselle Cranes (*Anthropoides virgo*) and Sarus Cranes (*Grus antigone*) as exotics. Cassowaries (*Causarius casuarius*) were imported to Europe from the sixteenth century. They were important as gifts and trade products in Southeast Asia and several Cassowaries reached the Netherlands with Dutch vessels, when others were brought to Japan. The island of Seram seems to have been a main exporter of Cassowaries for countries overseas.[23] An interesting example is also the Mute Swan (*Cygnus olor*), originally from the Caspian

Sea region, now part of the local fauna in most parts of Europe and elsewhere. Since medieval times until the twentieth century it was widely kept in moats, ponds and channels in parks and cities, although it has since become naturalized in some parts of Europe. It was used as food and in some areas it was regarded as a royal privilege to utilize its meat. In England the Act of Swans from 1482 formalised the ownership of the swans in the country.[24]

Other important ornamental birds in ponds are the waterfowl, mostly domestic geese and ducks, sometimes also captured native species from the wild, but with increasing trade also foreign ducks and geese. Mandarin Duck (*Aix galericulata*) has vast symbolic value in China and is well-known in Japanese folklore. Couples are often kept in temple ponds and villages in the Far East. The duck had reached Europe by the 1740s, when George Edwards observed a couple in the garden of Matthew Decker in Richmond, Surrey, England. There is no record that the species bred in Europe until 1858 in France. Today it is easy to care for and a very popular bird.[25]

Other swans, geese and ducks used in park and garden ponds are e.g. the Black Swan (*Cygnus atratus*), introduced to England in 1791 and somewhat later to France; Black-Necked Swan (*Cygnus melanocorypha*), introduced in 1846 to Europe and bred in the Netherlands in 1865; Barheaded Goose (*Anser indicus*), kept in Indian temples for centuries before it reached Europe; Redbreasted Goose (*Branta ruficollis*); Carolina Duck (*Aix sponsa*); Ruddy Shelduck (*Tadorna ferruginea*); Magellan Goose (*Chloephaga picta*), which bred at London Zoo in 1852, and many other species. Ornamental waterfowl remain continuously popular until today. The reason for this is probably the same as a century ago, when an author concluded: "Waterfowl invariably improve the appearance of any piece of water, even if they are only considered for artistic effect. They are inexpensive to feed and keep in health and always make beautiful and interesting pets if they can so be called", wrote C. Laidlay in 1914.[26]

The Ringneck or Common Pheasant (*Phasianius colchicus*), found in Inner Asia and westwards to the Black Sea coast and the Bosphorus, is mentioned as an ornamental bird in early Medieval times. It was kept in pheasantries in monastic and royal settings. Charlemagne mentions keeping pheasants in 802 CE. A pheasant guard is mentioned from Prague castle in 1625 and another one was employed by the Swedish royal court in 1638. The pheasant had both culinary and ornamental functions. In the nineteenth century, its value as a bird to be hunted increased and it was released into the fields in many parts of the northern hemisphere. Although commercial large-scale farms for release-and-hunt purposes are probably the most important for this species today, it is still kept as an ornamental bird as well. Especially Asian pure-bred subspecies are popular.[27]

The trade with the Far East in early modern times brought also other pheasants to Europe. The introduction of the Golden Pheasant (*Chrysolophus pictus*) from China could serve as an interesting example. Although one Golden Pheasant may have been brought to Rome in the reign of Claudius and presented as a Phoenix bird, this was probably an isolated case, as new specimens were not brought to Europe until modern times. The earliest record of pheasant presence in Europe dates to the 1730s, when a picture of a "red Pheasant from China" was published by English illustrator and naturalist Eleazar Albin in 1740. He wrote in a commentary to his painting: "I do not find this beautiful Bird described by any Author; it was in the Possession of the Honourable John Spencer Esquire, at his House in Windsor Park, where I went by his Order to draw it". According information on the illustration, it was drawn on 21 July 1735, so the bird obviously was present by then. Around ten years later the species was already common in England:

"These birds of late Years are frequently brought from China: I have seen several of them in the Possessions of our Nobility, and some curious Gentlemen", wrote George Edwards in 1747. The Duchess of Portland kept Golden Pheasants in her aviaries at Bulstrode Park in the mid-eighteenth century, and one cock and two hens are recorded from Ashby Castle in the 1770s. In the 1760s, Madame de Pompadour, mistress of Louis XV, kept these pheasants in her Hermitage near Trianon at Versailles. Buffon gives a rather detailed description of the Golden Pheasant in his *Histoire naturelle des oiseaux* (1771). Probably George Washington kept pheasants from Europe, gifts from General Lafayette around 1787, in his estate at Mount Vernon. The popularity of the Golden Pheasant increased during the nineteenth century, when it became more common in the bird trade and thus available for aviculturists and zoological gardens in Europe.[28]

Along with the Golden Pheasant, other Asian species also entered European aviculture during the eighteenth century. Many were certainly kept in China long before, like Reeves's Pheasant (*Syrmaticus reevesii*), which was not introduced to Europe until 1831. The Silver Pheasant (*Lophura nycthemera*) arrived about the same time as the Golden Pheasant and is mentioned by Eleazar Albin in 1737, but the Golden Pheasant remains one of the most popular ornamental aviary birds. It is now a thoroughly domesticated bird, available not only in its original coloration, but also in various other color breeds. It is also often interbred with the closely related Lady Amherst's Pheasant (*Chrysolophus amherstiae*), which reached Europe in 1828. The hybrids are obviously fertile and concern has arisen about the genetical purity of the captive birds.[29]

Figure 7: A cat sneaking upon a pet bird. Print from 1771

Watch birds

An interesting feature of aviculture is the use of captured birds as "guard" or "watch birds". This practice is described from Africa, Asia and Europe. Domestic geese in particular are well-known for protecting farms against intruders. The Chinese Goose is also regarded as an excellent watch-bird and some even claim that geese are better guards than watchdogs. There is a legend about geese kept on the Capitol, which betrayed the Gaul invaders in 390 BCE by their cackling, and thus saved Rome. Also Indian Peafowl (*Pavo cristatus*) have long been acting as watch birds in parks and larger gardens. They also warn the poultry when birds-of-prey are approaching. Guineafowl are known for raising the alarm if anyone enters the yard or farm where they are kept. They protect both the farm animals and the owners from intruders and predators. In Central Asia tamed Chukar (*Alectoris chukar*) serve the same purpose.[30]

Farmers have used also other non-domestic bird taxa for the same purpose. Captive Cranes (*Grus grus*) were kept in castles in Central Europe during the fifteenth and sixteenth centuries as night guards. With their sharp call they alerted the inhabitants, if there were any movements in the vicinity. Also Hungarians employed cranes as guards. Cranes of various taxa have been used as guards also in other parts of Eurasia, e.g. ancient China and among the Moghuls of India.[31]

In the Amazon (Guyana and Brazil) the Grey-winged Trumpeter (*Psophia crepitans*) and related species with loud calls are often tamed because of their ability to guard settlements from dangerous animals and humans. The German explorer Robert Hermann Schomburgk found the bird in the mid-nineteenth century in all villages inhabited by indigenous peoples in Guyana. Northern Screamer (*Chauna chavaria*) is famous for its extremely load cry and is therefore kept in a semi-tame state in the La Plata region. In Bolivia and Brazil the natives keep various species of Jakupempa (*Penelope superciliaris*), Chachalacas (*Ortalis*) and Currasows (*Crax* sp., *Mitu* sp.) for the same reason.[32]

The custom is also known in the far north in Greenland. The Great Northern Loon (*Gavia immer*) was caught alive and tied with a rope around its legs or neck and put somewhere high up on guard, for instance on a house-top or roof, as it made loud sounds when it saw someone approaching, and thus warned the inhabitants of coming visitors.[33]

More recently parrots have been used as guarding birds in apartments and store premises. Some species are very protective in captivity and are thus helping their owners to scare away unwanted visitors.

Fighting birds

Bird fighting is considered to be one of the oldest spectator sports in the world, of course impossible to prove, but fights between roosters of gallinaceous birds is a widespread practice in Eurasia. Probably the first aviculturists who actually bred non-domesticated birds were the owners of birds used for fighting. Ancient sources describe the custom in China and Persia. Although many

Figure 8: Mr. John G. Stephenson's famous outdoor aviary at the Clevelands at Bishop's Cleve in the 1880s.

species have been used, especially in Inner Asia, most famous are fights with roosters.

Cock-fights are still legal for instance in Japan, but banned in many other countries. The practice possibly originates in Asia and is today, despite being often illegal, still widely spread in South and Southeast Asia, parts of Europe and North and South America, especially Cuba and Mexico. Illegal cock fighting exists in the United States as well; the last state to ban this sport was Louisiana in 2007.

Cock-fighting has been appreciated not only by common people, but also by royalty and American presidents. In England it was once one of the most popular pub sports, but it became forbidden by law already in 1834. Not only poultry of various breeds have been used for bird fighting. Old breeds of fighting domestic geese exist in Germany (Steinbacher Kampfgans) and in Russia (Tula Goose).[34]

Quail (*Coturnix coturnix*) fighting is probably as old as cock-fighting and the Romans were fond of this sport. Octavianus is said to have kept Quails, which always won over those of Anthony. It is described from southern Europe, Central Asia and China. In Southeast Asia (Java and Sumatra), Barred Buttonquail (*Turnix suscitator*) was used for the same purpose and Chukar (*Alectoris chukar*) or Rock Partridge is widely used in Afghanistan. The fighting partridge is gently taken care of and the Afghan villagers train them carefully, feeding and fattening them. Grey (*Francolinus pondicerianus*) and Black Francolins (*Francolinus francolinus*) have been used in bird fighting in India; this is still practiced in some rural areas.[35]

Very competitive small finches can also be used for fighting rings, and in some Latin American countries especially the Saffron Finches (*Sicalis flaveola*), known for their aggressive nature, are

Figure 9: Crested Canary Bird

used in bird fights known as *rinhas* in Brazil. The fighting birds are placed in a large cage while the owners and the audience place bets on which bird will win the fight. When one bird gives up and tries to escape from its competitor, the fight is over. The small finches are often severely wounded and do not survive the fight.[36]

Chinese used the Dayal Thrush (*Copsychus saularis*) for bird fighting, as described in an interesting note from Hong Kong in the early 1930s:

When a combat is to be staged, cages containing the two cock-birds are placed one against the other so that their open doors admit free passage from one cage to the other. The birds, if in a pugnacious mood, sing wildly the whole time with open wings and extended tail. But should one of the birds be fearful to challenge his rival he will not burst into a song of defiance but display considerable uneasiness at his proximity, and in this case, as the result of a contest is a foregone conclusion, the fight does not take place. The actual fight, if it occurs, does not last longer than a few minutes. One bird finds his way into his opponent's cage and a scuffle takes place on the floor, in which both birds use their claws in a manner that suggests human wrestling.

Both combatants sing madly the whole time. A few feathers fly, but no blood is drawn. Soon one bird is chasing the other round the cage, and at this juncture the owner of the defeated bird intervenes. It is not considered advisable to fight a bird more than a few occasions each season. And after each fight, he will need a fortnight's rest. The Chinese say that some cock-birds if continually exhibited to each other but not allowed to fight will go blind with suppressed passion![37]

Songbirds and singing competitions

"What is most prized and admired in house birds is undoubtedly their songs", wrote German aviculturist Johann Matthäus Bechstein in 1795. Keeping songbirds in cages for entertainment and pleasure has a long tradition in the Mediterranean region, Central Europe and East Asia. [38]Songbirds and pet birds were kept already during Antiquity; Greeks are said to have kept nightingales, while Blue Rock Thrushes (*Monticola solitarius*) are mentioned among Roman favorites. Pliny the Elderly describes the trade in trained nightingales. They "fetch the prices that are given for slaves and indeed larger prices than were paid for armour-bearers in old days". He knew "of one bird, a white one it is true, which is nearly unprecedented, that was sold for 600,000 sesterces to be given as a present to the Emperor Claudius's consort Agrippina". Romans found their pet birds charming and amusing and when the pets died they mourned them, and sometimes they even wrote commemorating poems for them. Catullus's poem over his beloved Lesbia's dead sparrow, *Passer mortusest,* belongs to the classical world literature.[39]

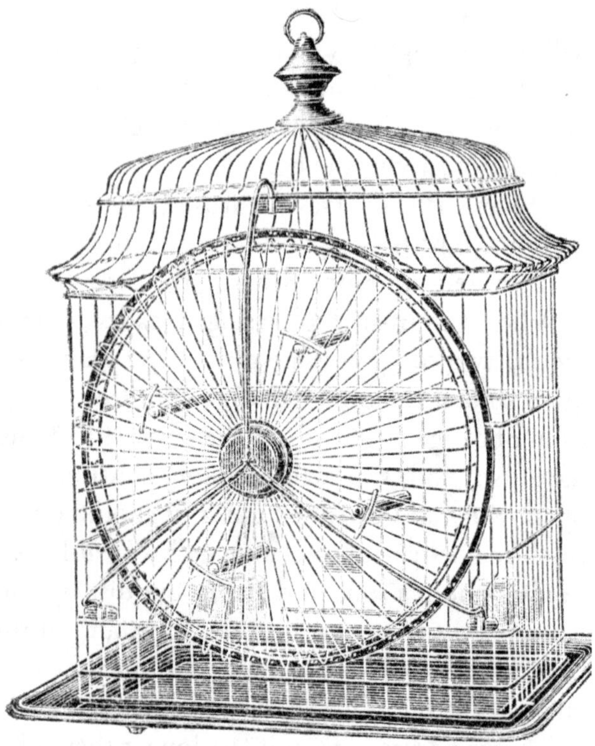

Figure 10: Nineteenth-century wheel cage for Canary Birds

Goldfinch (*Carduelis carduelis*), Eurasian Siskin (*Carduelis spinus*) and Linnet (*Carduelis cannabina*) have been popular songbirds in Europe at least since Medieval times. Books on songbird-keeping were published in early seventeenth century, e.g. Antonia Valli da Todi, *Il Canto degl'Augelli* (Rome 1601) and Giovanni Pietro Olina, *Uccelliera overo discorso della natura, e proprietà di diversi uccelli, e in particolare di que' che cantano, con il modo di prendergli, conoscergli, allevargli* (Rome 1622).[40]

Birds were sold for centuries by bird-catchers in street markets in Paris, Vienna, Nuremberg, Cologne, London, Stockholm etc., actually all over Europe. Moscow had a famous bird market which was visited by German and Swedish explorers in the eighteenth century. They could observe not only finches, but also Blue-throat (*Luscinia svecica*) and Orioles (*Oriolus oriolus*), which were sold to city people interested in keeping songbirds for pleasure. Also the entertaining Common Crossbill (*Loxia curvirostra*) was cheap and popular. Nineteenth-century Russian author Anton Chekhov described in 1883 a visit to the Sunday bird market on Truba Square, obviously a popular pastime for Muscovites for centuries:

Along one side of the square there stands a string of waggons. The waggons are loaded, not with hay, not with cabbages, nor with beans, but with goldfinches, siskins, larks, blackbirds and thrushes, blue tits, bullfinches. All of them are hopping about in rough, homemade cages, twittering and looking with envy at the free sparrows. The goldfinches cost five kopecks; the siskins are rather more expensive, while the value of the other birds is quite indeterminate.[41]

From the late Middle Ages and in the early modern age, itinerant bird-sellers transported songbirds in small cages on their backs in Europe. Goldfinches (*Carduelis carduelis*) and Serins (*Serinus serinus*) were captured in Carinthia and South Tyrol and sold on the markets in northwestern Europe. Linnets (*Carduelis cannabina*) and Bullfinches (*Pyrrhula pyrrhula*) were popular and sold everywhere.

Others were bred or trained in Tyrol and Thuringia. Carl Zeller's famous operetta *Der Vogelhändler* (The Bird Seller) is a romantic story about an eighteenth-century bird-seller arriving in Germany from Tyrol. Also in northern India there are ethnic communities specialised in bird-catching, such as the Baheliya in Uttar Pradesh and Mirshikar of Bihar.[42] Local native birds have been appreciated and kept as songbirds everywhere.

The hill tribes of Himalaya kept the Wedge-tailed Green Pigeon (*Treron sphenurus*) in cages for their song, the Japanese Bush-Warbler (*Cettia diphone*) belonged to the favorites among the Japanese, the Dayal Thrush (*Copsychus saularis*) could be found everywhere as a cage bird in southern China, Seedeaters (*Sporophila*) are popular in Costa Rica, the Mockingbird (*Mimus polyglottos*) was kept as a pet among Pima, Yaqui, Mayo and various Yuma-speaking tribes in southwestern United States, native finches (*Carduelis cannabina, C. carduels)* are still common cage birds in Spain and the Blackcap (*Sylvia atricapilla*) was once very appreciated in Central Europe.[43]

Figure 11: The Northern Cardinal (*Cardinalis cardinalis*) has a long history as a cage bird in Europe. Illustration from the mid-18th Century

The Canary bird (*Serinus canaria*) was

brought to Europe in the late fifteenth century; British physician William Turner mentions it in 1544. Swiss zoologist Conrad Gessner writes in his *Historia Animalium* (1555) that canary birds were sold everywhere and were appreciated for their song. Cesare Mancini, who wrote the first book on keeping cage birds in 1575 (*Ammaestramenti per allevare, pascere, e curare gli ucceli*) has a few data on the canary bird, and Italian ornithologist Ulisse Aldrovandi gives in 1599 some interesting details about the import of canary birds from its native islands.[44] It was domesticated because of its song, but with selective breeding it turned part-yellow, yellow, white and other colors.

In the late sixteenth and early seventeenth centuries it seems to have been distributed in aristocratic circles, but in the seventeenth century it was readily available in bird markets and a breeding center was developed in Tyrol. A French book by Jean-Claude Hervieux de Chanteloup, *Nouveau traité des serins de Canarie* (1713), which gives many interesting details about the state of canary bird culture in the early eighteenth century, became a best-seller and was translated into many languages: Dutch, German, English, Spanish, Danish and Swedish.

Tom de Roo exemplifies how the interest for song-birds developed in Belgium during early modern times and shows how the canary bird gradually replaced native birds among urban people in Antwerp and other cities. Songbirds were popular all over the civilized world: "There are few persons who have not at some period or other of their lives, nourished and cherished a feathered pet", wrote Henry Gardiner Adams in 1854.[45] Good song canaries were exported from the islands and carried by pedlars throughout Central Europe. They probably reached Scandinavia in the early seventeenth century. Today canaries are a popular songbird distributed all over the world and several song-breeds have developed.

The most famous is the German Roller, which developed in the Harz Mountains especially around Sankt Andreasberg. Other early song breeds are the Saxon Singer, the Belgium Malinois or Waterslager from the nineteenth century, the American Singer which traces its history to the Boston region in the 1940s, the Spanish Timbrado, which became a modern breed in the 1940s, and the Russian Song Canary. There is also a rather recent Irish Roller, which hardly exists outside Ireland.[46]

Training singers was essential for canary breeders and this was a sport which developed into perfection in Germany. Certain districts were specialized in training songbirds, not only canaries, but also bullfinches and other species. Eleazar Albin writes in 1737 that trained Bullfinches were very much esteemed in England. American bird-dealer George H. Holden described in 1888 how Bullfinches were trained in Hesse in Germany:

The Bullfinch is taken from the nest in the early spring, when fourteen days old and thoroughly tamed before being given his elementary lessons in whistling. By regularly feeding from the hand, he becomes very tame and strongly attached to his master, whom he soon begins to regard as a substitute for his mother. When taken from the nest, he is allowed liberty for two or three days in order to accustom him to his new mode of life, and afterwards put in a small wicker cage, so commonly seen in the bird-shops. This cage is set into a box of dimensions just allowing its admission, and having a small swinging-door in front, which is kept closed, except during lesson-hours. A professional trainer's house has, as its chief furniture, probably fifty or 75 of these study-boxes, placed in rows around the wall, as far apart as the space will admit. As the houses

Figure 12: Two peafowl cocks and a hen. Note the parrot sitting on its cage to the left. Engraving from *Historia de Gentibus Septentrionalibus* 1555

are usually one room ensuite structures, the space separating the boxes is limited. When a trainer has two or more rooms, the number of airs taught accords with the number of rooms; for only one air can be taught in a room, as the birds would get two songs mixed. But a bird having learned thoroughly one air is taken to another room, and taught a second piece.[47]

There are many ways to improve the perfection of a songbird. The cruel way of blinding birds, recommended by earlier birdsong trainers, was still practiced in some European countries in the early 20th century, is now banned everywhere. Many birds have a reputation of being excellent songbirds and especially in Asia many so-called softbills have been and are still used for that purpose. The Shama Thrush (*Copsychus malabaricus*) has been one of the most popular songbirds in India. In Calcutta (Kolkata) there was still in the early twentieth century a category of men, who made a regular business of feeding caged Shamas. For a small monthly sum they supplied a daily allowance of insects and other food.[48]

Traditional songbird competitions still take place in many parts of the world, Surinam, Singapore, Malaysia, China, Belgium and elsewhere. Song contests with Zebra Doves (*Geopelia striata*) have been popular in Indonesia. Their popularity among the Muslim Javanese is explained by the likening of its song to chants from the Koran. Filip Santens describes in his chapter the fascinating old tradition of keeping and training Chaffinch (*Fringilla coelebs*) in Belgium, a sport especially thriving in Flanders, but rather recently also in the Netherlands and Germany.[49]

Less well-known is the domestic Japanese Quail (*Coturnix japonica*), kept as a songbird by the Samurai in the pre-Meiji period. Such singing breeds existed until World War II.[50]

Parrots in a global world

Parrots have probably been kept by humans for millennia in Africa, Asia and America; at least Indian rulers had them in royal palaces over 2,000 years ago and through trade with Southeast Asia parrots reached China. Vernon J. Kisling discusses in his chapter their presence as captive animals in American high cultures before the arrival of Europeans, thus giving us a rare insight into early American aviculture. Parrots have long been part of European cultural history as well, although they originally do not exist in the wild on the continent. Feral parrot populations exist today in some cities, for instance groups of Ring-necked Parakeets (*Psittacula krameri*) live in twelve European countries, originating from cage birds which escaped in the 1960s and 1970s.[51]

The oldest evidence for European encounters with parrots is from 398 BCE when Greek physician Ctesias of Cnidos described in his *Indica* a bird which could speak an "Indian" language. According to most commentators this must have been a Plum-headed Parakeet (*Psittacula cyanocephala*). During the Indian campaign of Alexander the Great in 326–327 BCE the soldiers also came in contact with parrots and a few were obviously brought back to Europe. Ornithologists discuss if they were a kind of Ring-necked Parakeets (*Psittacula krameri manillensis*), still common in northern India, or the so-called Alexander Parakeet (*Psittacula eupatria nipalensis*). The parakeets were described by Aristotle, who never saw any live specimens and had to base his description on hearsay.[52]

During Greek and Roman times a rather extensive trade with Ring-necked Parakeets developed. Most were imported from India, although wealthy Romans probably also imported the African subspecies (*Psittacula kramerii kramerii*) of the same parakeet. Pliny the Elder, Solinus, Apuleius and others mention parrots; Pliny describes it as follows: "its whole body is green, only varied by a red circlet at the neck". The birds were kept in precious cages made of silver, ivory or tortoise-shell.[53] In his book *Amores* (20 BCE), Ovid includes one animal *epicedium*. This is a well-known poem lamenting the death of Corinna's parrot. In *Silvae* (92–96 CE) by Statius we find two animal *epicedia* (II.4 and II.5). One is a poem on the parrot owned by Melior, the poet's patron. Both poets have most likely read Catullus's already mentioned grave poem on Lesbia's sparrow. There are also a number of animal epitaphs in *The Greek Anthology* which probably were known to the Roman Golden and Silver Age poets.[54]

During Late Antiquity, however, the import of parrots seems to have ceased and not until the Middle Ages parrots were again found in Europe. In bestiaries from Medieval England parrots are mentioned in the twelfth century and from the next few centuries some good illustrations exists, mostly of Ring-necked Parakeets. Further Asian species also began to be known. Frederick II, Holy Roman Emperor, received a White Cockatoo (*Cacatua alba*) as a gift from the Sultan of Babylon in 1220.

At the end of the period the Southeast Asian Yellow-crested Cockatoo (*Cacatua sulphurea*) were present at Italian and Spanish royal courts and the earliest known illustration is found on Andrea Mantegna's painting "Madonna della Vittoria" from 1496. The first Senegal Parrot (*Poicephalus senegalus*) in Europe was reported in 1490.

Ring-necked Parakeets were still the most common species and are depicted for instance on Jan van Eyck's painting "Virgin and Child with Canon van der Paele" from 1434. In *Dialogus creaturum* from 1483, the first book printed in Sweden, there is a simple illustration of a parrot.[55] In later centuries e.g. the famous fictional pirate Long John Silver was always depicted with a parrot on his shoulder. We do know if pirates actually kept parrots as pets, but exotic birds were common as cargo and companions on ships. Sailors brought birds on board the vessels, especially parrots and other exotic species, as gifts for home and to trade with them. Also Japanese rulers appreciated the many exotic birds, brought by Dutch vessels from Southeast Asia.[56]

Parrots came from Asia to Europe with the East Indian trade, but several also arrived from South America and West Africa. Blauw Jaan in Amsterdam, who opened a combined wine cellar and menagerie on the Kloveniersburgwal, was a famous trader in exotic animals. He bought birds and other animals from ships in the harbor. The inn, which operated for almost a hundred years, sold parrots and other exotic birds to royalties in Europe, including the Swedish royal family. Some of these parrots were given to Carl Linnaeus, who described them in his *Systema naturae* (1758). He had among other species also a "talking" Grey Parrot.

Bird traders (*oiseleurs*) in Paris, who were members of an official guild, also imported parrots from Africa, South America and the Caribbean.[57] From the sixteenth to eighteenth centuries parrots were popular pets amongst aristocratic and royal families in Europe. Exotic beasts such as lions, monkeys and other animals including birds were regarded as status symbols. Many important

Figure 13: King Conrad the Younger hunting with falcons, 14th century manuscript Codex Manesse

art works feature parrots. Lazarus Roting created a painting in the early seventeenth century with among other species a Chestnut-fronted Macaw (*Ara severa*), a Jamaican Amazon, (*Amazona collaria*), a Peach-fronted Conure (*Aratinga aurea*) and a St. Thomas Conure (*Aratinga pertinax aeruginosus*). A painting by David Klöcker Ehrentrahl from 1670, showing some of King Charles's XI exotic animals, has a Yellow-crested Cockatoo, Hawk-headed Parrot (*Deroptyys accipitrinus*) and an *Aratinga*-species. Another painting by the same artist portrays macaws, including a Military Macaw (*Ara militaris*).[58]

African Grey Parrots have been common on the market for centuries, like macaws and Amazon parrots from South America. In the nineteenth century Australian parakeets and cockatoos became increasingly popular trade goods. Breeding parrots were, with a few exceptions, rather uncommon before the twentieth century.

French zoologist Buffon relates that a pair of Grey Parrots (*Psittacus erithacus*) bred successfully in the late eighteenth century for several years in Marmande in southern France. Blue and Yellow Macaw (*Ara ararauna*) is said to have bred in 1818 at Caen in France. Still one hundred years later the breeding records continue to be comparatively few and e.g. the Scarlet Macaw (*Ara macao*) did not breed until 1916. This changed after World War II and today most species can be bred in captivity. Only a few parrot species of the 348 known have never been successful, such as the New Guinean Pygmy Parrots (*Micropsitta spp.*), although some tried to breed them (e.g. Sten Bergman). Some very rare species are only kept within special breeding programs, such as the Kakapo (*Strigops habroptila*), Spix's Macaw (*Cyanopsitta spixii*) and the Orange-bellied Parrot (*Neophema chrysogaster*). The trade with captured parrots from the wild still continues outside Europe and North America. Researcher Rosie Low has given an excellent overview of the development of parrot-keeping during the 20th century in Europe, North America, Australia, New Zealand and Japan, and we can only refer further interest to her work.[59]

Imitators

The traditional habit of teaching birds to "talk", that is to mimic human speech, is an ancient custom known already in Antiquity. Several authors provide accounts of talking birds and it is clear that people found parrots comical. In his *Naturalis historia*, Pliny the Elderly writes: "Above all, birds imitate the human voice, parrots indeed actually talking. India sends us these birds [...]. It greets its masters, and repeats words given to it, being particularly sportive over the wine".[60] In his book on the history of animals, Aristotle writes that the parrot "is said to be human-tongued (and it becomes even more outrageous after drinking wine)". History can tell many interesting anecdotes about "talking" birds.

One of the most astonishing stories is the Prussian explorer Alexander von Humboldt (1769–1859), who in the early nineteenth century during his expeditions in the Orinoco region learned about a tribe, the Atures, who perished after being hunted down by indigenous rivals. Humboldt visited a nearby village: "An old parrot was shown at Maypures, of which the inhabitants said, and the fact is worthy of observation, that they did not understand what it said, because it spoke the language of the Atures". Humboldt recorded some words spoken by the parrot; this is the only remnant of the extinct Ature language.[61] Pedro Fernandes and Ingvar Svanberg give in their contribution a brief ethnographic insight into the now vanishing custom of teaching jays and other corvids to "talk" in Portugal. This very ancient habit is disappearing in Europe due to legal prohibitions.

Parrots are still the most popular "talkers", but also other birds can be trained to imitate whistling or speech. Hyatt Verrill claims that the Imperial Parrot (*Amazona imperialis*) of Dominica Island in the Caribbean "learns to talk within a few weeks after taken from its native forests, and even when captured fully grown he becomes perfectly gentle and affectionate within a week or two". Beside the Grey Parrot (*Psittacus erithacus*) and Moluccan Cockatoo (*Cacatua moluccensis*) also the Budgerigar (*Melopsittacus undulatus*) and Cockatiel (*Nymphicus hollandicus*) are often kept for their

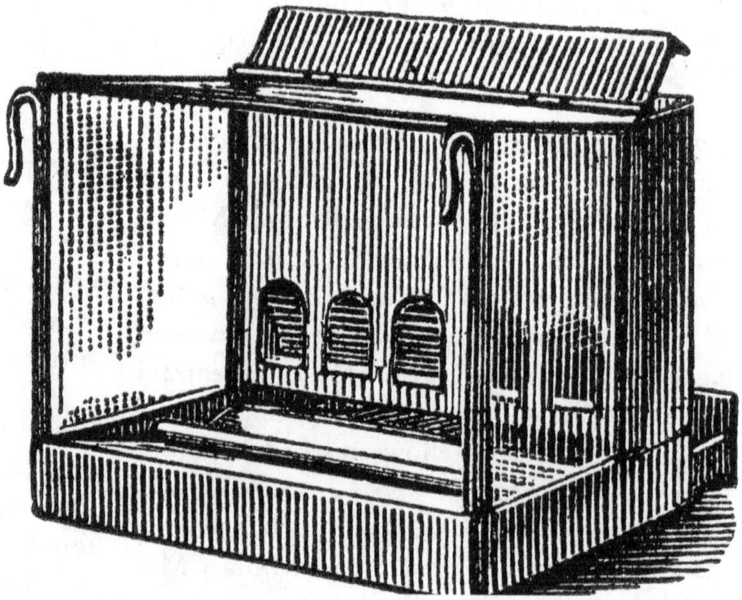

Figure 14: Seed feeder

ability to whistle or imitate phrases. Some of these birds perform in zoological gardens and shows. According to Guinness World Records there is Budgerigar which is said to have a vocabulary of 1,728 words. Keeping parrots for this ability is still a very popular way of keeping tame birds.[62]

Talking parrots and other birds have fired the imagination of poets and in many poems we find funny examples of talking birds or parrots revealing secrets, e.g. "On the Death of a Lady's Sparrow, in Imitation of Catullus, for his Lesbia's" by George Farquahar (1678–1707) and "On the Death of Lesbia's Green-bird" by Henry Needler (1690–1718). John Skelton's famous poem "Philip Sparrow" is another example. In a Swedish Baroque poem, "Complaint on the Death of a Very Accomplished Parrot", by the learned admiral Werner Rosenfelt (1639–1710), we find a dead parrot who speaks from the other side of the grave. Rosenfelt is not very well known today, but in his lifetime he won some recognition.

In the poem he laments the death of a beloved parrot with numerous allusions to the Roman authors Virgil, Ovid, and Statius. The epitaph consists of three minor parts – one in which the parrot tells us where it was born and where it died, one in which it provides information about its life, and one in which it tells us what happened to it after death:

> In India I was born and carried overseas
> On Värkö, more than 1,000 miles away, I died
> I have given my owners lots of funny moments,
> So that they will not forget me too soon
> Now I live amongst gods and goddesses
> Where I find more pleasures than I did on earth.[63]

Funerary poems on animals form a sub-genre of the standard epitaph written to commemorate a deceased person. As in the case of many other Baroque poems, the poem on the parrot is an inter-textual game, a specimen of sophisticated literary entertainment. However, despite Rosenfelt's copious use of Roman models, his poem should not be considered a collage of lines derived from other poetic works; it would be more correctly described as a well-balanced mixture of imitation and individuality.

Furthermore, the parrot's death provided Rosenfelt with an opportunity to compose a witty and humorous poem and the task he set for himself seems above all to be to raise his reader's spirits. His ambition was to teach (*docere*) and to amuse (*delectare*), as his poem had two groups of recipients. Adult readers were supposed to be amused and entertained and for children the purpose was to guide, comfort and inspire a good cheer. Poets and artist have sometimes also identified themselves with cage and aviary birds.

In his famous poem "Auguries of Innocence" (1803), William Blake writes: "A robin redbreast in a cage / Puts all heaven in a rage / A dovehouse filled with doves and pigeons / Shudders hell through all its regions". Leonardo da Vinci, perhaps globally the most famous artist, is reported often to have bought caged birds at the market in order to release them.[64]

International trade

Most birds on the market into the eighteenth century were locally available, captured species from the wild, such as siskins, finches, starlings and thrushes. Due to colonial expansion, more exotic bird species became available in Europe and e.g. Common Waxbill (*Estrilda astrild*) from Sub-Saharan Africa, Adavat or Strawberry Finch (*Amandava amandava*) from India and Virginian nightingale, that is the Northern Cardinal (*C. cardinalis*), became available in France and England for those who could afford it. French ornithologist Louis Jean Pierre Vieillot probably bred the last mentioned in the late eighteenth century. Bird traders in Paris stocked not only cardinals, but also amazons, parakeets, whydahs, grenadiers, muniahs, firefinches, waxbills etc. According to one survey, none less than 72 non-European bird species were offered in Central Europe in 1794.[65]

During the nineteenth century aviculture developed quickly in Europe. With better communications, steamships and railways, but also through increasing demand due to new tastes and fashions of the time, birds from all continents became common trade goods. Pet stores were established all over Europe during the second half of the century. In 1870, around 230 species were mentioned in the Central European bird trade; in 1920, lists of 1,450 species of interest for private aviculturalists were compiled. Most species mentioned already in C.E. Dyson's *Bird Keeping* from 1878, including doves, weavers, estrildids, finches, thrushes, waxbills, tanagers, tropials, bulbuls etc. from Africa, America, Asia, Australia and Europe, were still available in the 1970s.[66]

The songbird business mainly relied on captured birds. Locally trapped birds were much cheaper than exotic and many European species were kept by rich and common people alike. Today most of these finches, thrushes and small insectivorous birds are born in captivity, at least in Europe and North America and are more or less domesticated. Keeping songbirds in captivity continues to be very popular also in other parts of the world; China has probably by far the largest numbers of pet birds in the world, and still many of the Timaliids (Old world babblers), white-eyes, thrushes, mynahs and starlings etc. are captured from the wild. Thriving markets with wild birds can be found in countries like Indonesia and Brazil. Despite efforts to stop international bird trafficking, it is still a problem in many parts of the world.[67]

Religious rituals and releasing birds

Using birds for religious sacrifice is a practice known at least since Antiquity. The Bible tells that the Israelites sacrificed pigeons and turtle doves: "And to offer a sacrifice according to that which is said in the law of the Lord, A pair of turtle-doves, or two young pigeons", according to the King James version of Luke 2:24. Also quails were fit for sacrifice. The pigeons and doves were most probably raised in captivity; historian Oded Borowski emphasizes this, because hunting would damage the bird and prevented its fitness for sacrifice. When Samoa chieftains traveled they were always accompanied by tame Pacific Imperial Pigeons (*Ducula pacifica*). These birds were seen as messengers to the gods.[70]

In Buddhism-influenced regions in East and Southeast Asia the custom of religious release of birds is common; it is considered an act of compassion. Chinese sources record it several hundred years back in time and today it has become a huge, destructive industry for birds and biodiversity. Buddhist and some Taoist temples, organizations and also companies provide imported and captured native birds for release. Large numbers of Java Sparrows were released in temples in Singapore in the 1920s and today the custom is increasingly popular not only here and in Japan, but also in Thailand, Vietnam, Laos, China, Hong Kong and Taiwan. Huge quantities of birds are captured and traded with, not only Java Sparrows, but also Nuns, Munias, Spotted Doves, Bulbuls and Mynahs, and then released, often far away from their original habitats.

The ceremonial and religious release of birds is an increasing problem for local biological diversity. Alan Dundes has described the custom in Iran, but this custom is spreading fast; it exists also in

Turkey and many other places.

To release white pigeons in connection with weddings, birthdays and other celebrations such as graduation ceremonies has developed into a modern custom also in Australia, northwestern Europe and North America. Small companies or bird traders provide the pigeons for the events.

New companions

The canary must be regarded as the first real companion animal, domesticated only for being a pleasure for its owner. Today it is not only kept as a songbird, but there is a long tradition of keeping it as an ornamental bird with various breeds. Especially the British Isles have contributed with old breeds like the Lizard Canary, Norwich, Scotch Fancy, Yorkshire, and the now extinct almost mythical London Fancy, but there are Belgian (Bossu), Dutch (Nordholländer), French (Frisé Parisienne), German (Münchener, Rheinländer), Italian (Gibber Italicus), Japanese (Japan Hoso), North American (Columbus), Spanish (Melado Tiner-feño) and Swiss (Berner) type canaries as well. Some of these breeds are old, while others are more recent and offer many color mutations. The most famous is the red-factor canary, a product of a hybridization with the Black-headed Red Siskin (*Spinus cocullatus*). The canary hobby is probably the most widely spread section of aviculture and there are special clubs for many breeds on all continents.[71]

Early domesticated finch species are the Bengalese Finch (*Lonchura domestica)* and the Java Sparrow (*Padda oryzivoria*), both bred in Japan. They are dealt with more in detail by Ingvar Svanberg in a separate chapter. White Java Sparrows "thrive well in very small cages, as they are quite and deliberate and seldom mar or wear their sleek feathers. They breed readily in confinement and are careful and affectionate parents", wrote Hyatt Verrill in 1915.[72] During the twentieth century further species, kept in cages and aviaries, have been fully domesticated.

Most well-known is the Budgerigar (*Melopsittacus undulatus*), which was introduced to England from Australia. The first Budgerigars were brought to Europe by ornithologist and artist John Gould in 1840. "I believe I was one of the first who introduced living examples to this country, having succeeded in bringing home several on my return in 1840", wrote John Gould in 1865, and continued: "Since that period nearly every ship coming direct from the southern parts of Australia has added to the numbers of this bird in England".

Thousands of the small parakeets were exported from Australia for the international bird trade. "I have more than once seen two thousand at a time in a small room at a dealer's in Wapping", confirmed Gould. The first breeding in captivity is recorded from France in 1846 and Antwerp Zoo had a breeding in 1850 and very soon after it became a common, popular and prolific cage bird. August Renoir featured the bird in a painting from 1871 with a lady and her pet Budgerigar.

Mass breeding production developed in England and France and new color mutations were much sought after. *The Budgerigar Club,* founded in England in 1925, has been important for the development of the species as a domestic bird. Especially the Blue Budgerigar became very popular and was very much in demand in Japan in the inter-war period. Today there are many different colors and the bird is a widely spread hobby, being almost as fascinating as the canaries. Budgerigars have also been popular because of their abilities to imitate human voices.[73]

The Cockatiel (*Nymphicus hollandicus*) is also available in several color varieties. "These are beautiful birds of handsome colors, and should be kept by everyone fond of ornamental birds," noted Hyatt Verrill in 1915. It is a newcomer within aviculture and became popular after World War II. Details in Cockatiel history within aviculture is very little known.[74]

After the war several almost domesticated species can be added to the list. These species are today adapted to life in captivity, breed without problems and due to true selective breeding many color mutations exist, e.g. Chinese Button Quail (*Coturnix chinensis*) from Southeast Asia, Bobwhite (*Colinus virginianus*) from North America, Diamond Dove (*Geopelia cuneata*) from Australia, Ring-necked Parakeet (*Psittacula krameri*) from India, Crimson Rosella (*Platycercus elegans*) from Australia, Bourke's Parrot (*Neopsephotus bourkii*) from Australia, Turquoise Parrot (*Neophema pulchella*) from eastern Australia, Catherine Parakeet (*Bolborhynchus* lineola) from Central America, Fischer's Lovebird (*Agapornis fischeri*) from east-central Africa, Peach-faced Lovebird (*Agapornis roseicollis*) from southwest Africa, Masked Lovebird (*Agapornis personatus*) from Tanzania, Greenfinch (*Carduelis chloris*) and Golden Finch (*Carduelis carduelis*) from Europe, African Silverbill (*Lonchura cantans*) from central and eastern Africa, Gouldian Finch (*Erythrura gouldiae*) from northern Australia and the Zebra Finch (*Taeniopygia guttata*) from Australia.[75]

Many other quails, pigeons, doves, parrots, finches and some softbills, such as Pekin Robins (*Leiothrix lutea*) and Laughing Kookaburras (*Dacela noaeguineae*), are also routinely bred and have existed for many generations in captivity and can now be regarded as semi-domesticated birds or domesticates.

The future of aviculture

Many birds have become extinct in the last few hundred years. Hunting, human expansion, destruction of habitats, introduction of invasive pests and also legal and illegal bird trade are contributing factors. Some of the disappeared species have been kept in captivity, but at a time when breeding programs did not yet exist. Dodo (*Raphus cucullatus*) was brought from Mauritius to Europe, e.g. to London in 1638.

The Danish scientist Ole Worm kept a Great Auk (*Pinguinus impennis*) from the Faroe Islands as a pet in his Copenhagen yard in the 1650s. Cuba Macaws (*Ara tricolor*) were also kept in Europe, including Sweden, in the late eighteenth century. The last Mascarene Parrot (*Mascarinus mascarinus*) from Réunion died in captivity in a Bavarian royal menagerie during the early 1830s and the last Norfolk Kaka (*Nestor productus*) died in captivity in London in 1851. The Passenger Pigeon (*Ectopistes migratorius*) and the Carolina Parakeet (*Conuropsis carolinensis*), both species available in zoological gardens in the nineteenth century, are gone because no real effort was made to save them. Joseph Lindholm III tells us the sad story of the two species in captivity.[76]

To this date, the number of bird species in zoological gardens is decreasing. Birds that once were common are now becoming rare in bird collections. One reason is that many exotic birds have very little visual appeal, for instance gulls, easy to keep and once a well-known group of birds in captivity.[77] Also other species, previously common in the bird trade, are now lost to aviculture. Hummingbirds were available for private aviculturists a few decades ago in Europe, but now only a few zoological gardens keep them. Christoph Hinkelmann tells the story about hummingbirds among hobbyists in Europe before import was banned.

Birds are kept in different kind of contexts, but domestic birds in general have a tremendous value for the well-being of humanity. Companion birds for amusement continue to be of great importance and aviculture as a hobby can be carried on in most settings. Benedict Taylor tells in his chapter the fascinating story about bird-keeping among prisoners. The "Birdman of Alcatraz", Robert Franklin Stroud, became famous after the film with the same name, starring Burt Lancaster in 1962. As Taylor shows, it has not been uncommon to keep birds within closed institutions.

The future of aviculture depends on the aviculturists' ability to breed birds. Breeding rings are important for hobbyists, if they want to keep rare species for the future. Even common species will be gone if aviculturists lose an interest in them (many small doves and estrildids are becoming rare in Europe because of this). Zoological gardens today place the emphasis on rare birds and participate in breeding programs, with the goal of securing captive-bred populations of vanishing birds for the future. However, zoological gardens are also dependent on their visitors. Pheasants need large aviaries and although they are easy to keep and breed without problems, many zoological gardens do not have them any more and for instance Copenhagen Zoo, which used to have pheasants in a row of aviaries, has removed the collection. Joseph Lindholm III discusses in his chapter the importance of aviculture development in American zoological gardens through work with rare birds. Some breeding programs have been rather successful, for instance Bali Mynah (*Leucopsar rothschildi*) and Socorro Dove (*Zenaida graysoni*), extinct as wild birds, are today available in private collections. Large numbers of Bali Mynahs have been bred and in the last few years several specimens have been released on a small island close to Bali. The Hawaiian Goose (*Branta sandvicensis*), once very rare, has successfully bred and is being reintroduced to its native islands. It is also common in captivity and found in many private collections in Europe.

Maybe it will be possible to save the Guam Kingfisher (*Todiramphus cinnamominus cinnamominus*), which remains only as a captive bird in American zoos and breeding centers. Studbooks and breed registries are now maintained for many rare birds in zoological gardens.

Many pheasants, rare in their native countries, are quite common within aviculture. The Blue-throated Macaw (*Ara glaucogularis*) is easy to breed and captive-born specimens are available for common bird keepers. Only a few hundred remain in the wild in Bolivia. The knowledge developed by skilled aviculturists, both private enthusiasts and professional zoo employees, is very important for these breeding programs and, in the long perspective, for the survival of many bird species.[78]

Figure 15: Bird escaping from cage

Chapter 2. Aviculture in Ancient Old World Societies

Vernon N. Kisling, Jr.

Ancient bird-keeping developed initially from the convergence of two trends: the keeping of wild animals for utilitarian purposes and the evolution of societies into civilizations with attributes favorable to collecting. Essential to both trends were the relationships of individuals to wild animals in particular and to nature in general. The domestication of animals, which began about 10000 BCE, was an intensified effort at keeping wild animals. As a biological process, domestication requires the keeping of wild animals through many generations and is achieved only after significant changes are made in their behavior. During ancient times these changes could not have been anticipated initially, so the consequences of domestication were not known until after they had occurred.

Ancient societies evolved into urbanized, literate civilizations between 3000 and 1500 BCE. While the bulk of the population had little resources or time for leisure activities or developing physical and genetic attributes of bird species, royalty and a wealthy class of individuals, such as government officials, priests, merchants and landowners, prospered and took advantage of cultural luxuries that became available, including animals in the wild. For these privileged classes the urban lifestyle provided a relatively stable social environment, conducive to long-term endeavors such as collecting. Wealth to buy what they desired, leisure time to do what they wanted, the availability of luxury items and heightened aesthetic sensibilities provided the upper classes with the opportunity to build reserves and parks, plant gardens and collect animals.

Keeping birds is known from ancient civilizations in Mesopotamia along the Tigris and Euphrates Rivers in what is now Iraq, in Egypt along the Nile River, along the Indus River in India and Pakistan, and in China along the Yellow River (Huang He). There were other, less complex societies throughout Europe, Asia and Africa; however, most of them probably did not develop the complex social structures that were favorable to collecting. Over time a complex evolution of animal collections occurred, as the epicenter of power shifted from these regions and finally settled in Europe. Animal collections in Mesopotamia and Egypt became less prominent but nevertheless continued, while those in India disappeared with the Indus society and re-emerged with India's Indo-Aryan societies. Collections in China developed in isolation, while newer collections appeared in Greece, Rome, Persia, Arab regions and eventually Medieval Europe.

Mesopotamia

Mesopotamian societies developed as city-states along the Tigris and Euphrates Rivers. Natural sciences were less important than mathematics and astronomy, but they were of practical importance and consisted of observing and classifying animals, plants and minerals. The recorded practice of medicine began with the Sumerian, but was more fully developed during the Babylonian and Assyrian periods. Physicians practiced human medicine, while veterinarians known as "ox and ass doctors" practiced animal medicine. By 2500 BCE the natural world around Mesopotamia was broken down into domestic animals, wild animals, wild birds, fishes, insects, plants, trees, vegetables and minerals. The natural world and its wildlife were also becoming increasingly utilized. Trade in luxury items increased, and included among this trade were animals. Animals were also obtained through confiscations in conquered lands and as tribute or gifts from other societies. Early trade in wild exotic animals, including birds, was undoubtedly negligible, since it was only satisfying the passion of a small class of wealthy individuals; however, over time it grew as royalty and an increasing number of rich attempted to surpass the collections of their predecessors. Although all classes of society had gardens and fish ponds, royalty and the land owning class had shade gardens, ornamental gardens and parks. Sumerian shade and ornamental gardens were often small and were combined with vegetable plots and orchards. Smaller gardens and parks were strictly botanical, but the larger parks also contained animal collections.

Property holdings included domesticated animals, as well as wild ones. Animals listed included household pets, fish in ponds, birds in flight cages, falcons for sport, lions in cages and wild game in parks. Hunters used blunt arrows to stun wild animals, or traps (usually concealed pits) to catch the animals alive for pets, collections and trade. Some animals, particularly rare species, were valuable luxury items. Lions (*Panthera leo*) and other animals were kept in cages and pits for exhibit purposes to impress and entertain local guests and foreign dignitaries during the Ur III period (from around 2100 BCE). It is conceivable, therefore, that cages were constructed to hold other dangerous or rare species as well.

Bas-reliefs from Assyrian royal palaces show monkeys, antelopes, camels, elephants and other species brought to the kings as tribute. While they usually viewed nature from a practical perspective, the Sumerian, Babylonian and Assyrian kings were proud of their animal collections. These animals were symbols of power, wealth and authority and the kings were especially proud of rare specimens their subjects and foreign dignitaries sent after diligent searches and difficult transport. Animals often came from Asia via trade with the Indus society and from Africa via trade with the Egyptians. Royalty and wealthy individuals further constructed fish ponds that served also the economic purposes of keeping fish fresh for the table. They kept wild birds such as ibis, cranes, herons, peacocks and pelicans as pets and in flight cages. Initially these bird collections also served the economic purposes of being handy food sources and commodities for sale. Falconry and wild beast hunts were royal sports conducted in the wild and in royal parks. Park collections of elephants, wild bulls, lions, apes, Ostrich (*Struthio camelus*), deer, gazelle, ibex and other species were a combination menagerie precursor, hunting reserve and garden-park. Royal families in addition used the parks for entertaining guests and for personal pleasure.

Eventually, the ability to maintain animals and plants in large park areas was taken to a new level of sophistication with the recreation of entire habitats. Sennacherib (Assyria, 704–681 BCE) simulated a marsh environment of southern Babylonia to exhibit rarely seen marsh species from that region. He also re-created mountain habitats, one of which may have been the famous hanging garden thought to exist in Babylon. His successors, Esarhaddon (Assyria, 680–669 BCE) and Ashurbanipal (Assyria, 668–627 BCE), also provided mountain habitats that resembled the nearby Amanus Mountains. Within the cities, the monumental, pyramid-shaped, terraced buildings known as *ziggurats* sometimes had their terraces planted with trees, shrubs and vines to give a mountain-like appearance. The hanging garden of Babylon (now thought to have been in Nineveh) was probably one of these structures. One aspect of civilization was its specialized labour and this included work related to the keeping of wildlife. Fishermen, fowlers (bird catchers and keepers) and shepherds cared for the domestic stocks. Servants (animal keepers) captured animals in the royal park for release at the appropriate time and place for royal hunts. These animal keepers probably served other animal care functions as well. Veterinarians ("ox and ass doctors") dealt primarily with domestic and military livestock, and their fees were pre-set in King Hammurabi's Code of Laws (Babylonia, 1728–1686 BCE).

Animal collections were not a unique activity of the Mesopotamian societies. Neighboring Hittites also kept pets, domestic livestock and wild animals; however, their wild animal collections were less extensive. Their collections also served more than one purpose, which included secular uses such as food, hunting and pleasure; religious uses such as rituals that required animals; and symbolic uses (certain species, such as lions and eagles, were reserved for royalty as emblems of their power and authority). As with Mesopotamian collections, Hittite collections contained native and exotic species, including lions, tigers, leopards, wolves, foxes, bears, deer, wild goats, boars, bison, elephants, hares, mice, eagles, snakes, frogs, bees, ants, spiders and several other animals whose identification has not yet been determined. The Persian kings, who conquered Mesopotamia in 539 BCE, continued the Mesopotamian collections, using their own traditions of garden design and animal collecting. Greek rulers, conquering the Persians in 330 BCE, did the same. Collections in this region still existed in 363 CE when Roman armies conquered Babylonia and the Muslim Arab rulers who followed after the seventh century CE continued the traditions.[79]

Egypt

Egyptian civilization began with the unification of lower (the northern delta region of the Nile River) and upper (the southern riverine region of the Nile River) Egypt into the Old Kingdom (2700–2200 BCE). This was a period of belief in magic and superstition and achievements in astronomy, geography, biology and technology, but zoology and botany apparently did not become well developed, even though native and exotic animals and plants were known to the Egyptians, domestication of wild animals and plants was one of their most important activities and the keeping of animals and plants was one of their great passions. Representations of animals and plants were pervasive throughout ancient Egyptian culture. Many Egyptian gods assumed animal forms, while other kinds of animals were worshiped without being transformed into gods. Symbols for the lower and upper regions of the Nile River valley were flowers (the papyrus and the flowering rush respectively). Animals and plants were graphically rendered as much as any other subject. Domestication of a large variety of birds, ruminants and carnivores was attempted.

Physically, the extent of the Egyptian natural world was not large. The Nile valley was relatively narrow, with deserts bordering it on both sides. It had an almost harborless Mediterranean coast to the north and the tropics of Africa to the south. While the valley was a pleasant place to live, it was also monotonous with a limited variety of wildlife, few trees and few wild flowers. Animal husbandry was primarily concerned with cattle, but the Egyptians also experimented with other species. In addition to several breeds of cattle, there were breeds of sheep, goats, donkeys and pigs. Horses and mules came into use between the Middle and New Kingdoms, and the camel during the Ptolemaic period. Egyptian domestication attempts included many different kinds of native wild ruminants and carnivores. These animals were fattened on bread dough, as were many kinds of birds. A variety of birds were kept in domestic flocks, particularly geese and ducks, but also swans, doves, ibis, cranes, herons, and other marsh and water birds. Bees were kept as well, both in the desert and in private gardens, for their honey.

Animal-related jobs in Egyptian society included herdsmen, fowlers (bird catchers), bee keepers and honey gatherers and veterinarians. No doubt there were also animal keepers, but there is as yet no specific information known about the individuals who maintained animal collections. Veterinarians were physicians, who cared for cattle and whose practice was based on human medicine (no separate term existed for animal doctors). There are no records to clearly tell us how valued veterinarians, or any of these animal related positions, were in Egyptian society.

Fowlers came from the lower classes, as bird catching was considered one of the dirtiest jobs. Bird hunting, on the other hand, was a sport of the upper classes. Much of the country formerly covered by marshes and tropical forests became arable land. At the same time old river beds remained and the greatest delight of the Egyptians was to row in a light boat between the tufts of papyrus reeds, to start the wild birds and then knock them down with a throw-stick.

Pictures of all periods exist representing these expeditions, and we have but to glance at them in order to realize how much the Egyptians loved these wild districts. Both bird hunting and spear fishing, as opposed to commercial bird catching and fishing, were sports of the wealthy. Other sports included harpooning hippopotamus and crocodiles, as well as desert hunts for gazelle, antelope, hyena, jackal, fox, leopard, lion and numerous small mammals.

Research indicates that wealthy Egyptians kept menageries containing animals taken by the lasso and dogs in the desert, as well as those brought into Egypt by way of commerce or as tribute and gifts. From neighboring deserts they obtained lions and leopards, hyena, gazelle, ibex, hare, and other animals. From countries to the south came baboons and giraffe, and from Syria bears and elephants. Egyptians particularly liked to tame and eventually domesticate as many of these species as possible. Bird cages in the houses of royalty and wealthy citizens became popular in the New Kingdom, in addition to the more common bird houses that had always been used for keeping fowl for food, tribute and sacrifice. Egyptians took special delight in displaying wild birds from the marshes, birds of prey and imported exotic species. Fish ponds and bee hives were also maintained, often complementing their gardens.

Pools became micro-habitats containing fish, birds and aquatic plants. Larger houses and palaces had rooms opening onto gardens in such a way that painted gardens on the walls and floor blended with the real garden just outside the rooms. Paintings of plants and birds, along with live birds in cages, minimized the distinction between the house and the garden, as the garden was entered from these rooms.

Trading expeditions were made to obtain the needed resources, exotic materials and animals. These expeditions were made to the Divine Land in the east extending up through Palestine, Syria and Mesopotamia for wood, gold, silver, precious stones, horses, bears, elephants, fish and cattle. They extended to Nubia in the south via the Nile River, for gold, ebony, ivory, acacia wood, precious stones, dogs, cattle, ostrich feathers and eggs, panthers, panther skins, giraffes and monkeys. They also extended to the land of Punt further south in Africa via the Red Sea for incense, myrrh, other plants and exotic African animals. Little is known about these trading expeditions other than those to the land of Punt. When the Greek Ptolemaic dynasties (323–30 BCE) replaced the Egyptian New Kingdom dynasties, the city of Alexandria at the Mediterranean coast boasted many great amenities. One of these included the largest and most varied animal collection in the ancient world. When the Romans took over Egypt in 30 BCE, this animal collection continued and was probably the source of supply for African animals used in the Roman spectacles.[80]

Figure 16: Roman mosaic depicting a cock-fight. Santangelo, Naples

India

A civilization at the Indus River existed around 2500 BCE–1500 BCE, centered around two prin¬cipal cities, Harappa and Mohenjo-Daro, in what is now Pakistan. Little is known about its zoo¬logical activities; however, animals were domes¬ticated, most importantly the Elephant (*Elephas maximus*). Contact existed with Mesopotamia, as evidenced by seals used on traded commodities. These seals depicted wild animals, such as Elephants, Tigers (*Panthera tigris*), Rhinocer¬oses (*Rhinoceros unicornis*), Buffalo (*Bubalus bubalis*), antelope and Gharials (*Gavialis gangeticus*). Animals such as Peacocks (*Pavo cristatus*) and apes, as well as animal products including ivory, were exported. The Indus society had a wealthy class of people who enjoyed gold, silver, jewels and other expensive items. Because of this wealth, their artistic depiction of wild animals, and the contacts they had with other ancient societies (which did have animal collections), it is probable that animal collections of some kind were part of the society, although there is no direct evidence yet that any structures were used for animal collections.

There was little continuity between the ancient Indus civilization and the later Indo-Aryan societies which prospered throughout India between 1500 BCE and 1500 CE. Indo-Aryan animal collections are much better known due to greater documentation and the fact that these collections existed when Europeans arrived. Indo-Aryan societies, along with their animal collections, may be representative of the kinds of collections that existed throughout Asia throughout this same period. Observations of the earliest European explorers and merchants to reach Asia indicate that there were royal animal collections throughout the continent, but they provide little useful information. While these observations were made during the sixteenth and seventeenth centuries, they mention collections that were probably continuations of older royal collections. Considering what is known about collections in India and China, these Asian collections may date back to ancient times.[81]

China

Like Egypt, China's history has been marked with alternating periods of large states and fragmentation. Periods of large states represented stable times under influential rulers. The first is considered the Shang (1500–1000 BCE) along the Yellow River. It was during such relatively peaceful periods that significant animal collections often developed. Certain states in history are better known for their animal collections and gardens, import of exotic species and the study of natural sciences; nevertheless, it seems they all had animal collections.

Ancient Chinese attitudes toward wild animals and nature vacillated from ethical concern to uncaring use, but for the most part unfriendly and fearful. They viewed animals and landscapes they did not control as menacing and alien; as much, if not more so, than did the Mesopotamians or Egyptians. These attitudes moderated in later dynasties, but never completely subsided. Intellectual interest in the natural sciences was minimal except for plants, animals and minerals used for medicinal purposes. Nevertheless, wild animals, especially exotic species, held the interest of rulers and the wealthy classes, as it did in other ancient societies. Royal animal collections served both practical (food and ritual use) and aesthetic purposes. Starting with the founder of the Shang dynasty, the rulers built animal reserves; however, it was Wen Wang, founder of the Zhou Dynasty, who built the first well-known animal reserve. This reserve, along with similar royal and baronial parks (parks owned by the wealthy class) of the Zhou period (about 1000–200 BCE), were large walled-in natural areas that required their own staff of administrators, keepers and veterinarians.

Han Dynasty (around 200 BCE–220 CE) rulers continued this tradition of large royal parks. Emperor Wu Di enlarged an established reserve around 140 BCE. The reserve was a large wilderness area, but also had some 70 palace buildings and water gardens. It contained exhibits of gems and minerals, native and exotic plants and animals. Animals in the reserve included swans, geese, ducks, water birds, shore birds, Great Bustards (*Otis tarda*), herons, cormorants, turtles, Alligators (*Alligator sinensis*), sturgeons, various aquatic species, camels, asses, mules, horses, yak, deer, elephants, rhinoceroses and possibly Giant Pandas (*Ailuropoda melanoleuca*). Animals were obtained locally, as well as from Mesopotamia, India and other regions of Asia. In addition to keeping animals in their parks, Han emperors kept birds, bears, tigers and other mammals in palace annexes, in what appear to have been private menageries kept for their own personal pleasure. These animals, as well as dangerous animals in the parks, were kept in cages. After the Han Dynasty, the Qin, Tang and Song Dynasties (around 265–1279) continued the fashion of extravagant parks.[82]

Greece and Rome

Greek and Roman societies (1100 BCE–476 CE) existed along side those of Mesopotamia and Egypt and eventually overshadowed and conquered them. Greek intellectual inquiries went beyond what could be learned simply through observation, the factor limiting the growth of knowledge in other societies up to that time. However, Greek curiosity was still practical and descriptive. Although they had an appreciation for nature, they were intent on controlling nature; creating what Cicero later called "another nature" out of the wilderness that surrounded them.

A primary concern of the Greek city-states was the creation of this second nature, an agricultural and man-made environment. Associations between certain gods and trees were particularly strong, with places of worship originally taking place in sacred groves. Roman natural science was derived from Greek science and made few independent, original contributions. Nevertheless, Roman efforts to copy texts that were later lost and compile what was known into encyclopedic tomes, carried ancient knowledge about animals forward to the Medieval period.

Like the Greeks, Roman attitudes toward nature were practical and utilitarian. While Romans enjoyed the beauty of nature and preferred rural life over urban life, they felt the land and its natural resources should be used productively.

Extensive agriculture, deforestation, mining, quarrying and hunting modified the environment significantly. The early Greeks (up through 323 BCE) depleted many of their natural resources and had to look for resources outside their islands, e.g. by colonizing the eastern Mediterranean coast and the Black Sea. Agriculture and deforestation in particular modified the environment. As a result, the Greeks took an interest in forestry; controlling the cutting, exporting and importing of lumber, and experimenting with replanting some areas. Mining and quarrying also presented environmental problems, as did hunting, which brought some animal populations to extinction while decimating others. Grazing domestic stocks took its toll on the countryside and the wild animals that lived there.

Greek interests, travel and trade provided favorable conditions for developing animal collections; however, the ruling city-states did not have enough wealth or influence to develop large ones. Pets included various kinds of birds, monkeys, weasels (probably the Polecat (*Mustela putorius*), domesticated like the ferret to catch vermin), hedgehogs and harmless snakes. Temple collections maintained animals for processions, during which wild animals appeared in cages, while tame ones walked on leases or pulled vehicles. Showmen and professional animal trainers exhibited wild animals for entertainment. Some animals, such as bears and lions were common, while others such as tigers were rare. Bears and lions were common because they could still be found in Greece and its neighboring areas at the time. Rare animals did not exist locally and were not often seen within Greece. Whether an animal was considered native or exotic depended on where it was captured. An animal from another land was considered exotic, even if it was the same species as one found locally.

During the Hellenistic Greek period (323–27 BCE), from the death of Alexander the Great to the reign of the Roman Empire, the number of exotic animals brought into Greece increased, but this does not seem to have influenced the development of collections. Alexander's campaign into Persia and elsewhere in Asia was the most significant Greek expansion of power and brought the Greeks into contact with little known lands and animals. Alexander had many of these animals sent back for Aristotle to study, but there is as yet no known information on whether the animals remained alive for an appreciable period of time, or if they were maintained in collections. Hellenistic monarchies established after Alexander's death in Europe, Egypt and Asia maintained animal collections, but there is little information to indicate if these were very extensive, except for the collection at Alexandria in Egypt. Ptolemy II (283–246 BCE) established the largest and most varied of these Hellenistic collections at Alexandria. Egyptian Ptolemies were the wealthiest, most powerful and most influential of the Hellenistic monarchies. After taking over the Egyptian dynasties, Ptolemaic rulers established many excellent cultural institutions at Alexandria, making the city one of the centers of ancient scholarship. Unfortunately, while ancient scholars and travelers often mentioned the animal collection, no details were provided. This collection still existed when Egypt became a Roman province in 30 BCE.

The Roman Republic (around 509–27 BCE) began as a city-state in Rome and eventually occupied the entire Italian peninsula. Early Romans were content with their native soil and wildlife, having no desire to conquer other territories or to import exotic wildlife. During this time hunting and baiting native wildlife were popular. In baiting, animals (usually dogs) were used to attack another animal (such as a bear) that was tethered. This form of entertainment remained popular through the Medieval period. Deer, wild goats, boars, bears and bulls were often hunted and used in baiting. The first known exotics in the Republic were four Indian Elephants captured in the battle at Heraclea against King Pyrrhus, and later marched in the procession celebrating the victory (280 BCE). Early Republic hunts and processions evolved into increasingly elaborate spectacles which included religious ceremonies, triumphant marches, special events and the *venationes* (public

Figure 17: Madonna with the child and a Ringneck Parakeet (*Psittacula krameri*). Engraving by Martin Schongauer in the 1470s

games with animal vs. animal and human vs. animal combats). Extravagant spectacles began in the late Republic period, but did not reach their famous notoriety until the Imperial period (27 BCE–476 CE). Historians and many others have written extensively about these spectacles because of the large number of animals displayed and killed; however, little has been written about the collections from which these animals came.

In addition to these rather well-known uses of animals, more humane and aesthetic uses of animals existed among the wealthy class, civil administrators and emperors. Both native and exotic wild animals could be found in villa gardens, ponds, bird enclosures, cages, large parks and hunting reserves. The largest of these collections belonged to the emperors, and these collections grew as the Roman Empire spread throughout Europe, Africa and Asia. Expansions brought the Romans into contact with new species of exotic wildlife and with existing foreign collections. As has been previously noted, Ptolemy II's collection still existed when the Romans overtook Alexandria in 30 BCE. Roman armies also encountered a large collection in Babylonia when conquering it in 363 CE.

Public entertainment included itinerant acts with performing animals, but interest was directed to the more exciting processions and public games which the emperors, provincial administrators and sometimes wealthy citizens provided. After the initial appearance of elephants in the third century BCE, exotics began replacing native species in spectacles. Some of these early exotic species included lions, leopards and ostriches. Over time, a great many birds were seen, including ostriches, eagles, cranes, storks, herons, ibises, swans, pelicans, Indian parrots, Peacocks, Guineafowl (*Numida meleagris*), pheasants, partridges, quail, geese, ducks, pigeons, doves, crows, ravens, and numerous songbirds. Many of these were kept in cages as pets.

The emperors surpassed the Republic's grandeur in every way, including private collections and public spectacles. The spectacles were presented throughout the Empire; the most extravagant were those the emperors provided at Rome. Emperors and administrators had the best access to wild animals, but the upper classes had some access as well. The ability to obtain wild animals, especially exotic species from distant lands, depended on an individual's wealth and influence, and it was exotic animals that provided the highest status. *Vivarium* is one of several Latin terms that have given rise to modern terminology. Although popular in the nineteenth century, *vivarium* is seldom used today, while other terms such as *aviary*, are still used.

Various professions were associated with keeping wild animals, including dealers, performers, animal slayers (those who fought animals in public games), trainers, keepers and veterinarians. In addition to the state-owned stockyards, collections were maintained in a variety of small enclosures, large parks, hunting reserves, flight cages and fish ponds. Ordinary citizens commonly maintained freshwater fish ponds, while saltwater fish ponds (which were expensive to create and maintain) were a hobby of the nobility. Large, dangerous animals such as lions, leopards and bears were sometimes kept in cages or as pets in the houses of nobles and in the imperial palaces.

Rare species, such as tigers, cheetah, zebra, giraffe, rhinoceroses, hippopotamuses and crocodiles were usually only found in the imperial collections. Others, such as elephants, were only exhibited during the spectacles, and no more than one or two elephants are known to have been privately owned. As the Western Roman Empire declined, so did the public spectacles. However, they continued in the Eastern (Byzantine) Roman Empire. The spectacles of Constantinople were not the extravagant affairs they had been in Rome, and in time they declined as well. Disintegration of the Roman infrastructure and empire included the loss of animal collections; however, monarchs, monasteries and municipalities continued to keep wild animals in post-Roman Europe. These European collections continued through the Middle Ages to emerge as menageries in the Renaissance period.[83]

Persia and Arabia

Alexander the Great did not conquer Persia, now Iran, until 331 BCE, but the Greeks already knew of their *paradeisos*, the wonderful gardens of the Persians, which existed since at least 546 BCE. These gardens have been later defined as re-creations of the Garden of Eden, a state of supreme bliss, an ideal place. Essential elements of these gardens were geometric designs, water, trees and flowers. Animals were not an essential element, but were sometimes included, especially in the larger park-like gardens. The influence of these gardens was widespread, impressing both those who conquered Persian territories, and those whom the Persians conquered. These gardens also influenced Islamic garden design as it spread throughout Asia, northern Africa and Spain after 622 CE. Egypt became a province of the Islamic Caliphate (641) and the ruler required tribute from its southern regions, including camels, elephants, giraffes and other animals. Animals were also captured on the Caliph's hunting trips and military excursions in these areas. These animals were kept in the Caliph's animal collection at Cairo, occasionally to be used in processions or presented as gifts to foreign dignitaries in Europe and Asia.

During this time, Arabic science was probably the most advanced in the world, especially in astronomy, mathematics, alchemy and medicine, but natural sciences were not considered important. Despite this lack of interest, animal collections were part of the royal courts throughout the Muslim world. Abderrahman III (912–961) established an animal park as part of his new city, Zahra (north of Cordoba, Spain), in 936. Here he kept animals in cages and fenced enclosures. Large collections also existed in Baghdad and Cairo, as well as in Byzantine Constantinople. These collections continued through the Middle Ages and into the European Renaissance period and animals were exchanged between rulers independently of faith. Ottoman Sultans continued the collections after conquering Constantinople in 1453, exchanges between rulers continuing through the nineteenth century.[84]

Medieval Europe

The Medieval period (476–1453) in the western parts of Europe extended from the decline of the

Western Roman Empire through the emergence of the European Renaissance with its accompanying age of exploration and global influence. Provincial animal collections were dismantled or abandoned when the Romans withdrew, but monarchs, monasteries and municipalities filled the vacuum resulting from the collapse of the empire.

Rulers throughout Europe along with a wealthy aristocracy provided a continuity and structure to civilian life and government, and monasteries became the spiritual and political centers of the Church, influencing the community. The most important of these monarchies, monasteries and municipalities had animal collections like the previous Roman administrators and rulers.

Charles the Great (Charlemagne, 742–814), founder of the Holy Roman Empire, had royal collections at several of his estates. These collections included elephants, lions, bears, camels, monkeys and birds, especially falcons. During the Middle Ages falconry was a popular sport for royalty. Some of these animals were sent to Charles the Great from collections held by other monarchs in Europe and elsewhere, as such gifts were a form of recognition and signs of friendship.

When William the Conqueror ruled Britain (from 1066) he seized the existing game reserves, forming the first British hunting forests and parks. At his manor, Woodstock, he started a collection of exotic animals.

His son, Henry I, enclosed Woodstock and enlarged the collection, which included lions, leopards, lynx, camels and an owl considered to be rare. In 1235, during the reign of Henry III, animals from this collection were moved to the Tower of London. Frederick II (1194–1250) of the Holy Roman Empire had an extensive collection at his Palermo residence in Italy. He kept a particularly good collection of falcons and was an authority on falconry and bird biology. Based on his experiences, he wrote an authoritative text on the subject, *Über die Kunst, mit Vögeln zu Jagen* (*On the Art of Hunting with Birds*, or *The Art of Falconry*), which has been reprinted many times to the present day. Frederick II also had an animal collection that included elephants, giraffes and white bears.

Some of his animals were obtained in exchanges with other European monarchs and with Muslim rulers in India, from whom he acquired an elephant, Egypt, with whom he exchanged a white bear for a giraffe, Spain and Byzantine Constantinople. Louis IX of France (1214–1270) brought back an elephant when he returned from the Crusades. Philip VI (1293–1350) created an area at his Château de Louvre for lions and leopards. Charles V (1337–1380) moved this collection to St. Pol and enlarged it. He particularly liked birds and kept many of them in gold and silver cages in every room of his Château. He also kept caged Nightingales (*Luscinia megarhynchos*) in his garden trees and had a large flight cage for other birds. Many of the French nobility had animal collections as well. A collection of René, Count of Anjou (1440s), was one of the largest. His estate included a lion house, a small mammal house, flight cages for birds, ponds for water birds, ostriches, camels and elephants.

The Medici rulers of Florence, starting in 1434, maintained animal collections for many generations. Some of the Popes had collections as well, which were housed at the Vatican. An early collection of Pope Benedict XII (1285–1342) contained ostriches. Pope Leo X (1475–1521) kept tropical birds, lions, leopards and other cats and several mammals. His prize possession was an elephant named Hanno, which the King of Portugal, Dom Manuel I (Emanuel I, 1469–1521), had given to him. Collections of the Italian princes were the first to flourish as the Renaissance began to sweep across Europe. By the fifteenth century, animal parks, as well as the earliest botanical gardens, could be found throughout Italy. In the north of Europe, Sweyn II, King of Denmark (1047–1075), was known to have an animal collection. William IV, Count of Holland (1350s), kept lions, bears and falcons on his estate at The Hague.

About the same time, the Dukes of Guelre had a collection at their castle, Rozendaal, near Arnhem. Prosperous Dutch citizens also maintained collections which whenever possible included royal animals (that is, those considered emblematic symbols) such as lions and eagles.

In addition to the monarchs, monks and nuns at monasteries usually had a keen interest in gardening and small animal husbandry. These activities had practical purposes, but sometimes served ornamental or aesthetic purposes as well. Monks at St. Gallen, Switzerland (early to mid-800s) maintained one of the better known animal collections, which consisted of animal houses, outdoor paddocks and keepers. Municipalities also provided their citizens and visitors with views of animals, which were kept in moats no longer needed for defensive purposes, as well as in pits and cages. Municipalities often kept wild animals that appeared on their coats-of-arms, such as eagles, lions and hoofed stock; today the city of Bern in Switzerland still keeps bears.

Throughout Europe the population at large, while not allowed to view collections of the privileged, saw animals when they appeared in itinerant animal acts during fairs. Travel accounts also added to the information about exotic animals: Marco Polo's travels to the Far East around 1271–1295 made the Yuan (Mongol) Dynasty and its emperor Kublai Khan known in Europe.

As a state servant, Marco Polo had ample opportunity to observe the emperor's estates and to travel throughout the country. As was popular with earlier emperors in China, the largest collections existed in the royal parks. The Great Khan's royal park was rich with meadows watered by many rivulets, where a variety of animals had their pastures.

Hawks and other birds were employed in the chase and their pens were also on the grounds. The number of these birds was more than 200. The Great Khan also had a large collection of falcons for falconry, requiring 10,000 falconers to care for the birds. Marco Polo also saw the great parks of regional rulers, which were stocked with animals, but unfortunately he says very little about them.

In turn, Grand Eunuch Zheng He made China's presence better known in Asia and Africa with his Indian Ocean expeditions (1405–1433). These trading expeditions were short-lived, however, because court officials did not hold the merchants, who encouraged the commercial ventures, in high esteem and the expedition costs were felt to be an unnecessary expense to the state. Zheng He was the admiral of these expedition flotillas, which involved 62 ships, over 100 auxiliary vessels and almost 30,000 persons. The ships called at ports in Indochina, Indonesia, India, Ceylon and the eastern coast of Africa.

Included in the fleet were "horse-ships", normally used to transport horses and domestic stock, but they carried wildlife obtained during the expeditions. Spectacular exotic species obtained during these expeditions included giraffe, lions, rhinoceros, oryx, leopards, ostriches and zebra.

Europe emerged from the Middle Ages into a continent of nation-states with increasing power, wealth and influence during the Renaissance. Relatively small and scattered royal, monastic and municipal collections from the Medieval period began to increase in size and numbers.

These collections were to become known as menageries and with collections throughout the world, eventually evolved into zoological parks, aquariums and aviaries.[85]

Figure 18: A white gyrfalcon attacks a tundra swan (*Cygnus columbianus*). Scroll painting on silk by Yin Xie, Ming dynasty (15th– 17th centuries)

Chapter 3. Icelandic Trade with Gyrfalcons

From Medieval times to the Modern era

Sigurður Ægisson

Falconry, or hawking, is the art of training birds of prey for the taking of wild game. Its origin is lost in the shadows of antiquity, but it is often thought to have begun in the Near, Middle or Far East, and to be at least 4,000 years old, as there is some evidence that trained falcons were flown at prey in China around 2000 BCE.

This is a point of dispute, though, as the first artistic images come not from there but from Asia Minor. Several carvings from around 1500 BCE show a large bird on the fist of a human figure. Grasped in the same fist is the figure of a hare (presumably the quarry) held by the back legs. The earliest record, on the other hand, is said to be an Assyrian (northern Iraq of today) bas-relief dated 722–705 BCE. It shows a hunting scene where a bearded man, carrying a falcon, is apparently in the process of throwing it off his right wrist.

According to the Greek physician and historian Ctesias of Cnidus, who for 17 years around 400 BCE served at the court of Artaxerxes II Mnemon, "the pygmies", a mysterious people inhabiting some part of Inner Asia, hunted the hare and the fox not with dogs, but with ravens, kites, crows and eagles. From his description and silence regarding the Persians, falconry apparently was not known in Persia, strangely enough, as this region is cited by many as the cradle of falconry, with a background of some 8,000–10,000 years.

The first person to use a raptor in this way may have seen in it an opportunity of catching fresh meat, perhaps after a stay beside a nest, taking a share of what the birds brought to their chicks. At least this was to become one of the main functions of the bird, especially before the invention of guns; it was partly a method of hunting.

Another idea involves the role of birds of prey as guides to the spirit world in shamanistic cults.

Falconry in Europe

How and when falconry reached Europe is unknown, but there are sources claiming it was known already in the time of the Roman Emperor Vespasian (9–79 CE) and even earlier, as there are some references to Julius Caesar (100–44 BCE) using falcons to destroy pigeons carrying messages. It may have been known in a crude form to the ancient Greeks at the time of Aristotle (384–322 BCE), who makes two references to it in Thrace.

Others say that true falconry appears to be first mentioned by Julius Firmicus Maternus, the Sicilian lawyer, Christian Latin writer, and notable astrologer. He states in one of his books, from around 330 CE, that persons born when Venus is in Aquarius will be addicted to hunting and hawking, and to the breeding of hawks and hounds for these exercises. The growing number of references to falconry in western Europe from the fifth century onwards, suggests the increase in interest may have been due to barbarian influence. The earliest known European pictorial documentation is a floor mosaic at Argos in Greece, dated at around 500 CE.

Information on hawking, as well as the oldest data on falconry among the Germanic tribes who overran western Europe, is found in their early laws, which date from the sixth century, called the Salic Laws. Called the Salian Franks or Salii, a subgroup of the early Franks, they date from the sixth century. These laws and others, such as the Burgundian Laws (500–505 CE) and the

Laws of the Ripuarian Franks (530–570 CE), show that falconry was a well-established form of hunting. In Denmark sixth-century documents record that Rolf Krake and his men, on a visit to King Adils in Uppsala, each carried a falcon on his shoulder, and archaeological evidence in Sweden seems to confirm the introduction of falconry at that point in time. By the seventh century, the sport was popular enough in France for trained hawks to gain value, and was already familiar in Ireland by late Celtic times, in the seventh century. By the next centur,y or probably earlier, it had reached England. Icelandic sagas mention falconry in Norway during the Viking Age (793/800–1050/1066/1100).

This sport became part of the Medieval education and most kings and members of the aristocracy took an interest in it. In many courts the office of Royal Falconer was created. In Wales. the Master of the Hawks, or Penhebogyd, held fourth place at court and was honored with certain privileges. "But," as one source puts it, "this proud distinction was not without its drawbacks, as this high official was only allowed to take three draughts from his horn, lest his brain should not be as clear as it ought to be, and the precious birds might be neglected." The stock of the Icelandic Gyrfalcon was for centuries the property of the king of Denmark and he often used the falcons as diplomatic cement, to curry favor with other heads of state.

After 1000 CE, vast sums were spent on falconry establishments and no bird was a greater favorite in regal and imperial circles than the Gyrfalcon, (*Falco rusticolus*), especially a female; it was much bigger and stronger than the male, and could overpower prey as large as the Capercaillie (*Tetrao urogallus*), Snowy Owl (*Bubo scandiacus*), heron (*Ardea* sp.), swan (*Cygnus* sp.) and even the Common Crane (*Grus grus*). The Gyrfalcon's legendary status spanned civilizations as well as continents, its call echoing down the corridors of the Alhambra palace and fortress complex of the Moorish rulers of Granada in southern Spain, as well as in the palaces of the Mongol emperor Kublai Khan, who was said to have 200 of them; there, as elsewhere, they were most commonly used in hunting, but they were also known to be reliable and fast messengers for the delivery of letters.

The color of the birds was, however, important. A "ger-fauk that is milke white," as described in the fourteenth-century English romance *Guy of Warwick*, was the goal of every king. In 1396, during a Crusade, Ottoman Sultan Bayezid I captured the son of Philip the Bold, duke of Burgundy. The duke offered 200,000 gold ducats for the return of his offspring. The sultan demanded instead, and received, something even more valuable: Twelve white Gyrfalcons. According to Charles d'Arcussia, the author of *La Fauconnerie* in 1598,

Figure 19: Two medieval falconers. Detail of two falconers from Frederick II's work *De arte venandi cum avibus*, 1240s

to own a falcon of that type resulted in disappointment with all other hawks.

Various circumstances in the eighteenth and nineteenth centuries, such as the development of firearms, the enactment of land reforms and the evolution of an urban, industrial society, combined to bring about the gradual decline of falconry in Europe.[86]

When did it begin?

It is not certain when the export of Gyrfalcons from Iceland began, but there is some evidence to suggest it may have been around the year 950, or a little later. The locals probably knew by their association with foreigners that important (and wealthy) persons in England, France, Germany and even further afield found these birds highly desirable. Possibly the Icelanders saw in Gyrfalcons a valuable trading commodity – if they had not known this before.[87] Among others, the brothers Þórólfur and Egill Skallagrímsson from the farm Borg at Mýrar were in the service of Athelstan the Victorious, king of England, during 936–938.[88]

After the peace treaty between England and Wales was signed in 927, every year the latter had every year to hand over to the king of England: "20 pounds of gold, 300 of silver, to add 25,000 oxen, besides as many dogs as he chose, which could discover with their keen scent the dens and lurking places of wild beasts, and birds which were trained to make a prey of other birds in the air."[89] From this we may deduce that the king practiced hunting with these birds, as did his predecessors for at least two centuries, and that Icelandic Gyrfalcons would have been at times a topic of conversation.[90] Relatively few species of birds of prey were used by the English kings; the main varieties before 1307 were Gyrfalcon, Peregrine (*Falco peregrinus*), Lanner (*F. biarmicus*), Goshawk (*Accipiter gentilis*), and sparrowhawk (*A. nisus*); the saker (*Falco cherrug*), Hobby (*F. subbuteo*), and Merlin (*F. columbarius*) being used far less frequently. The Gyrfalcon was the most highly valued, here as elsewhere.[91]

Grágás, the Law Book of the Icelandic Commonwealth (930–1262/1264), stipulates that it is forbidden to hunt Gyrfalcons, swans and geese on land belonging to others.[92] This indicates that the capturing of Gyrfalcons was seen as a privilege and that the birds were regarded as a trade commodity.[93] It is likely that this provision was made law not long after the establishment of the Icelandic parliament, around the mid-tenth century (when laws were not written down, but instead memorized by an elected Lawspeaker).

There is good reason, and sources point to this direction, to believe that it occurred at the instigation of the previously mentioned chieftain Egill Skallagrímsson. He is also the first Icelandic poet, as far as we know, to use the word 'haukur' (hawk, meaning Gyrfalcon) and discuss the handling of this bird by the aristocracy, as an inspiration for poetic 'kennings' and figures of speech, to denote the word "hand".[94] Also, at this time the Icelanders had a large fleet of excellent ships and voyages between Iceland and neighboring countries were a simple matter.[95] In short, Icelandic landowners had the sole right to hunt falcons on their own land and could enjoy that bonus just as freely as they did the other benefits and bounty of their estate. Their counterparts in other countries, however, did not have the same right of possession and had to bow to the higher claims of their rulers in the business of both hunting birds and selling them, as indeed they had to do in other matters.[96]

Precious birds

Moving into the eleventh century, it was claimed at the parliament that Gyrfalcons from Iceland would be a suitable, handsome gift for

Ólafur Haraldsson, King of Norway.[97] An Icelandic Gyrfalcon as a present from a king is dealt with in an epic poem as early as 1024.[98] During that century Norwegians began to monopolize Icelandic foreign trade and continued to do so for the next three centuries.[99] Emma Ford relates the following story in her book *Gyrfalcon*:[100]

"During the siege of Acre in Syria, in AD 1104, the crusading king, Philip Augustus of France lost a white Greenland Gyr from her perch when her leash broke. The falcon flew off to the city walls, whence she refused to be recalled. Eventually the falcon was captured by Saladin's people and taken into the city. Hearing of his falcon's recovery and desperate for her safe return, Philip sent a succession of envoys to Saladin. All were dismissed. Finally, Philip offered the vast sum of 1,000 gold ducats. The gesture was in vain. Saladin spurned the offer and kept the Gyrfalcon."

This excerpt shows the esteem in which the noble bird of prey was held; it is possible that the bird in question came from or through Iceland. The oldest known sources describing the export of these birds is, however, from Lincolnshire in England and found in the financial accounts compiled by Henry II's financial adviser and treasurer. In 1169, a man named as Hemelinus Decanus is said to owe the king one Icelandic Gyrfalcon ("Girfalconem Islandensem") and during the next few years this same debt is referred to. In the account register for the year 1177, however, it is revealed that the man had paid for the bird,[101] the ship having probably sailed first to Norway and then south with the precious cargo.[102] A little later, or around 1185–1186, the court priest to Henry II, Welshman Giraldus Cambrensis (Gerald de Berry), compiled his work *Topographia Hibernica* in Dublin, where he stayed at the request of the king as the attendant to Prince John. He states that in Iceland there are excellent, large Gyrfalcons which are being exported.[103] Here it might be noted that both the king and his son John were avid enthusiasts of falconry and followed this pursuit with great competitive spirit.[104]

Figure 20: Stephen, King of England 1135–1154, feeds a white gyrfalcon. From the Chronicle of England, by Peter de Langtoft, ca 1307–1327

In Iceland the Church, at the time Roman Catholic and the owner of large areas of land, had the same right of possession as other landowners and was the sole hunter of the Gyrfalcon within its own boundaries, according to the before mentioned article in *Grágás Book of Law*. Also the Episcopal See at Niðarós (now Trondheim in Norway) had the same rights as any other landowner. Páll Jónsson, bishop of Skálholt from 1195 to 1211, sent the birds as gifts to friends abroad. In other countries the sole right to hunt falcons was possessed by the rulers; this also applied to

Norway, where the king owned all the Goshawks and falcons which nested in the mountains. Although it was banned to hunt Gyrfalcons on land owned by others in Iceland, this activity was probably free to anyone to practice in common grazing areas and in the mountains.[105]

In the wake of the Crusades in the twelfth and thirteenth centuries, rulers in Europe began to take an increasing interest in falconry[106] and in the thirteenth century at least 17 instances of Gyrfalcon export from Iceland had their destination in England.[107] For example, in 1223 or 1224 the old Norwegian King Hákon IV (Hákonarson) sent his emissaries to Henry III, King of England, with six Gyrfalcons and the promise of more birds, when his men made the next trip to England with birds from Iceland. In return Hákon asked that the king of England protect Norwegian traders visiting England from illegitimate procedures, and English traders to Norway would be protected in the same way.

One year later, Hákon once again sent an emissary to Henry III and explained to him that two years previously, he had sent men to Iceland to hunt birds for the king. These men had been obliged to put up with unbelievable hunger and cold while sailing the Arctic Ocean. Indeed, the fowlers had just recently returned to their homeland with their quarry.

Therefore he could now present the king with 13 Gyrfalcons – ten grey and three white – and ask that these be received in the spirit in which they were given. And the letter goes on: "... if you place such value on this offering as did your father and predecessors, who were said to regard Iceland Gyrfalcons as more precious than gold and silver, then you will not hesitate to hold this last gift in due esteem, since we had, as before, to go to such lengths and overcome countless difficulties to secure these birds."[108]

King Hákon IV sent Gyrfalcons as gifts also to other rulers, e.g. the Sultan of Tunis and Frederick II of Hohenstaufen,[109] the powerful Holy Roman Emperor in Sicily, who was also a first rate naturalist. Frederick's book, *De Arte Venandi cum Avibus* (Art of Falconry), which is assigned to the last part of his life around 1241–1250, took thirty years to compile. The book distils the knowledge of falconry of many cultures, and immediately became one of the basic sources of reference dealing with the subject. In one place it states: "Gerfalcons are fledged in or near the most distant parts of the seventh climatic zone... Some of them are brooded on the high cliffs of the Hyperborean territory, particularly on a certain island lying between Norway and Greenland, called in Teutonic speech Iceland ("Yslandia") ... These falcons are the best birds for hunting."[110]

These words had great influence on the other European rulers and even further afield and did much to kindle and intensify the fires of interest in Gyrfalcons.[111] The Emperor was also the overlord of the town of Lübeck, in northern Germany, and twelve Gyrfalcons from Iceland were a kind of annual duty the town had to pay him. In 1378 it was the business of Lübeck merchants to bring Gyrfalcons alive from

Figure 21: The falconry garden (Falkonergaarden) in Frederiksberg, Copenhagen. Drawing by J. L. Ridter, 1900

northern territories for sale, through Venice, to Alexandria in Egypt.[112]

German historian Klaus Friedland has very interesting information about these matters, e.g. how big a role the export of Gyrfalcons played in Icelandic economy. In an article, "The Hanseatic League and Hanse Towns in the Early Penetration of the North", which appeared in *Arctic* in 1964, he says:

The trade in Icelandic hawks could be perceived as relevant only to the pleasures of the courts, and not to the penetration of the North. This would be only half the truth. Hawks, tercels, formales – and whatever else they were called in the late Middle Ages – are an indication of the approximate value of goods imported to Iceland, i.e. flour, timber, and wax. As these goods are mentioned fairly often in the literature of medieval economic history, without sufficient information concerning trade routes and destination, they are not a clear indication of the average demand for staple imports of the Iceland settlers. Hawks, on the other hand, provide such indications. The price for 12 hawks… was nearly 350 aurei; this was the value of approximately 50 tons of cereal grains, which would mean the livelihood of 100–200 people a year. Given that the 12-falcon duty was only a part of the annual Lübeck hawk import total, and given that the Lübeckers were not the only Hanseatic merchants who traded in Icelandic hawks, it appears that a considerable proportion of the Icelandic population secured their livelihood by importing food on the Hanse trade route.[113]

Greenland

This could be the opportunity to mention something else. To continental Europeans, Greenland was unknown until the tenth century, when people from Iceland settled on the southwestern coast. This part of Greenland was apparently unpopulated at that time. According to estimates, the first settlers numbered around 500. This became a significant colony, existing as two settlement areas – the larger East settlement and the smaller West settlement – and it traded in ivory from walrus tusks with Europe, as well as exporting rope, sheep, seals and cattle hides.

On the other hand, it depended on Europe (mainly Iceland and Norway) for iron tools, wood especially for boat building (although this it may also have obtained from coastal Labrador), supplemental foods and religious and social contacts. Ruins of 330 farms have been found, as well as several other buildings. At its peak, the East settlement population is thought to have numbered around 2,000–2,500 and the West around 500, although higher figures have been guessed at, or up to 10,000 in total.[114]

In 1126, a diocese was founded at Garðar (now Igaliku). It was subject to the Norwegian archdiocese of Niðarós (Trondheim). In 1261, the population accepted the rule of the Norwegian king, although it continued to have its own laws. Trading ships from Iceland and Norway (from the late thirteenth century all ships were forced by law to sail directly to Norway) traveled to Greenland and would sometimes winter there. The Norse settlements along the southwestern coast eventually disappeared in the fifteenth or perhaps early sixteenth century.[115]

A few scholars have expressed the opinion or at least suggested that the White Gyrfalcon (the *candicans* "type") was caught in Greenland, not Iceland. There is little solid evidence to support such claims (see Emperor Frederick's book), due to a lack of sources or perhaps because it is not true.[116] The author of *Speculum regale* (The King's Mirror), which is thought to date from around 1260, describes animal life in Greenland thus: "There are… many large hawks in the land, which in other countries would be counted very precious – White Gyrfalcons, and they are more numerous there than in any other country; but the natives do not know how to make any use of them."[117]

English records do not mention falcons from Greenland; they refer only to "White Gyrfalcons", but they often mention Gyrfalcons from Iceland.[118] Arabic sources from the thirteenth century refer to the "island of the white hunting falcons", which lies from the west to the east, and is about seven days' travel in length and a four days' journey in width. The sources obviously point to Iceland, "and from it are obtained the white falcons ... [for] the Sultan of Egypt, who pays from his treasury 1,000 dinars for them."[119]

Certainly it is not impossible[120] that vessels sailing between Greenland and Europe transported Gyrfalcons as well as other types of valuable cargo, or that ships brought them from Greenland to Iceland to be collected together in some kind of aviary. After all it was only a journey of four days between the two countries, although this was not without danger once the vessel had rounded Hvarf Point, heading north to the East and West Settlements, due to sea ice and submerged skerries.

The trip continued with its proceeds to Europe when conditions permitted.[121] According to the ancient source *Grænlendinga þáttur* (Greenland Saga), early in the twelfth century there were many foreign traders present in Greenland (about 1135–1136, e.g. on three to four vessels) and Icelanders numbered among those visitors.[122]

Some facts, however, appear to contradict this claim, among other things the complete lack of information by contemporaries. Another strange fact is that while foreign kings were said to have been presented with Polar Bears (*Ursus maritimus*) both during this and other historical periods, no mention whatsoever is made of anyone else receiving the gift of falcon from Greenland.[123] There could be some kind of conspiracy here, as in the case of the tusk of Narwhals (*Monodon monoceros*). To hide the origin (until the seventeenth century the tusk was talked of as the horn of the unicorn) meant to keep prices up; this is however unlikely, as the two commodities are too dissimilar to warrant comparison.[124]

It may also be mentioned here that in spite of an abundance of Arctic Foxes (*Vulpes lagopus*) along Greenland coastlines, the skins of these creatures, which can be white or blue or at times a mixture of the two, were never referred to as goods worthy of export. Such a possibility did not appear to have occurred to the traders until the sixteenth century and was brought to attention for the first time as a scheme of Erik Walkendorf, Archbishop, to kindle the interest of his royal master and friend Christian II in the Greenland settlements, and not least in the excellent commodities to be found there.[125]

In the previously mentioned letter written in 1225 by Hákon IV to Henry III, the Gyrfalcons were said to be captured in Iceland; had they been brought all the way from Greenland, it would have been even better for King Hákon. He would certainly not have concealed this piece of information or failed to emphasize how much he had endured to collect these treasures.

Few researchers take seriously the account in *Króka-Refur Saga,* which is supposed to occur in the tenth and eleventh centuries, but is in fact an exaggerated account from the fourteenth century. Sveinn Úlfsson, King of Denmark had been presented with, among other things, a gift of fifty Gyrfalcons, of which fifteen were white and all of them emanating from Greenland. Possibly we actually hear an echo from the previously mentioned *Speculum regale* (The King's Mirror), but it is still an indication of the value of the birds and other assets, which could be be found there.[127] In the same way, *Grænlandslýsing* (Description of Greenland, after 1360), attributed to Norwegian priest Ívar Bárðarson, says towards the end of the manuscript that White Gyrfalcons are among the many treasures found here, meaning perhaps that they are one of the many trading commodities used by the Greenlanders.[128]

However, this part seems to be an addition from a later time;[129] the original text, probably Norwegian, is lost, but the Danish translation from the sixteenth century has survived along with many other manuscripts.[130]A few Gyrfalcon bones have been found among osseous material excavated at Norse ruins in the East settlement,[131] but what it means exactly is impossible to say; the bird(s) could simply have been used as food. Or as Eugene Potapov and Richard Sale put it in their book *The Gyrfalcon*: "Gyrfalcons are known to have been eaten by the Vikings… The excavated fragments do not appear to be wholly indicative of the Viking menu, though they do clearly show that the Gyrfalcon was not a taboo bird for the Norse settlers."[132]

The Gyrfalcon population in northeastern Greenland (*candicans* "type" birds) is thought to track the four-year cycle of their lemming prey. This population is migratory, main wintering grounds are situated further south in Greenland and important winter prey include Rock Ptarmigan (*Lagopus lagopus*). The Greenland Rock Ptarmigan population is also cyclic, peaking every ten years, but not in synchrony with the Icelandic ptarmigan population.[133] According to Danish ornithologist Finn Salomonsen, in summer the gyrfalcon frequents particularly the interior of the country, but in migration times and in winter it is commonly seen in the coastal areas. The migration in the northern parts of the eastern coast (north of Scoresby Sound) takes place mainly from late August to the end of September, the latest record in Hold-with-Hope Land being 6 October, Germania Land 1 November and Franz Joseph Fjord 27 November; the two latest are thought to have been exceptionally late birds.[134]

In Scoresby Sound the migration lasts from early or mid-September to late October or early November. In the first eighteen days of September in 1937 between 200 and 300 white Gyrfalcons were observed passing Scoresby Sound, where already a few Gyrfalcons had begun their winter sojourn. In Angmagssalik District, migrants arrive from September to December and a large number spend the winter there.

A number carry on to Iceland, where they are regular winter visitors.[135] Salomonsen adds that in years when ptarmigans are very scarce, the greater part of the Gyrfalcons vacate Angmagssalik District about New Year when the fjords freeze over.[136] He does not say where they go. It is difficult to assess whether the gyrfalcons come directly from Scoresby Sound to Iceland, or later from the area around Angmagssalik. Researchers Potapov and Sale appear to be of the former opinion.[137]

In the Middle Ages it was believed that the white falcons were chiefly carried to Iceland by the sea ice.[138] In latter times, however, this notion has been viewed with scepticism.[139] Is it not possible that during conditions as those described around Angmagssalik, a large proportion of the birds headed across the Denmark Strait, instead of migrating south? In any case, during the winters, when white and "half-white" Gyrfalcons (cf. later; those are in fact also thought to have had, at least for the most part, their habitat in Greenland) were most numerous in Iceland, sea ice happened to have drifted to its shores, or the country experienced frequent spells of heavy snowfall with extreme cold. A multitude of sea birds of many varieties are generally to be found at the ice edge. One source describes the climatic conditions in 1744 and 1745 as follows: "The winter was rather mild, but frosts in the spring. Just before Christmas, sea ice drifted close to the shores and [at the beginning of 1745] merged in a frozen expanse with land-based pack ice. Early February was greeted with a harsh wintry spell with frosts of unprecedented severity. The sea ice pressed against the entire north coast, drifting into every fjord, and virtually the entire island was surrounded by ice, even the southernmost part. The ice brought ashore 4–5 Polar Bears in the north and 5 bears were killed in the south-eastern district of Skaftafellssýsla".

In 1745, 21 white Gyrfalcons were captured and 11 half-white ones. The year 1751, in which a new record was set when 42 white Gyrfalcons were caught and 11 half-white birds, is thus described in the same source: "The period after Christmas [1750] brought the most severe frosts lasting well into February.

The northeast of the country suffered the greatest hardship with farming stock perishing *en masse* and many districts were deserted by all their inhabitants. People in the west also had to endure spells of biting frost." In 1752, when 18 white Gyrfalcons and 12 half-white were captured, the story goes: "The weather became harsher at Easter with heavy falls of snow. The winter was accompanied by such severe freezing that frost cracks were discovered in grassy valleys during the following summer, some of them 40 fathoms (c. 250 feet or c. 75 meters) or more in length. These conditions caused mishaps in many locations, loss of livestock and widespread hunger in numerous districts."[140]

A total of 31 non-breeding white Gyrfalcons were seen in Iceland during the period 1979–1992 (1–4 annually), and all observations were made in autumn or winter (October–April), except for one from May and one from June. Most of the records came from the west or north.[141] They continue to come; two were seen on 6 December 2008 and 12 January 2009 in the Reykjavík area.

Fighting over the falcons

Iceland was an independent state for more than three centuries, but after decades of conflict, as several clans fought for power, causing havoc, turmoil, and chaos, it entered into a treaty, establishing a union with the King of Norway in 1262–1264.[142] In reality not much changed, but a new law book (*Járnsíða*) compiled for the Icelanders in 1271–1274, was adopted. However, it was deemed unsuitable, since it had its foundation in Norwegian law. The law book contained no provisions for the hunting of Gyrfalcons and the laws of the Commonwealth (*Grágás*) remained unchanged. In 1279 Árni Þorláksson, Bishop of Skálholt, was commissioned by the Archbishop, either as a safety measure or emanating from suspicion engendered by experience, to keep a strong hold on the privileges of the Church when dealing with royal officials regarding this matter.[143] Eventually the king had another law book compiled, which became valid in 1281 (*Jónsbók*). This event gave rise to acrimonious disputes between royal authorities and the Church with regard to the right to Gyrfalcon hunting. The king wanted unlimited free rights in this respect, which the Church regarded as a violation of its ancient prerogative. The conclusion of the matter was that the king was granted permission to hunt "on any man's land,... in return for payment, with the exclusion of estates belonging to the church." Thus the royal plans of appropriating the hunting of Gyrfalcons without compensation had come to nothing. A law amendment dating from 1314 stated that anyone living on a farm with land attached was entitled to two thirds of the remuneration and the hunter to the remaining third.[145] The king's hunting prerogative was further curtailed as a growing number of farmlands were acquired by the Church and the monasteries.[146]

Iceland remained under Norwegian kingship until 1380, when the death of Olav IV extinguished the male royal line. Norway and Iceland became part of the Kalmar Union with Sweden and Denmark; Denmark became the dominant power. The King of Norway had, as may be expected, chiefly sought Gyrfalcons in Iceland and in Norway, but after the trade was taken over by the king of Denmark, his main sources were Iceland and Sweden.[147] Through this period, the entire fifteenth century, or even longer, the history of the hunting of Gyrfalcons is to a great extent lost in the mists of time. But one thing is certain: Gyrfalcons were in great demand at the courts of Europe and the demand for them seems to have been supplied by the English, Dutch and German sailors who made their voyage to Iceland every summer to fish and trade. Whether the birds were trapped or taken from the nests, by foreign falconers or the local people, is not known.[148] Maybe both methods were used.[149] A number of white Gyrfalcons from Greenland spent the winter there, as already mentioned, mainly from December to April, a period of difficult sailing conditions. If foreign trappers were active then they had to stay during winter, which was not considered attractive because of food shortage; nor was this allowed. The capture of birds may have been practiced by the local population far back in history,[150] but this was to become the norm later.[151] No doubt an indication of the way people felt about this noble bird is the fact that it is sometimes drawn on contemporary maps, e.g. in 1539 on *Carta Marina* by Olaus Magnus, where it occupies the top of the Tröllaskagi peninsula, as it later came to be named, and below is written "FALCO ALBI", i.e. "white falcon".[152]

In 1538 Christian III of Denmark began to impose the Lutheran creed on his Catholic subjects in the middle of the North Atlantic, and in 1551 Iceland officially became Lutheran and all the lands of the Church came into royal ownership,[153] thereby giving rise to an entirely new situation. The kings of Denmark gradually came to understand what a significant resource they possessed in the Icelandic Gyrfalcon. During the reign of Frederick II (1559–1588), a regular institution of falconry was established, where falcons were tamed, trained and looked after by special servants who had the required skills.

All this was ruled over by a Grand Master of Falconry, Niels Ulfstand of Barsebäck, who took office on 13 October 1571.[154] The project, however, appears not to have been as successful as expected.[155] Foreign falcon hunters were given permission to capture birds (among the first was an Englishman, Henrik Getle, on 4 November 1580) in return for a specific toll. During the reign of Christian IV (1588–1648), his men were sent to Iceland specifically to purchase Gyrfalcons, especially of the white variety.

At the same time, Icelanders who possessed the necessary skill to capture such treasures, were invited to sell them to the king and in return a royal official would provide a financial grant to the seller. Furthermore, the king gave his friends from abroad permission to buy Gyrfalcons in Iceland exempted from the toll.[156]

In 1649, Frederick III (1648–1670) for the first time, sent his own men to hunt for Gyrfalcons. This turned out to be a troublesome expedition for various reasons.[157] In 1662 Crown Prince Christian, later Christian V (1670–1699), was sent on a trip around Europe. In France, he stayed for a considerable time at the court of young Louis XIV. Two years later, in 1664, inspired by what he had learned, he established an elaborate mews in his newly built "Falkonergaarden" in Copenhagen. The main purpose was to receive and distribute Iceland-caught Gyrfalcons as diplomatic gifts among the courts of Europe.[158] Around 1670, however, the practice of sending men from Denmark to Iceland to hunt gyrfalcons was discontinued. A new plan was adopted and a Master of the Hunt was appointed; a master-falconer, usually a foreigner.[159] The whole country was now divided into districts and to each an official trapper was assigned. No one was allowed to hunt in districts belonging to others. This became a lucrative profession in which some Icelanders were directly involved and known as "falcon captors". There were originally ten districts, but the number fluctuated; nor were they always of the same size, one was divided and others enlarged according to the occasion.[160] The royal representative, the itinerant Master of Falconry, in Icelandic sources usually referred to as the "traveling falconer", sailed to Iceland every summer with his servants and received the captured Gyrfalcons at Bessastaðir. From 1690 he undertook these voyages in a specially equipped vessel, the so-called "falcon ship". He returned to Denmark with his proceeds during the first half of August.[161]

The Icelandic trapping method

A remarkable narrative exists, dating from 1752, by Danish Niels Horrebow, who lived in Bessastaðir during 1749–1751, outlining the way the purchases were conducted along with other events. The following quote serves as an example (based on an Icelandic translation from 1966):

> As for the color of the Gyrfalcons, they are white, half-white and grey, but all are of the same species; people have even found all the color varieties in chicks occupying the same nest, as is generally known in Iceland. I can say for sure that almost every Gyrfalcon's nest in Iceland is known, for the falcon captors search them out, each in his own district so that they may plan their traps accordingly and lay out their nets near to the nets where the Gyrfalcons are most likely to be enmeshed. Apart from those nesting Gyrfalcons, occasional visitors arrive from Greenland, most of them white. The trappers call them 'flying Gyrfalcons' since they do not nest in Iceland. It is a matter of general knowledge that Icelandic Gyrfalcons are the best and most efficient of all hunting falcons. Whereas Norwegian Gyrfalcons only last one year for hunting, the Icelandic Gyrfalcons carry on for 12 years or more, not to mention the fact that they are both larger and more precious than the others...
>
> The itinerant master of falconry comes... every year accompanied by one or two attendants to Iceland on the ship from Copenhagen and stays at Bessastaðir where the king's falcon aviary is located. But he does not hunt a single Gyrfalcon. In every district in Iceland there are falconers who have been granted permission by the regional governor, each in his district where no one else is allowed to hunt Gyrfalcons. These are all Icelanders and they make a handsome profit from the activity when things go well. About Midsummer's Eve those men ride to Bessastaðir with their catch. Each horseman can carry 10–12 falcons. The birds are fitted with hoods reaching down below eye level and are tied to a crossbar fixed to a pole which the man holds in his right hand in the manner of a battle flag, letting its lower end rest in the stirrup. The master's task is to assess the falcons, accept the useful ones and reject the others. Then he takes care of their transportation to Copenhagen. The fowlers are given a receipt for the birds they deliver and then the treasurer pays them 15 Danish rd. (rigsdaler) for a white falcon and 10 rd. for a half-white bird. In addition they sometimes receive 2–4 rd. as a prize for the capture. The price of grey falcons has been 5 rd., but a few years ago the king graciously increased this to 7 rd.

Horrebow describes the hunt itself as follows:

> Two poles are driven into the ground a short

distance apart. A Rock Ptarmigan or Domestic Pigeon [*Columba livia*] is tied to one of them, or, if neither of those is available, a hen [*Gallus gallus*] will serve the purpose, with a string six or eight feet in length which is attached to the bird's leg so that it may flutter a little in the air to help the Gyrfalcon to spot it. Another piece of string, about 80 fathoms long (c. 500 feet or c. 150 meters) is also tied to the bird's leg. This string is led through a hole in the other pole so that the hunter can pull the bird between the poles if he so wishes. A net is attached to this pole, shaped like a scoop and stretched in a semi-circle with a diameter of about 6 feet. The scoop is then raised on to its edge and reaches between the poles when lowered. A string is attached to the upper end of the scoop, similar in length to the former which the hunter pulls to bring down the scoop over the falcon, in the same fashion as he uses the other string to pull the ptarmigan between the poles, the string being pulled through a hole in the first pole. The equipment is placed near a falcon's nest or where the trappers are expecting a flying Gyrfalcon.

When the falcon approaches and sees a ptarmigan or pigeon fluttering near the ground, it first circles in the air over the location, making a few rounds to check whether any danger is present. Then it swoops down and strikes the bird, often with such force that its head flies off as if cut with the stroke of a knife. When the falcon has struck the bird, it takes to the air again, unless desperately hungry, to make sure it can settle down peacefully and enjoy its prey. In the meanwhile the trapper pulls the bird towards the other pole, bringing it under the net. The falcon does not realise this and when it returns to gorge on its quarry the falcon hunter pulls the string bringing the scoop down over the falcon which thereby has been imprisoned in a cage. The hunter carefully removes the bird from the net, making sure that no feather is broken, neither in the wings nor in the tail. His assistants then pulls a hood over the falcon's head, blocking its vision. While the trappers wait for the Gyrfalcon, they remain motionless, hiding behind large rocks or lie flat on the ground at a distance of about 50–80 fathoms [300–500 feet or 90–150 meters], so that the falcon suspects no danger from them, even if it takes note of their presence. When the falcon vessel is ready, the royal itinerant falconer has as many cattle slaughtered as deemed sufficient to feed the Gyrfalcons for two weeks. As much livestock as possible, both cattle and sheep, is also placed on board and slaughtered as the need arises during the voyage to Copenhagen.[162]

If the Gyrfalcons fall ill, the meat is mixed with eggs and fish oil, or else dipped in milk:[163] "On board the vessel, the hoods are kept over the falcons' eyes as on shore. The poles they perch on form two rows lengthwise and lie amidships under the deck, lined with sod and small pieces of turf over which there is a cladding of rough homespun woolen cloth. And to all this are attached small cushions of Icelandic woolen cloth, stuffed with hay. Taut strings are placed a tight intervals between the poles and the laths in the ship's hold as a foothold for the birds so they will not fall far if they lose their footing when the ship rolls in rough seas."[164]

There are also accounts of the freighting of Gyrfalcons from Iceland to Denmark, their food, and their transportation to foreign courts from the period 1746–1755; it cost 16,736 rd.,[165] and from 1760–1769 the corresponding sum amounted to 18,093 rd. In 1765 it is reported that 460 cubits (around 690 feet or 210 meters) of homespun woollen cloth were purchased for use as an underlay for the falcons during their voyage from Iceland.[166] A description from 1785 by Skúli Magnússon, treasurer, is worth investigating further;

he was in charge of the royal finances in Iceland, chief of police in Reykjavík and lived on the island of Viðey off the shoreline:

> Icelandic Gyrfalcons all belong to the same species, only varying in color which is described as grey or white, the white further subdivided into white and half-white. The grey variety is by far the most common and the white birds are much rarer. In one district there could thus be 5 Gyrfalcon nests or more, each containing 3 eggs with only one white Gyrfalcon emerging from this total number, but more commonly none.
>
> The color does not appear to be completely grey, where an equal number of white and brown feathers would be necessary; for there are more brown ones which appear grey in hue. Neither is the white Gyrfalcon completely white (although it develops in that direction through time), but mixed with brown, although white feathers are predominant. But if the white feathers are more abundant, and white and brown equally conspicuous, the Gyrfalcons are half-white. The correct and surest distinction between the two colorations may be determined by looking at the beak and the talons which are black on grey Gyrfalcons, but white in the case of white Gyrfalcons. This threefold division only applies in the Icelandic countryside, but neither here nor abroad, since the Gyrfalcons are received here and sent to foreign courts as either white or grey Gyrfalcons.
>
> A fully acceptable grey falcon must not be more than a year old, or two years at most, and a white falcon must not be above three years of age. The female is larger than the male and therefore more sought after. The wing tip and tail feathers of acceptable falcons must be intact, neither broken nor missing, something that often occurs when they are transported by falcon trappers on long overland journeys. The Gyrfalcon should also have a healthy appetite, stand straight on its legs, not let its wings droop, nor tilt the head, for anything amiss in those aspects is symptomatic of internal maladies which cannot always be easily eradicated. The usual recipe for their diet is mixing their meat with cotton oil and egg yolks. Suspicion of disease causes the king's itinerant falcon master to pronounce no Gyrfalcon as valid and acceptable until it has remained in captivity for 8 days. The Gyrfalcon's daily diet consists of boneless beef or mutton from which all sinews and fat has been carefully removed (although ptarmigans, poultry and pigeons are more suitable).
>
> The Gyrfalcon's age can be fairly accurately estimated until the fourth year, for one and two year old falcons have bluish-white feet, in the third and fourth year their feet are a little yellowish, turning complete yellow from then on... The hunt does not begin until 1 March and the birds are delivered to the falcon aviary on 24 June when the falcon trappers receive 20 rd. in krona currency for a white Gyrfalcon, 14 rd. for a half-white one and 7 rd. for a grey.[167]

To facilitate an estimate of current prices at the time, a sheep cost 1 rd., a cow 7 rd. and a horse 8 rd.[168] The best Gyrfalcon hunting regions were the districts of Barðastrandarsýslur, Húnavatns-sýslur, Snæfellsnessýsla, Mýrasýsla, Borgarfjar-ðarsýsla, Gullbringusýsla, Kjósarsýsla, as well as Rangár-vallarsýsla, Ísafjarðarsýslur, and the Westman Islands; these are the western and southern regions of the country. The northeast had the fewest falcon trappers, probably due to transportation problems, since it was a long way to Bessastaðir across rough terrain and thus a considerable risk of damage or injury to the birds, which would discourage falcon hunters.[169]

There are place names in the east and north which indicate falcon trapping and capture. In the index of places in Suður-Múlasýsla district (stored, as other sources listed in the following entries, in the archives of the Árni Magnússon Institute for Icelandic Studies, Department of Name Studies) is the islet of Fálka-Jónshólmi, named after an eighteenth century falcon hunter, who used it for storing falcons (possibly chicks he had captured in nests). The index of places at Keldunes in the district of Norður-Þingeyjarsýsla (northeast Iceland) contains the following entry: "There is a falcon shed a short distance to the south and west of the farm Kelduneskot where it is said that, once upon a time, falcons used to be tamed and the remains of stone and turf walls are still to be seen." The index of places at Reykjahlíð in Mývatnssveit district (also northeast) states: "There are old remains of a falcon shed in the west part of Syðri-Höfði, from the times of the falcon trappers."

Figure 22: Icelandic stamp, ssued in 1960, with a face value of 25 Icelandic crowns

Place names have also been preserved from the most popular trapping districts. On the farmland of Fremri-Fitjar in Fremri-Torfusta-ðahreppur, district of Vestur-Húnavatnssýsla (northwest Iceland) is the place name Fálkabyrgi (Falcon shelter), which is not further explained, and at the farm Lönd in Miðneshreppur, district of Gullbringusýsla (southwest) contains the following reference: "Below Akrar farm was the small croft Fálkhús (Falcon house) where there is now a sheep pen." The farm Kot in the district of Rangárvellir (south) makes the following remark: "Svartahraun lava field extends from the small stream Selsundslækur into the edge of the lands of Kot. It is now completely eroded, although some vegetation is beginning to emerge again. Towards the middle of the lava field is the hill Valhúsahóll (Falcon house hillock)." Then there is Fálkakofi (Falcon shed) at Gunnarsholt farm, Rangárvellir district (south), on a hill beside the Hólhesthús, but there is no further information regarding the name. At Akurtraðir farm, in the district of Eyrarsveit (southwest) states that in "the middle of Gerðuhamrar rocks where they are highest is a small hut called Fálkabyrgi (Falcon shelter) where, in olden times falcons were kept before they were sent out of the country."[170]

The falcon aviary at Bessastaðir

Captured Gyrfalcons needed careful attention and it would early have been considered necessary to keep them in special quarters. The most famous of such buildings was at the Bessastaðir estate. It is not clear when it was first built, perhaps it existed already in 1662, when the king proclaimed his monopoly on falcon capture and sales. The itinerant falconer and his attendants also used to stay in this house. The premises were built of single stone-masonry walls with the addition of double timber cladding.[171] Later on it was thought more convenient for the falcon aviary to be situated in Reykjavík and it was moved there in 1763, according to most sources,[172] but some indicate that it was temporarily placed at Seltjarnarnes peninsula on a hill, which is now known as Valhúsahæð (Falcon house hill) and transferred to its final location only some years later.[173] This is uncertain; perhaps this specific Valhús (Falcon house) was only a temporary storage facility for Gyrfalcons like all of those referred to above.

It was the duty of the king's tenants to carry the Gyrfalcons free of charge to the island of Örfirisey, a distance of about one kilometer, when the ship came to fetch them. In addition the tenants had to provide hay for the livestock that was intended as food for the birds on the journey to Copenhagen. In 1760 for example, twelve bulls aged three winters and older were collected on Mosfellsheiði moor for the Gyrfalcons, and in 1761 police commissioner Brynjófur Sigurðsson was ordered to have 24 bulls aged three winters and older herded southward to Gufunes point as fodder for the falcons. That same summer 106 falcons were sent to Copenhagen, so that each ten of them had been provided with traveling provisions of two fully grown bulls as well as poultry and mutton which served the purpose of delicacies.[174] Yet in times of hardship it was no simple matter to come by the necessary provisions. In 1702 only very few farm animals could be obtained in the countryside near Bessastaðir. so they had to be fetched from the district of Árnessýsla in the east and the uplands of Borgarfjörður among others. In 1759 the treasurer had to resort to harsh methods to force farmers in four districts (Gullbringu-, Kjósar-, Árnes- and Borgarfjarðarsýsla) to deliver the required number of cattle for the falcons.[175] The bulls were usually stored at Geldinganes until they were transported on board the falcon ship.[176]

In 1764 it is believed that well over 200 Gyrfalcons were kept in the falcon aviary during the eight days they were being fed under the direction of the falcon master, until a final assessment could take place to ensure that no unhealthy birds were being purchased. This suggests that the falcon aviary must have been a sizeable building. Although it was not a tall structure, being all on the same level, it was of a large floor area like the Falkonergaard in Copenhagen. A number of poles were placed along the entire length of the building on which the falcons perched. Small hay cushions with covers of homespun wool were attached to the poles, into which the falcons could sink their talons and thus stand more securely than they would have on the bare poles.[177]

Falconry as a sport was at this time declining in Europe and the supply of Gyrfalcons from Iceland began to exceed demand. In 1766 an annual limit of 100 grey birds was set for export to Denmark, but all the half-white and white birds that could be captured were allowed to be exported. By the early 1770s this had been diminished to 60–70 for all colors and by 1785 the limit stood at 30.[178] Although Gyrfalcon capture in Iceland is referred to occasionally in historical sources and some figures are indicated, no records exist, with the exception of most part of 1664–1709 and 1731–1806. These figures are not accurate nor fully consistent from one book to another. In 1664, 48 birds were captured, but for the following three years where records exist – 1673, 1674 and 1676 – the only available information relates to cost (549 rd., 506 rd. and 366 rd.), and later on the figures are as follows: 1690: 62+, 1691: 98, 1692: 82, 1693: 107 and 1694: 59+. All birds appear to have been grey; at least no categorization is indicated until after this period. In the following years the overwhelming majority of birds are grey, with very few half-white specimens, and white falcons are even fewer. As far as the half-white birds are concerned, the years 1745 and 1751 are exceptional (as already mentioned with regard to the severe winters) when eleven of these were caught each year, 1752: 12 and 1766: 10. Correspondingly, white gyrfalcon-captures were in two digit figures for the following years: 1733: 10, 1734: 11, 1737: 12, 1745: 21, 1751: 42, 1752: 18, 1753: 22, 1754: 10, 1756: 16, 1761: 16, 1766: 12, 1770: 15 and 1784: 14.[179]

A total of 4,848 gyrfalcons, 4,318 grey, 374 white and 158 half-white, were exported from Iceland to Denmark during 1731–1793, i.e. on average 78.2 falcons per year.[180] The largest number was caught in 1764, a total of 211 falcons, 205 grey, 4 white and 2 half-white. The feed purchased in Iceland for their maintenance and the journey abroad consisted of 72 steers, 339 sheep and 65 lambs.[181] In Copenhagen the gyrfalcons were divided as follows: 50 to the French King, 30 to the German Emperor, 60 to the King of Portugal, 20 to the Landgrave of Hessen, two to the French Ambassador and three were kept by the Danish King; the remaining birds were killed.[182] No fewer than fifty different courts received gifts of Gyrfalcons from the Danish kings in the eighteenth century.[183]

Only very few birds were caught during the last years of operation of the falcon aviary in Reykjavík. Only three were exported in 1803 at an expense of 2,500 or 2,600 rd. In 1806 the falcon ship visited Iceland for the last time, rented at a cost of almost 4,000 rd. The proceeds were 19 falcons, all grey.[184] The last courts to enquire for Icelandic falcons and receive them were the Portuguese and the Russian.[185] A year later, 1807, the last two falcons at the royal mews in Copenhagen died of hunger during the bombardment of the city by the English fleet.[186] Shortly afterwards, in 1810, the local falcon aviary was closed.[187]

Epilogue

Although the export of Gyrfalcons, which began in the Middle Ages, was now formally at an end, some individuals completely unconnected with the Danish King continued to secure gyrfalcons from Iceland well into the nineteenth century. In 1869 John and James Barr were sent to Iceland by the Maharajah Dhuleep Singh for the purpose of obtaining Gyrfalcons; they captured and took to India no fewer than 33 birds. Around the same time, the Germans sent men to Iceland for the same reason, but this time the birds were to be used to attack and destroy French carrier pigeons in Paris during the war between the two countries in 1870–1871. Gyrfalcons were also of interest to certain foreign parties for a totally different reason, which required the massive slaughter of the birds in the interest of science, where vast numbers of eggs and dead falcons were sent to different countries.[188]

In 1903, Christian IX decreed that the heraldic shield of Iceland should boast a "white Icelandic falcon, on a blue background, perching, and turned to the left" instead of the previous flat image of a cod. The Gyrfalcon was not a protected species, it never had been, and it is worth mentioning that the Danish king and his Icelandic farmer subjects never shared a common understanding of the qualities or worth of the Gyrfalcon. Some important personages and their descendants had displayed the bird as early as around 1300 on their family crests and/or seals, e.g. Haukur Erlendsson (1265–1334) law-man, Loftur Guttormsson (1375–1432), Jón Vídalín (1666–1720) bishop, Páll Vídalín (1667–1727) law-man, Finnur Jónsson (1704–1789) bishop, Hannes Finnson (1739–1796) bishop, Ólafur Stephensen (1731–1812) regional governor and Magnús Stephensen (1762–1833) chief justice.

With increased tendencies towards nationalism and the struggle for independence in the nineteenth century, certain individuals carried this torch onward. Sigurður Guðmundsson (1833–1874) wanted the Icelandic heraldic shield to display a white falcon with spread wings against a blue background. He made a drawing to this effect, which was carried in public by 1860 and kept on top of a flagpole in 1873. A year later the formal adoption of the gyrfalcon as part of the Icelandic flag was given serious consideration. In 1885 the matter was brought up again and now the number of falcons had grown from one to three, all of them white. The idea was never accepted.[189]

When the Parliament House was completed in Reykjavík in 1880–1881, many wanted to plant a shield displaying a white falcon above the balcony window beside the insignia of the state, replacing the heraldic shield of that time, the split cod against a red background. This was not accepted due to opposition from the king's representative and the Danish architect. Eventually the shield was mounted in 1904 and remained in place until 1911.[190]

Often the falcon was looked upon by Icelanders as a pest and threat to poultry and livestock. In 1619 James I of England wrote to Christian IV that he had heard that the Icelanders were destroying Gyrfalcons' eggs and killing the young birds, and begged the Danish king to put a stop to the practice. King Christian immediately became alarmed and instructed the Governor of Iceland to enquire into the matter and to report upon it.[191]

When the formal export of the falcon was discontinued at the beginning of the nineteenth century, the old hatred for the "feathered predator" raised its head once more, to such a degree that in 1885 a law was passed for the first time, declaring that it could be killed at any time. As mentioned above, in the same year ideas including the gyrfalcon in the national flag were suggested but not accepted. The majority of parliament members were farmers or from farming backgrounds, and they were looking for their own or their friends' interests and livelihood, e.g. the nesting population of the common eider (*Somateria mollissima*).[192] The opinion was not unanimous; some wanted to show the bird sufficient respect to give it protection, in spite of its predatory nature and the damage it caused.[193]

After the Gyrfalcon finally found its way to the Icelandic heraldic shield, it appeared on various banknotes, first in 1907. Sometimes the representation was identical to that of the heraldic shield, but the bird was also pictured in the company of "the Lady of the Mountain", which used to be and still remains a poetic image of Iceland.[194] In 1919 conservation laws at last came into force, luckily, as the falcon was supposed to be on the brink of extinction.[195]

After changes were made to Iceland's heraldic shield that same year and the crest displaying the country's four guardian spirits took its place, in 1920 there was a decree proclaiming that a special Icelandic royal flag should be designed and on this should be an Icelandic falcon. This particular flag was used by the Danish king Christian X when he visited Iceland in 1921.[196]

In the same year, an Icelandic Order of Merit was established, which was given the title "The Icelandic Falcon" (later often called "The Icelandic Order of the Falcon"), which was at first divided into three classes and later into five. This medal comprises a cross with center front, a blue stained shield on which is superimposed a silver falcon with wings lifted in preparation for flight. This was and still is the highest possible honor and recognition of the Icelandic nation.[198]

In 1930 the conservation laws were lifted again and people began hunting the falcon both for stuffing and live for export.[199] However, in 1940 after excessive hunting and slaughter of the bird, the conservation laws were put back in place and have remained so until this day.[200] At the beginning of the 1960s, foreigners, especially Germans, began to visit Iceland once more, and now mainly for the purpose of smuggling Gyrfalcon eggs and chicks out of the country.

These were sold to Arabic oil magnates for the sport of hunting; they were prepared to pay astronomical sums of money for the birds. The practice continued intermittently for the next three decades or so, but stringent rules were applied to this form of smuggling, and many smugglers found themselves being punished by the courts of law.[201]

Increased surveillance of nesting grounds also meant that this activity gradually fell away until it ceased altogether. Today the Icelandic Gyrfalcon has a relatively safe existence. The population is said to be around 250–400 nesting pairs, or around 1,000–2,500 birds over the winter months.[202]

Acknowledgement

The author thanks the following persons for their invaluable help and support in the composition of this article: Áslaug Óttarsdóttir director of the Sólheimar Library, Reykjavík, Anna Sveinsdóttir director of the library and archives of the Icelandic Institute of Natural History, Reykjavík, Kristín Konráðsdóttir director of the University of Akureyri Library, the staff of the Municipal Library of Akureyri, the staff of the Municipal Library of Siglufjörður, Daníel Bergmann nature photographer and bird specialist, Jón Baldur Hlíðberg artist, Gunnlaugur Pétursson engineer and bird specialist, Edward B. Rickson bird specialist, Guðmundur Ó. Ingvarsson cartographer, Örnólfur Thorsson secretary general of the Office of the President of Iceland, and last but not least Rafn Kjartansson and Frances Jane Milne McQueen, translators into the English language.

Chapter 4. Parrot Aviculture in Ancient New World Societies

Vernon N. Kisling, Jr.

Ancient societies in the Americas were fairly isolated from the rest of the world, but not from each other. A long history of established urban settlements created extensive road systems, which facilitated organized and structured trading systems throughout the Americas. Trade items included live birds, which were kept as pets, in animal collections and were bred in aviaries. In addition, Native American societies in South, Central and North America used feathers from many species of wild birds, some of which were obtained locally, while others were obtained through the trade network. A few of these bird species, most notably the Quetzal (*Pharomachrus mocinno*), and Scarlet Macaw (*Ara macao*) and other parrots, were important enough to be used in sacred ceremonies and sacrifices and their feathers were used on sacred artifacts.

Parrots of the Neotropical region – the zoo-geographic region of Central and South America and the Caribbean – include some 148 species. These parrot taxa include the macaws, which are the largest of the parrots, have the longest tail feathers and some of the most spectacular colors. Other Neotropical parrot taxa include many smaller parrots, parrotlets and parakeets. Many of these parrots are green with some mixture of other colors, while a few are primarily blue. However, the Scarlet Macaw, is a colorful mixture of red, blue and yellow feathers, especially prized by many societies in the Americas.[203]

South American aviculture

The Incas of South America (1100s–1535) had a refined knowledge of astronomy and various aspects of nature, particularly medicinal plants. Their religion combined features of animism, fetishism and the worship of nature gods; the highly organized state religion worshiped the sun god, but also other faiths were tolerated. Rituals included the sacrifice of humans and animals. The Incas kept animal collections and gardens, both real and artificial. The royal palace at Cuzco contained an artificial collection in gold and silver. Chronicler Garcilaso de la Vega noted that the royal palace hosted herbs and flowers of various kinds, small plants, large trees, animals (large and small, tame and wild), including snakes, lizards, snails, butterflies and birds. Each of these plants and animals were in imitations of their natural surroundings. In this extensive royal collection there were birds of all kinds, some perched on the trees as if they were singing, while others were flying or sucking honey from flowers. There were also deer and stags, "lions," "tigers," birds and other animals, each being set in its natural surroundings to give greater similitude. This palace collection was imitated in estates of the provincial lords as well as in temples. The Temple of the Sun was covered with many gold and silver figures, including men, women, waterfowl and other birds, "tigers," bears, "lions," foxes, hounds, mountain cats, deer, Guanacos (*Lama guanicoe*), Vicunas (*Vicugna vicugna*) and domestic "sheep." These were placed around the walls in spaces and niches. In addition, figures imitated herbs and plants that grow on buildings and were placed on the walls so that they seemed to have grown on the spot. Scattered over the walls were lizards, butterflies, mice and snakes, which seemed to be running up and down the walls through the plants.

There were also real gardens filled with orchards, groves of trees, flower beds and sweet smelling herbs, native to Peru, and real animal collections. The Royal Inca, as well as provincial lords, conducted hunts and captured a variety of wild animals, along with Llamas (*Lama glama*), and Vicunas that were caught for the domesticated herds. Locally captured wild animals were kept in collections maintained at a variety of locations and other wild animals were obtained through trade. The collections also included animals that regional lords sent to the Royal Inca as tribute or taxation. Some of these animals were kept at the royal court, while others stayed in various districts of the capitol; the districts were named for the kinds of animals kept, such as the *amarus* district (where large snakes were kept), the *pumacurcu* and *pumapchupan* districts (with large cats and bears), and the *surihualla* (the ostrich field as it tends to be referred to, but it should be known as the rhea field).

During the most important festivals of the year, especially the festival to honor the sun known as *raymi*, many macaws, vultures and condors were presented to the Inca priests. Various tribes brought these birds and other animals, found locally in their regions, to the capitol of the Inca Empire, Cuzco, where the animals were kept and used as needed during festivals; birds in districts outside the city as mentioned above. Also gold figurines in the shape of animals, including many kinds of birds, were offered during these festivals.

South American birds sacred to the Incas and other tribes were the Condor (*Vultur gryphus*) and some parrots. Food birds included rheas, tinamous, ducks, geese, curassows and guans. Pet birds included rheas, curassows, guans, parrots, parakeets, macaws, toucans and songbirds such as orioles and thrushes. Birds used for their feathers were numerous, but of most importance were the macaws, parrots, toucans and rheas. Bird feathers were used for many kinds of decoration. Most often this included feather-work on headdresses, garments such as cloaks, caps, belts, arm bands, bandoliers and warrior artifacts. In addition to parrot feathers, feathers were obtained from herons, storks, flamingos, hummingbirds and hawks. Some artefacts were elaborate, requiring highly skilled artisans and large amounts of feathers. Headdress found among the Chiquitoans and Mojo of Bolivia consisted of 300 parrot tail feathers as well as other feathers. Color was important for many visual and symbolic reasons; preferred colors were often obtained naturally, making certain bird species more important than others, but some colors were produced artificially. Many South American tribes used a technique known as *tapirage,* by which they were able to change the color of the feathers produced by some parrots. The technique varied, but usually involved the use a chemical substance from frogs (blood or skin secretion, sometimes mixed with a plant dye, sometimes using the fat of certain fish), which was smeared on the skin where green feathers were recently plucked. This caused new feathers, through several moults, to acquire a yellow or orange color.

Even in recent times (based on expeditions and studies conducted in the early 1900s), many South American tribes kept and use parrots. The Mochicas of Peru had parrots as pets; the Aymara of Bolivia used imported green parrot feathers in many different ceremonial dance costumes; the Araucanians of Chile routinely captured parrots for their feathers; various tribes of Ecuador had an established trade in parrots among themselves; when at war, the Cañari of Ecuador and Pasto of Colombia decorated themselves with parrot feathers; every Tupinamba village of Brazil kept parrots as pets and plucked their feathers yearly for feather-work ornaments; boys of the Tenetehara of Brazil carried wands having wooden handles containing 30-40 Scarlet Macaw tail feathers; the Tapirapé of Brazil attached red and blue feathers from the Scarlet Macaw to their bows; the Caraja of Brazil routinely captured parrots for their feathers, as well as young birds for trade; the Yuruna of Brazil had headdresses and other ornaments made out of various parrot and macaw feathers; the Mundurucú of Brazil made extensive use of feather-work and were particularly fond of red, blue, yellow and green feathers from parrots; a single Apiacá village in Brazil contained 80 tame macaws in thatched stick cages; the Mojo of Bolivia kept tame parrots and plucked them on a regular basis; all the tribes in the Montaña region (covering adjacent areas of Bolivia, Ecuador and Peru) had tame parrots, and tribes of the Uaupés-Caquetá region (in the Colombian-Venezuelan area) commonly kept parrots as pets.[204]

Central American (Mesoamerican) aviculture

The urban societies in Central America included the Olmec and Zapotec (around 1100 BCE–300 CE), Maya (300s–900s), Toltec (900s–1200s) and Aztec (1200s-1521) civilizations. The Zapotec worshiped the god *Xaquija,* who took the form of a macaw and was the first Zapotec deity to receive a temple. The sacred macaw was believed to fly into the temple and become an oracle, answering questions from the priests. As a representative of the sun and fire, the macaw's feather colors were important representations of these things, as well as directions dictated by the rising and setting of the sun.[205]

Mayans had gardens, fish ponds and pets, including parrots and other birds, monkeys, Coatis (*Nasua narica*) and Kinkajous (*Potos flavus*). They hunted wild animals, were familiar with wild animals and plants, and domesticated a few species. Folk systematics of their descendants, the Tzeltal society, indicates a knowledge of animals and plants that no doubt existed in the past as well. Long-distance trade, tribute and taxes were used to obtain exotic luxury items, including feathers, jaguar hides and teeth, shark teeth, stingray spines and shells. While it is difficult to know the bird-keeping habits in this and other early Mesoamerican societies, it is clear that the Aztecs continued their traditions. We have a better knowledge of Aztec activities because of exploration and colonial records, rather than being dependent on archaeological evidence alone. [206]Aztec society possessed a good understanding of animals, plants and minerals. Metals, gemstones, silver, gold and other materials, in addition to their practical uses, were made into animal figurines resembling animals with which the people were familiar. Aztec live collections were supplemented with well made and detailed figurines. These replicas amazed Cortés, because he found that every living thing on land or sea (as far as he knew) was represented in gold, silver, precious stones or feather-work, and was so realistic as to be the thing itself.

Montezuma had several residences, each of which contained gardens, fish ponds and birds. His most elaborate residence was at the new capital, Tenochtitlán. This aggregation of several thousand buildings had, among other things, a bird house, a wild animal house (with mammals, birds of prey and reptiles) and a collection of deformed humans. The animal and bird houses were contained within extensive gardens of flowering

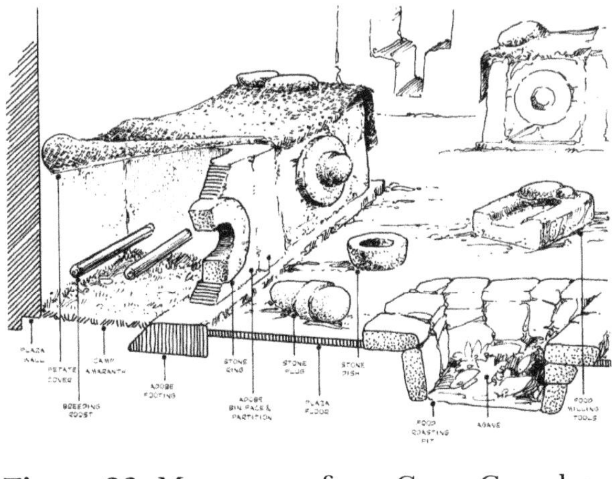

Figure 23: Macaw pen from Casas Grandes, Mexico

plants, fragrant shrubs and medicinal plants. These gardens also had freshwater and saltwater ponds filled with fish and waterfowl. Describing the bird house, Cortés observed ten pools of water in which were kept every kind of waterfowl known from the region. Fresh water was provided for the river birds and salt water for those animals from the sea. The water itself was frequently changed to keep it pure. Every species of bird was provided with its own natural food, whether fish, worms, maize or smaller cereals. Impressed with this collection, Cortés noted that it was the task of 300 men to look after these birds (early aviculturalists). Others were employed in ministering to those that were ill (early veterinarians).[207]

A separate collection contained mammals, birds of prey and reptiles, perhaps because these animals required a building with sturdy cages. This house contained a large courtyard, paved with decorative flagstones in the manner of a chessboard. The palace had cages about nine feet (three meters) high and eighteen feet (six meters) round, each of which was half covered with tiles and half by a wooden trellis. The cage areas contained birds of prey, including every kind Cortés was familiar with from Spain, as well as many others that were new to him. The size of this collection required a work force equal to the one needed at the bird house. Another 300 men looked after these animals and birds. Bernal Díaz del Castillo, who traveled with Cortés, also described these aviaries and mentioned that they kept and bred quetzals and many kinds of parrots, all of which had to be imported from the tropical regions to central Mexico. He also mentions that these birds were used to obtain feathers and that they were regularly plucked. While the feathers from several birds were used to create headdresses, garments, and ceremonial objects, the most sacred artefacts were made from green feathers, in particular those of the Quetzal.[208]

The primary purpose of keeping and breeding

Figure 24: A parrot is perched on a rafter in Canadian Indian dwelling, 1558

quetzals, macaws and other parrots was to have a readily available and continuous supply of feathers. It also provided live birds for sacrifice when needed. Feather-work was not only a significant part of Mesoamerican culture, but also an important part of the economy for these societies, from ancient times up through the age of European exploration. Continuing a tradition from earlier societies, the Aztecs elevated their artisans of luxury goods to exclusive guild status. This included feather-workers known as *amanteca*. The guilds resided in separate residential districts and had control over their members. Some were public artisans who made feather-work for the rulers while others were private artisans who made feather-work for the wealthy and elite classes. The business was a family affair with the craftsmanship passed down from parents to sons and daughters.[209]

Feather merchants also enjoyed an elevated status and were found in market areas separate from other merchants. Some merchants sold completed feather-work, while others sold just the feathers. Those who sold feathers were further grouped into those who sold precious and those who sold ordinary feathers. In Central America the most highly appreciated feathers were the brilliantly green feathers of the quetzal, especially the tail feathers that could reach twenty inches in length. Other valuable feathers came from the cotinga, trogon, Roseate Spoonbill (*Platalea ajaja*), Yellow-headed Parrot (*Amazona oratrix*) and the long-tailed macaws. Lesser, ordinary feathers came from other local birds and had many uses among the general population. Merchants, known as *pochteca*, comprised the long distance trade outside the urban centers. They also traveled outside the empire and sometimes obtained products directly from the source, rather than trading with other merchants. These merchants were able to go abroad, because they and their activities were considered neutral and they traveled with immunity among other societies. When obtaining resources such as feathers directly from the wild, the merchants however took personal risks, since they were encroaching on the supply of other merchants.[210]

An interesting facet of the long-distance trade and the keeping of live exotic birds in aviaries and gardens, is the effect this had on the distribution of some species. It seems some birds escaped and adapted to their new environments. Such an exotic introduction has been documented for the Great-tailed Grackle (*Quiscalus mexicanus mexicanus*). A native of the Mexican coastal lowlands, this species is now found in the central highlands of Mexico. While not mentioned as a source of feathers (although feathers from many different species were used), the Aztecs apparently considered this grackle to be strange, rare or precious. The Aztec Emperor Auitzotl (ruled Mexico 1486–1502) required this bird to be brought to him for his gardens. Many escaped into the wild and because they were considered of importance no one was allowed to kill them. As a result, they multiplied and now inhabit their new environment around Mexico City.[211]

Southwestern North American aviculture

The southwestern region of North America incorporates a cultural and geographical area that is now known as New Mexico, Arizona, southwest Colorado, southern Utah, western and southern Texas and northern Mexico. It is the region north of Mesoamerica and south of the Plains, Plateau and Great Basin regions of the United States. This region was culturally influenced by Mesoamerican traditions, rather than from the north. It was also the northern periphery of the exotic trade which filtered up from Central America, although there is evidence this trade continued to move east to the Mississippi River. None of this happened, however, without modifications. These cultural and ceremonial modifications can also be seen in the use of bird species for feather-work.

While several native bird species were used to obtain feathers, the most important bird feathers came from the Scarlet Macaw with its red, yellow and blue feathers, and to a lesser extent the Military Macaw (*Ara militaris*) with its green feathers. Even less frequent use was made of the Thick-billed Parrot (*Rhynchopsitta pachyrhyncha*) with its green feathers. Green feathers were more valued by Mesoamerican societies and less so by the northern Mexico and southwestern United States societies, which preferred the red, yellow and blue feathers of the Scarlet Macaw. Although the Military Macaw and Thick-billed Parrot are native to northern Mexico, the Scarlet Macaw is not native to any southwestern North American region and had to be imported a considerable distance from the more southern regions of Mexico.

Since ancient times the natural habitats of many macaw and parrot species have diminished. Eighteen parrot taxa have become extinct since 1600 throughout the Americas and West Indies. Eight macaw species no longer exist in the West Indies as they did, when Columbus visited that region. Probably several other macaw and parrot taxa have disappeared which we do not know about, and these birds were no doubt used by tribes and sighted by early explorers. West Indian species were more susceptible to extinction because of their island habitats; however, South and Central American macaw and parrot habitats have also been reduced or destroyed.

It is estimated that Scarlet Macaws reached southwestern North America during the 500s through long distance trade from the Tamaulipas-Huasteca area, along the lowland forests of the Mexican Gulf coast, where the Huastecs were known to breed macaws in captivity. Live birds were caught while still very young and transported, fed and cared for by the merchants so that they arrived at various destinations alive and healthy.

This trade was maintained up to the European exploration period in the 1600s–1700s. From the earliest explorations to later explorations in the 1700s and 1800s, parrots and feather-work were observed throughout the entire southwestern North American region, as well as the Mesoamerican, South American and West Indian areas. Columbus was presented with macaws and brought sixty of them for the King and Queen of Spain upon returning from his first voyage in 1493. These birds created a great deal of interest because of their long tail feathers and bright colors, unlike the smaller, short-tailed parrots Europeans were used to from Asia and Africa. Throughout the 1500s and 1600s, macaws and parrots were popular birds for import to Europe in menageries, private collections and as pets. Cabeza de Vaca, after being shipwrecked with three companions (1528-1536), was the first European to wander from Florida across the Gulf of Mexico through Texas and northeastern Mexico. Upon entering what is now Mexico, they were among the first Europeans to encounter merchants with live macaws and macaw feathers.[212]

Following the introduction of Scarlet Macaws in the 500s, a trade center was established at Casas Grandes in the Paquimé region of northern Mexico (probably during the 1100s or 1200s). In addition to other trade items, Casa Grandes specialized in the breeding of Scarlet Macaws in order to supplement the trade in these birds. It was also a breeding center for Military Macaws (even though these birds were found locally) and turkeys (these were domesticated). Other minor trade centers were established over time as well; however, no evidence of bird breeding has been found at any of these lesser trading centers. Over 300 Scarlet Macaws and 80 Military Macaws have been excavated from the Casas Grandes site. The site contains numerous breeding cages, a design replicated wherever macaws were bred. Bird keepers at Casas Grandes, who could be considered aviculturalists, cared for these macaws and lived around the central breeding center, known as the House of Macaws. Macaws were kept in adobe nesting cages, as well as enclosed courtyards.[213]

About the time Casas Grandes was established in the Paquimé area, parrots and their feathers appeared in the Hohokam region of southern Arizona and the Mimbres region of southern New Mexico. From here they spread into the Anasazi and Rio Grande areas further north. From Casas Grandes numerous trade routes took its traders throughout northern Mexico, southwestern United States and even east to the Mississippi River. The La Salle expedition in 1682 followed the Mississippi River south from Canada and found evidence of an interregional trade exchange from Mexico through southern Texas to the Natchez area and into Arkansas. Trade in macaws may have only reached the larger communities, but the birds were seen and known throughout the region and appeared as depictions on pottery in many villages. Some of these larger communities had bird keepers, both male and female, to care for the birds and these individuals are also depicted on pottery with the implements they used to care for and transport the birds.[214]

While it seems the merchants trading in live birds were able to properly care for these birds and deliver them in good health, almost half the birds excavated in the southwestern United States had pathological bone conditions. This was caused primarily from a lack of proper care and dietary deficiencies. It seems that the knowledge of the merchants may not have been passed on to those who bought the birds. Among some individuals who kept and bred these macaws, there was a lack of knowledge about biological and dietary needs, or an inability to provide what was needed for the birds.

Zooarchaeologists have been able to determine the sites were where macaw breeding occurred, because the cages used to raise macaws have the same design features as those found at Casas Grandes. Also the extent of this breeding activity can be determined due to the number of cages present at the sites. The cages can be distinguished from turkey pens used to breed and raise the native turkey (there were two subspecies of the Common Turkey (*Meleagris gallopavo*), and the Ocellated Turkey (*Meleagris ocellata*), an effort resulting in the domestication of these turkeys. However, macaws were not domesticated (there were no physical changes to the birds over time) and so the care of macaws and other parrots can be considered a purely avicultural effort. Macaw breeding cages were adobe boxes lined up in rows with each box being approximately three feet (one meter) square and one and a half feet (half meter) high.

The floor had a covering of amaranth (a native herb from Mexico) and the open top was covered with a flexible *petate* (a long woven palm mat similar to the bed mats used by the Indians). The backside of the cage was usually against an adobe wall and the front of the cage had a stone ring with a stone plug. Macaw cages at Casas Grandes were concentrated in one building, the Macaw House, while the turkey pens were concentrated in a different building, the House of the Dead, and other wild birds in yet another building, the House of the Well.[215]

Military Macaws, thick-billed parrots and other smaller, short-tailed parrots were kept as well, but their green feathers were less significant than the red, yellow and blue feathers of the Scarlet Macaw. These three colors were symbolic, each color representing a different direction. The feathers were also associated with the Sun deity, summer and supernatural deities connected with the development of water control for agriculture, including Quetzalcoatl (a patron carried over from Meso-american culture), the Pueblo Corn Maidens and the Lord of Crops. Scarlet Macaw feathers were used to decorate prayer sticks, insignia of priestly office, garments, headdresses and warrior costumes. To give an idea of the extent to which enormous numbers of feathers were needed, one ancient Anasazi artefact (probably used as a sash) contained over 2,000 Scarlet Macaw feathers.[216]

Other birds were of sacred and sacrificial interest, in particular turkeys and eagles, but further native species were of lesser interest. While the remains of most birds are excavated from trash dump sites, the bodies of Scarlet Macaws were carefully sacrificed and ceremoniously buried in sacred locations. Many were sacrificed at about one year of age, when they become fully feathered, as part of the Spring Equinox ceremony. The birds were plucked, carefully wrapped and buried under the floor of *kivas* or special rooms of the house, and sometimes buried with humans. Macaw clans were devoted to these birds and artisans specialized in their feather-work.[217]

Even as late as 1880, a survey by John Bourke among the southwestern tribes reveals a continuing use of parrots. The Zuñis had a parrot clan and live parrots; the Acoma had a parrot clan; the Laguna had a parrot clan; the San Felipe kept parrots in cages; the Sandia had bundles of parrot feathers; the Isleta had a parrot clan and live parrots; the San Juan had abundant parrot feathers; the Santa Clara had down and feathers from parrots in all their houses as well as a parrot clan; the San Ildefonso had parrot feathers; the Pojoaque had parrot feathers, and the Nambe had parrot feathers in all their *pueblos*. In all cases these parrots were most likely Scarlet Macaws, since Bourke use the word *huacamayo* which usually referred to macaws. In addition, modern Hopi consider the colors of parrot feathers sacred and use them on the Parrot Kachina mask. Macaw symbols are also used in the Flute Ceremony.[218]

While many birds were of interest to the tribes and societies of the Americas, few were esteemed more than the quetzal and parrots, in particular the Scarlet Macaw. From ancient times up to the time of European exploration, and in many cases up to the present time, parrots and macaws were of such importance that they were expected as tribute by the more powerful societies, as trade throughout the region and as gifts to highly esteemed individuals.

These individuals included European explorers, who were often presented with macaws as presents, and these birds were brought back to European menageries. While many tribes kept individual birds in cages and sometimes bred them, rulers of the more powerful and wealthy societies maintained elaborate aviaries with keepers and veterinarians. New World aviaries of the Aztecs and Incas were even more extensive than those found in Old World Europe at the time. All of that has been lost; however, the fascination of macaws and parrots continue to captivate visitors to American zoological parks and aviaries.

Chapter 5. Chaffinch Sport in Flanders

Filip Santens

While driving along Flanders roads on a Sunday morn¬ing between 1 April and 31 August, one will observe, at least once, the Flemish fincheneers (*vinkeniers*). Sitting in a long row, they listen careful¬ly for up to an hour to the song of a bird in a cage. Every time the bird in the cage sings a very specific song, a tally mark in chalk is made on a long wooden stick. The bird in question is a male Common Chaffinch (*Fringilla coelebs*), and fincheneers organize contests to find out who owns the best singing Chaffinch. The contest only takes one hour and the Chaffinch with the highest number of calls wins the com¬petition.219 Drivers who do not know about the fincheneers often shake their heads or even curse them, because the road is blocked once again.

In this chapter I will explain the fincheneers' history and motivations. The first part highlights the Chaffinch itself and the singing competition. The second discusses the history of the development of the chaffinch sport, and it also shows how this sport survived into the twentieth century. The third part shows the way the chaffinch sport has adapted to new laws with the support of the General Chaffinch Sitting Federation. It also shows how fincheneers evolved from chaffinch catchers to Chaffinch breeders. We conclude by looking at neighboring countries, because the Chaffinch sport is also popular in other European regions.

Chaffinch sitting

In popular speech the Common Chaffinch, is called the "bachelor's chaffinch". The bird is widespread throughout Europe, North Africa and Asia Minor. Locally the Chaffinch is also known as *Boekvink* ("Beech Finch", "Debt Finch", "Calf Finch" or *Boomgaardvink* ("Orchard Finch"). In some areas people call it "Blue Head" or "Warbler". In outer appearance Chaffinches look alike and show little or no difference; yet it is possible to distinguish different groups of Chaffinches. Especially the different ways of singing make them different from each other. For those still uncertain it might be important to know that only male Chaffinches sing.[220]

Before being able to participate in a singing competition, a fincheneer has to turn his or her Chaffinch into a prize winning bird. First the bird needs to learn a correct Flemish way of singing. Then the bird is assigned a fixed location and is trained intensively. Only then is the Chaffinch ready to enter contests. For persons not familiar with the Chaffinch sport it might be a little confusing. In plain words it boils down to this: the Chaffinch needs to learn to sing a song ending in "wie". A wild-caught Chaffinch often sings a song that ends in "weeuw" or "beeuw", but captive-bred Chaffinch¬es also have to be taught the correct way of singing. Young Chaffinches imitate the singing behavior of the other birds and every fincheneer has a master, i.e. an older bird which

Figure 25: Early photo of fincheneers in Flanders, ca 1900

knows the correct way of singing. This might be the male parent of the captive-bred Chaffinch, but not necessarily. There is a complex regulation that constitutes the requirements of correct Chaffinch song. During contests specially trained judges, called singing judges, make sure that only legitimate calls are counted. In case of doubt the bird in question is judged by the singing judges and they decide whether a bird is fit for competition or not.

The decision of letting only birds with the right way of singing enter into the contest was made long ago. Only local wild-caught Chaffinches participated in the contests. Apparently their dialect ended in "wie". Later on fincheneers started catching migratory birds as well, though they sang in a different dialect. It was decided that the "wie"-singers were Flemish Chaffinches and that all other birds were foreign and had to learn the Flemish way of singing before being able to enter a competition. These foreign Chaffinches were commonly denoted as Walloon Chaffinches. In Chaffinch sport, the term Walloon has nothing to do with the French speaking part of Belgium, Wallonia (viz. the Walloon provinces of Belgium), since the name originates from Wales and means in general "everything that is not from Flanders".

In Flanders it was legal to catch wild Chaf¬finches until 2002, when a law was passed establishing a 0 quota; now captive Chaffinches can only be captive-bred. A Chaffinch lives for about seven to eight years in captivity; in exceptional cases the birds live for fifteen years or even more. This suggests that some old wild-caught birds can, at the time of writing, still be participating in con¬tests. During the contest every Chaffinch sits in a specially designed cage. Old Chaffinch cages indicate that fincheneers used to participate in contests with open barred cages. Later on blinderd cages were used for the Chaffinches. In the past, people were convinced that a chaffinch cage had to be as firm and dark as possible. Therefore only the front of the cage was covered with milk glass, the rest being completely in wood. Tradition changed, however, and today there are specially designed cages that make utmost use of glass, sometimes also at the back of the cage.

Figure 26: Finch sittings in the mid-1900s

Keeping Chaffinches in Flanders is strictly regulated. The law states that wild-caught birds have to be officially ringed and registered in an inventory for birds that were not hatched in captivity. Even home-bred Chaffinches need a closed ring and they have to be registered in a separate inventory. Moreover, since 1982, every bird needs an individual index card that accompanies the bird wherever it goes. Every Chaffinch which participates in a contest has to have a federation card. The card declares the name of the bird, its ring number, the name of its owner and the associations of which the owner is a member.

One of the most important parts of Chaffinch sitting is without question the contest itself. Most competitions are held on Saturday or Sunday mornings between nine and ten o'clock. Registration normally starts about one hour before. Fincheneers show their federation card and sometimes also a contest card and pay the inscription fee of 1.50 Euros. After this procedure a number is drawn from a number box or bag. This number is written down on the contest card or attached to the federation card. Afterwards all cards are returned to the owner. Now the owner can freely check out the surroundings of the contest, or immediately put the bird with his/her number in a long row on the road or street, which is kept traffic-free during the contest.

The distance between the cages is set at about 2.40 meters. In the eighteenth century, the distance between two cages was ten feet, which equals the modern distance. The different numbers are chalked on the street along the side. Sometimes local authorities give permission to paint the numbers on the road; in that case they only need to be painted once a year. Usually the fincheneer goes to his or her spot about ten minutes before the contest begins. The cage is stabilized to prevent rocking. One minute before the start of the contest the organizer of the contest shouts: "Birds on the ground", and then: "Move over from left to the right".

The executive committee of the association supervises the contest. One minute after that first signal, there is a second, which announces the beginning of the contest. The organizer then shouts: "All that is good singing, score!" From that moment it becomes wonderfully silent, a fascinating moment for an outsider. One hour after the second signal there is a third one announcing, the end of the contest: "Sticks on the ground!" Now the sticks can only be touched to register the number of songs. As quickly as possible, the results are put down on the contests cards by the executive members.

The normal duration of a contest is one hour, but in case of bad weather the contest can end earlier. About forty minutes after the contest the results have to be publicly shown or proclaimed at the local headquarters of the organizing committee. Once the results are public, participants have ten minutes to make their complaints. The winner receives a small amount of money, varying from 2.50 to 5 Euros. The Chaffinch with the smallest number of songs gets about 1.50 Euros. Every winner receives a flower. The award is of symbolic nature, since all organizing committees are non-profit organizations. All contests are announced in the *Aviboblad*, the local magazine of the Flemish fincheneers, and after a couple of weeks the results of the contests are published.

History

The oldest records of Chaffinches are usually texts describing the bird itself, either for its singing qualities or as a bird caught to be eaten, similar to Starlings (*Sturnus vulgaris*), Thrushes (*Turdus* sp.), Blackbirds (*Turdus merula*) and other small birds which were captured for food. During the period of the Ancient Régime only the king or sovereign and the nobility had the privilege of hunting and catching birds.[221] Before 1789, the area today called Belgium was divided into counties and duchies (e.g. County of Flanders, Duchy of Brabant etc.), ruled by counts or dukes. These counties and duchies were further divided into smaller entities. The head of such a dominion was a lord who acted as a local ruler. One of his typical privileges was the exclusive right to hunt birds and catch fish; common people were not allowed to fish in public waters or hunt, to keep pigeons, breed falcons or catch small birds without permission.[222] The fact that local folk did not always acknowledge these noble privileges is proven by the numerous regulations declared by the sovereigns.

While the privileges prevented large scale bird catching in the countryside, the towns escaped this legislation. As far as one mile outside the city limits people were free to catch birds and trade with them. Very soon bird markets were organized in towns and at the markets different kinds of birds – Chaffinches included – and small wild animals were sold as culinary delicacies. In Bruges there was such a bird market as early as the fifteenth century. Several guild hall orders, e.g. from 1490, regulated the sales in the markets and enumerated the merchandise offered, e.g. swans, peacocks, partridges, herons, hares and rabbits, geese, chickens, capons, chaffinches etc. On 17 October 1573 a woman was punished because she had sold chaffinches outside market hours. In the same year a similar sentence was imposed for selling quail (*Coturnix coturnix*).[223] From the sixteenth century onward retailers or traders of small birds and bird catchers would unite into guilds and crafts. In history books they are referred to as *vynckenaers* (fincheneers).

In his work *Langhen Adieu* from the *Testament Rhetorical* (1560) rhetorician Eduard de Dene greeted the *vynckenaers* (fincheneers) in Bruges.[224] Probably de Dene referred to the members of the old trade with small fowl. Sixteenth-century records are preserved for the organization. It was also called the Guild of Saint Marcus or the Guild of the Small Fowl within Bruges (*ghilde van Sinte Marcus ghenaemt de ghilde van het cleen ghevoghelte binnen Brugghe*).[225] In 1634 the craft was granted a statute whose content has not survived, but in 1661 this statute was referred to when penalizing a guild member for catching birds out of season, i.e. during the particular time of the year in which bird capture was prohibited (the so-called closed time). This shows that the Saint Marcus guild was a guild of bird catchers, in which undoubtedly chaffinches and other birds were caught with nets, bird snares and lime-twigs as was usual in those days.

Before 1629 there was a guild in Ghent under the protection of Saint Gilles and sustained by the *vrye vynckenaers van de stadt Ghent* (Free Fincheneers of the City of Ghent). The guild members entitled themselves as *gulde-broeders van Sint-Gillis geseyd vinckenaers, onderhouden met de slagnetten binnen Ghendt* (guild Members of Saint-Gilles, i.e. Fincheneers, Clap-nets in the city of Ghent).[226] In 1629 they requested the city council to pass a draft of a rule book for internal order, containing only a few articles. This rule book would thus become law, and was enforced upon its members. Among the items treated were the acceptance of new guild members and penalizing. It is not known whether the city of Ghent consented to this request.[227]

The guild of Saint-Gilles had the exclusive right, within the legal city limits, to catch birds either with nets, clap-nets or other devices. In 1673 the guild asked King Charles II for the privilege of bird catching within a three-mile range of the city center. Apparently the original bird catching area had become so limited that the existence and preservation of the guild was in danger. To back up their claim towards people outside the city, the people of Ghent said this measure was necessary in order to protect their crops (since birds were harming their fields). As the demand was not limited to the legislative area of Ghent only, the Secret Council made inquiries in Brussels – by means of research executed by the Council of Flanders in

Figure 27: Fincheneers sitting in a long row along the road with their birds in small cages, 2009

Ghent – and at three local city councils which lay within the three-mile radius. The council of Oudenburg and the Country of Aalst were in favor, even though it was not in line with the feudal prerogatives of some sovereigns. Only the council of Dendermonde was sceptical about this monopoly for the Ghent fincheneers (*vynckeniers*).

On 16 December 1673, Charles II decreed a patent for the Guild of Saint-Gilles of Ghent, through which it obtained the exclusive right to catch birds within a three-mile radius outside the Ghent city center, with exception of the county of Dendermonde. The text, containing fifteen articles, is still kept in the archives of the guild in the town records of Ghent and at the university library the parchment manuscripts are preserved, which is an authentic copy of the patent drawn up in the guild headquarters of the fincheneers in Ghent.[228]

Chaffinches and other birds were allowed to be caught from the day of Saint Magdalene (20 July) until Shrove Tuesday (in February or March).[229] Those who caught birds outside this period were punished with ten sixpences for every captured bird. All clap-nets and other catching equipment were forfeited, too. It was permissable to trap in well-defined areas, with or without call birds, but the catcher had to maintain a distance to make sure that the call of the call bird was not heard by other bird catchers. The guild was disbanded in 1795, when French law allowed free bird catch in Belgium as well. At the end of the sixteenth century, these bird catching guilds turned into real fincheneers.

Renowned Ypres choir tax officer and chronicle writer Augustijn van Hernighem, who carefully noted all things taking place in his town, has left a description of one of the first Chaffinch contests in 1595: "On the first day of May in the backyard or on the premises of the cloth fuller, there was a competition to find out who was the King of Birds. In his garden there was a secluded area against a hedge, and this area was lined with a cord, so that nobody could get close to the birds. And they were caught for a person who wanted to compete to become the bird singing champion. The songs of both were carved in a wooden stick, and they started at the stroke of six up until seven o'clock; and Anthones Cottyn was Caesar, with the same birds he had also been champion with one year earlier. After all that, they went to the church of Saint-Pieter and they hung their bird cages close to the altar".[230]

From this account it is clear that Ypres probably already had a fincheneer guild in 1593, because one could only become Caesar after becoming "King" twice in a guild. There is no doubt that the modern Chaffinch sport has its origin in these early bird catch guilds. To capture birds, the catchers used call birds. In order to keep them for the next catching season, call birds were kept imprisoned during the closed season, i.e. the time during which bird catching was not allowed. Most likely someone discovered that the chaffinches had a good fighting spirit, making it possible to conduct singing contests. Driven by human curiosity about whose Chaffinch could sing the fastest and the most, the bird catchers organized contests.

In the seventeenth century genuine fincheneers guilds came into being. The principal goal of these guilds was no longer the catching of birds for food, but organizing Chaffinch singing contests. To supply the necessary numbers of Chaffinches, the guilds were dependent on bird trapping. Due to very strict regulations in the countryside, it is likely that before 1800 only the towns had Chaffinch singing competitions. Early records of fincheneer guilds at the time of the Ancient Régime show that there were no uniform competition rules. Every guild competed according to its own statutes and regulations. Yet there are some similarities in the way the competition was organized. The chaffinches were either put on the ground in a single line (Ypres) or hung above the ground (Kortrijk and Bruges). The latter method in which the bird cage was nailed to a pole at a height of about 1.5 meters is still used today in the French-speaking south part of Belgium. Normally the contests took place between six and seven in the morning and before or after it the fincheneers usually went to Mass. We assume that the guilds only organized one Chaffinch sitting contest a year. The winner was subsequently called out as the "king of the guild". In Ypres every person was allowed to add up the numbers of songs of two Chaffinches (so-called plot-sittings). In Kortrijk, however, it seems rather to have been battles between birds ("libertine sittings").[231]

We are today unable to determine whether the contests were conducted with blinded Chaffinches or seeing ones. From the nineteenth century onward the blinding of Chaffinches was a common practice, but it is impossible to find out when this barbaric practice of blinding Chaffinches or putting them in a blindered cage took place for the first time. An ancient superstition was that Chaffinches sang better when they were in a dark space or when they were literally blinded. In his 1636 *D'uitstekende Digtkunstige Werken*, Dutch poet Jan Zoet mentions that "if the Chaffinch is blinded, it sings the best".[232] In an eighteenth-century dictionary, the great French natural historian Georges Louis Leclerc Buffon wrote: "The Chaffinch is subject to having pain in its eyes and to even become blind. Therefore, one has to give it plenty to drink during several days: the juice of the leaves of Swiss chards, mixed with a bit of water and sugar. One also has to place a little twig of a fig tree so that it can sit on it and rub its eyes against it. In addition to that one also gives the Chaffinch a bit of melon seeds".[233] By the end of the eighteenth century, the blinding of Chaffinches became a general rule, as noted in 1781: "There are the blinded Chaffinches that have been blinded to sing better".[234] Despite this, we are of the impression that the first Chaffinch sitting guilds did not blind their Chaffinches.

Nineteenth century

In 1794 the French invaded Flanders for a second time within a few years. This time they succeeded in chasing away the Austrians. As a consequence of the enactment of unity on 1 October 1795, France annexed the Austrian Netherlands. The French abolished feudal privileges in accordance with the ideals of the French Revolution. The privileges included hunting rights and now everybody was allowed to catch birds freely, without any restrictions. Permission was needed only for large hunting events. This sudden change is probably the reason why so many fincheneer associations suddenly emerged, especially in the countryside. Only on 29 March 1873, was bird capture regulated again by the General Administration.Trapping was only allowed during the period of partridge hunting from July to December. The use of bird lime was also prohibited and it was not permitted to catch birds when there was snow.[235]In 1882, the law of 1846 concerning the opening and closing of the hunting and bird catching seasons was radically reformed. As a result, catching or shooting thrushes, larks and other singing birds outside legally determined periods, were banned.[236] As during the Ancient Régime, there was no uniformity in the organization of contests between different fincheneer guilds in the nineteenth century. Yet, as before, we can discern some general principles. It is certain that in the nineteenth century only blinded chaffinches were used for singing; the blinding started in the eighteenth century and was applied exclusively throughout the next century.

Before blinding the Chaffinch, the bird was put in a dark room. There it could get used to eating and drinking in the dark. Then the treatment could be started: A thin, glowing iron wire with twisted endings was used for this practice. It was held very close to the eyes of the bird. The natural reaction of the bird was to squint its eyes, and because of the heat, the scorched edges of the eyelids stuck together.[237] When in the twentieth century this practice was rightfully abolished – first by the guilds themselves and later by law – owners often succeeded in giving back sight to blinded chaffinches by rubbing their eyes with grease.

There were two ways of "playing" with the Chaffinches: the plot-sittings and the libertine-sittings; the libertine-sitting is still legitimate today. Every fincheneer participated individually in the contest. The birds were placed in a row and the bird with the most calls won the contest. With the plot-sittings things were entirely different, as associations competed with each other. Birds were registered in groups of four. The songs of the birds were added up and the bunch which had most songs won the contest. The participating associations received between three and eight awards. Two additional awards were provided for, a silver metal plate – often some kind of silver remembrance medal – for the bird with the most songs and a special prize for the association which came from the furthest region. In order to obtain the latter prize, the participating association needed a certificate from the city council where it had its headquarters. This certificate had to list the names of all members of that association.[238]

In particular, the plot-sittings were truly popular in the nineteenth century and libertine-sittings were scarce. Only after World War I did this kind of sittings become popular. Because of the plot-sittings, the chaffinches were property of the associations. Amateurs pretending to have a good singing bird could only sell the bird with the permission of the association. It was considered a great honor if an association won a contest. As in the Ancient Régime, most guilds organized only one chaffinch sitting competition a year, but some associations also held contests at a more regular basis and the fincheneers often walked many miles to participate in a contest. It was not uncommon that associations from the north of France participated in contests in the province Oostrozebeke of West Flanders and vice versa.[239] The cages were either put in a row on the ground or hung against a wall. The amateurs brought their own tally stick (ruler) to the contests. In most associations only the Flemish way of singing could be scored with a tally mark in chalk on the stick. Not all associations obeyed this rule; some also tallied Walloon songs.

Many East and West Flemish towns had their own Chaffinch sitting associations in the nineteenth century. Some of those associations still exist today and can look back at a history of more than a hundred years. Such are: *Verheugd in de Zang* (Joyful in Singing) from Aarsele (founded in 1838), *De Lustige Zangers* (The Merry Singers) from Ardooie (founded in 1897), *Niet Rijk maar Recht* (Not Rich but Fair) from Meulebeke (founded in 1849) and *De Paradijsvogel* (The Paradise Bird) from Veurne (founded in 1840).

Twentieth-century competitions

At the start of the twentieth century the first bloom of the Chaffinch sport gradually came to an end for different reasons. A source from 1907 reveals the following: "Up to this moment the pastime of blinded Chaffinches was a true popular entertainment, not only for the lower classes, but also for the bourgeoisie and the noblemen. Yet times are changing: ever since the means of transport have become so numerous and assessable, all pastimes that used to unite the people in one town gradually decreased. Today people have to go to different places and they have to look for pleasure elsewhere. In this way the number of fincheneers gradually decreased. In some places there are still guilds, but where are those days from the past?"[240]

Among the craftsmen, tailors and shoemakers there were still some amateurs, and from time to time contests. A single bird competed on its own against others, the reason why the proposition was called libertine. One bird, one deposit and each participant on his/her own: "Is not the same principle omnipresent in community life in which everyone puts his/her own interests above those of others?"

Diminished interest in Chaffinch sittings was not the only cause of the fall. Once the realization of the cruelty of blinding the Chaffinches had grown among fincheneers, some radicals used this opportunity to criticize the entire sport. A Torhout weekly newspaper spread an angry response to the blinding of birds and bird catching in general: "After that there are the brutes, or to put it differently, the Chaffinch catchers, who against all odds stretch nets across public roads and catch the Chaffinches that are useful as caterpillar eaters in your orchards. They come home and what do these brutes do? They scorch the eyes of those ignorant and innocent birds."[241] The author of this article was apparently misinformed about the way chaffinches were blinded. Not their eyes, but only their eyelids were scorched, as was described above and by no means their entire eyes were burnt. The same weekly journal published a similar indictment a few years later:

> But, unfortunately, ancient barbarian practices do not only keep on existing in the countryside but also in the towns. We are talking about the blinding of Chaffinches to make them sing. Maybe that cruelty is committed mostly in the towns, where there are chaffinches associations, which organize contests for Chaffinch singing. People still believe that Chaffinches, often enclosed in a small cage of only a couple of centimeters – where are the times that these birds could fly freely wherever they wanted – only want to sing when they are blinded, because then they believe that they are still singing aloud freely in the glorious open forest. In very much the same way they are then deluded into not thinking about their miserable imprisonment. Well, this is a misconception and the blinding of Chaffinches, the burning out of their eyes of the magnificent singing birds, is a truly fruitless cruelty. A seeing Chaffinch sings as good as a blinded one.[242]

This author also mentions the commonly spread misconception that the eyes were literally burned. In 1906, J. Yserbyt, among other things, master of agricultural sciences and nature lover, reasoned against the blinding of Chaffinches and came up with the idea of letting seeing Chaffinches sing in blindered cages: "For the birds, singing beautifully is a big but not always enviable gift, because often it leads to their ruin. Such is the case for the Orchard Finch. Everywhere these Chaffinches are caught for the summer contests… Prices for such prize-winning Chaffinches vary from 20 to 200 Fr. The fiercest among these feathered artists call about 700 to 1000 times in one hour… We assume that the Chaffinches call faster in the dark, but what if we put them in darkened cages during the contest, wouldn't one achieve the same goal as with the cruel blinding of those poor animals that we love and cherish so dearly as we do with happy mother nature, sloped by God's mild sunlight?"[243]

Some fincheneers did abandon the Chaffinch sitting sport because of the blinding. Sylvain Wittouck from Hulste tried to convince his friend in 1907: "Those people who blind the Orchard Finch commit a barbarian and cruel act. But the

Chaffinch singing contest lover resorts to this act to increase the winning chances of his champions. This inhuman practice led us to forsake the Chaffinch pastime and its contests".[244] Meanwhile in 1902 a convention was held at Paris, which became very important for the Chaffinch sport. For the first time internationally approved treaties protected birds which were useful for agriculture. Fourteen countries, among them Belgium, signed this agreement. The Belgian ratification followed in 1907.[245]

Yet World War I caused an initial first interruption. Not only the former epicenter of the chaffinch sport, West Flanders and the north of France, was in the middle of the action, but, also, the Germans were obviously against bird catching and even more to the blinding of Chaffinches. In several towns they wrote ordinances against such practices; for instance in Ghent in 1915 the catching of birds in general, and the blinding of chaffinches in particular, was forbidden.[246]

This was the last straw and many fincheneers, put off by the harsh measures, sold their blinded Chaffinches and switched to competing with seeing Chaffinches. On 15 May 1916 Camiel Erffelynch from Ghent experimented with an specially designed cage. It was a veiled cage for a seeing Chaffinch. One year later, on 13 May 1917, he displayed this new method of competing to a group of fincheneers; the outcome was satisfying.[247] After 1918 the Chaffinch sport revived in some places. The German measures were revoked and although there were contests with seeing chaffinches, blinded Chaffinches sometimes reappeared, but not for long. Growing pressure from among others Belgian Queen Elisabeth, who had experienced the misery of soldiers blinded by gas attacks in the Yser Salient, forced the Belgian government to act and brought about the prohibition of Chaffinch blinding throughout the world.

On 23 October 1921 the blinding of Chaffinches with the purpose of competing in singing contests was strictly prohibited by law. The law also stipulated that Chaffinches could no longer be caught for food; the birds could only be kept as cage birds.[248] Singing contests with blinded Chaffinches were still allowed. In 1923 there was a big protest against these competitions by blind war victims.[249] Their demonstration reached its goal and the law of 10 September 1924 prohibited the transport of blinded Chaffinches.[250] This meant the end of competitions with blinded Chaffinches. A new law of 22 March 1929 forbade also the keeping of chaffinches which were deprived of their eyesight.[251]

Meanwhile bird catching also was restrained. In 1906 the permitted period for bird catching was narrowed down. Catching birds was only allowed between 15 September and 15 November. [252]From 1922 onwards all bird catchers had to be in the possession of a letter of leave, taxed with 30 francs and signed by the district commissioner. [253]In 1932 this bird catching period was cut down even further to 1 October to 15 November.[254] This law was probably enforced under the pressure of bird protection societies, who in 1932 organized a protest against bird catching in Brussels. An earlier petition against bird capture contained 668,251 signatures.[255] The Chaffinch sitting itself had also become more difficult because of royal decrees and laws. As early as 1906, anyone who wanted to participate in bird singing competitions had to acquire a document that proved bird ownership from the local town hall.[256] The collection of this document was made even more difficult by a new law of 25 October 1929.[257]

The General Chaffinch Sitting Federation

One Sunday morning in 1933, Florimond Verstraete was having a drink – as usual – in the tavern of the Old Town Hall in Rumbeke. Suddenly, a few men with Chaffinch cages in their hands entered the tavern. Florimond Verstraete, also known as Lieutenant Verstraete and officer from World War I, chicory (endive) farmer in times of peace, recognized these lads as Chaffinch sitters. In his youth Lieutenant Verstraete had attended Chaffinch sitting competitions in his home town Vinkem. Afterwards he had lost all contact with the Chaffinch singing sport.[258] Highly interested he joined the fincheneers. They were Emiel Barry, Henri Demeester and Soten Deroo. They had just returned from a chaffinch singing competition in Kachtem. The Lieutenant displayed particular interest in their cages, and in no time he was busy talking to the fincheneers. They spoke with pride about their pastime and the difficulties they faced in keeping the Chaffinch sport alive.

From that moment Florimond Verstraete displayed an unparalleled enthusiasm for the sport. In 1934 he organized a major competition in Rumbeke; 347 participants from all over Flanders attended. With friends he traveled through out the Flemish provinces and talked to members of old Chaffinch sitting associations, encouraging their revival. He made possible the founding of several new associations and by 1937 there were already 69 associations founded or revived by Lieutenant Verstraete or his fellowship. His biggest achievement, though, was the publishing of *De Vinkenbode* (The Chaffinch Journal) in April 1935, in close cooperation with printer August Tanghe. For the first time fincheneers had their own journal, in which they could announce singing contests and publish results. Until then they had resorted to privately printed posters or asking favors of local newspapers. However, it remained difficult to reach all fincheneers as they often lived far away from each other and were dispersed all over Flanders.

Through the activities of some diligent fincheneers, among others Gaston Veys from Deinze, and with the support of Lieutenant Verstraete, today's Chaffinch sitting slowly came into being. Their dream was to establish a universal organization that could address and end the existing abuses and anomalies. In practice, there was often tampering before, during, and after the contest. For example, the numbers were not always scored correctly; some birds sang for different owners; the difference between good singing and bad singing was not always clear; the contests did not always start on time; Chaffinch sitters might change places during the scoring. Even the scoring itself was not always done properly; the rules were often misapplied; the authorized awards were sometimes missing etc.

Many fincheneers shared Veys' ideas. On 24 August 1935 a preliminary meeting took place to prepare the foundation of a new association, but ended in failure. A couple of proposals, such as the creation of a book with regulations, an obligatory small folding chair, and the introduction of a membership card, were launched. Yet the year 1935 passed without any of Veys' ideas brought into practice. Meanwhile printer Tangle sold his share in *De Vinkenbode* to Lieutenant Verstraete; he had lost faith in the survival of the weekly journal. On 1 October 1935, the first issue of *De Vinkenier* (The Fincheneer) was published. The successor of *De Vinkenbode* kept on being printed with Tanghe, but the journal was now owned and produced by Lieutenant Verstraete.

Demand grew to establish some kind of arbitrary committee to make sure that all Chaffinch sitting contests obeyed the rules and proceeded in a fair way. Once again some thought it might be a good idea to start a general association to coordinate all existing associations. The conference in Rumbeke on 5 March 1937 was a great leap forward. Before noon, representatives of different associations gathered and discussed the uniform contest regulations. They especially took into account the different propositions put forward during regional meetings and articles that had appeared in *De Vinkenier*. The participants approved certain rules and regulations which are still applied today, for example, the distance between two cages had to be eight feet or 2.4 meters. The person scoring the number of bird calls had to be seated at least one meter from the bird. Only quietly singing Chaffinches could be scored from a distance of 0.5 meter. The use of cards with the name of the amateur fincheneer, the bird, and the stamp of its association was highly recommended. Those fore-runners to the general federation card were not compulsory, though.

Another important thing agreed upon was three distinct signals at the beginning of every contest. When the first signal sounded, the birds had to be put on the ground. On the second signal the bird signers moved up to the right, and on the third signal the actual scoring of the Chaffinch calls started. The signals had to be given within two minutes before the beginning of every contest. In anticipation of the foundation of a special Chaffinch singing commission, a Mr. Erffelynk from Ghent was appointed as advisor. Every question concerning valid or invalid singing was to be brought to him. Still, the year 1938 passed without any remarkable events. The uniform contest rule book existed, but was not obligatory. The vicissitude of the singing had been discussed, but there were still troubles and people did not talk any more about a general coordinating organization or a federation. Lieutenant Verstraete felt the threat of the coming war. Since he was a reserve officer, he would be called upon if troops were mobilized. He feared for his life and was concerned about the survival of his life's work, viz. *De Vinkenier*, and therefore started looking for successors, appearing to be prepared to give up his privileged position as general leader of the fincheneers in favor of a universal coordinating federation. On the eve of World War II he founded *Algemene Vlaamsche Vinkeniersbond* (A.V.V.B., General Flemish Fincheneers' Federation).

In regional meetings held during the winter of 1938–1939 in Meulebeke and Pittem, the Federation attained its stable shape. Gaston Veys presented a number of possible candidate members, but the federation was only established in 1939. Lieutenant Verstraete took care of the financial side and associations could join the general federation for a small fee, which however meant that they had to agree with the 1938 rules and regulations concerning Chaffinch singing contests. They could also ask the chief council to act as judge in case of dispute. The long awaited coordinating organization was finally in place, albeit still in its infancy. The autumn of 1939 was dominated by mobilization and numerous secretaries and board members, among them Lieutenant Verstraete, were called to serve their country. The Chaffinch sport was put on a hold and champion ship celebrations postponed. *De Vinkenier* of 24 November published a form to be filled out by associations wanting to join the chief council. They had to send back this form accompanied by 20 francs. On 11 February 1940, 77 associations had responded.

While German troops occupied Poland and threatened to invade Belgium for the second time in thirty years, the fincheneers held their annual conference in Rumbeke on 12 April 1940. It was a historic conference, since the Federation was finally officially founded. No less that 126 associations had become members. The Federation's purpose was to improve mutual cooperation between the associations and to look after the interests of its associations and their members.

The chief council, elected one year earlier, headed the new association. Clearly the war threat brought about drastic changes. Originally, Lieutenant Verstraete only intended founding the chief council as some kind of arbitrary committee, but to guarantee the survival of the sport, he transformed the chief council into a coordinating organization. Two weeks later the former members of the chief council gathered in the pub *Oud Stadhuis* in Rumbeke. All the main tasks of the coordinating organ were defined, but before the guidelines and regulations were put into practice, the Germans invaded Belgium. The last *De Vinkenier* was published on 10 May 1940 and the fincheneer season, only just started, came to an abrupt end.

Exactly one year later fincheneers resumed activities. On 9 May 1941 *De Vinkenier* was published again. Every time the journal had to be shown to and approved by the *Kreiskommandatur* (regional war command) in Roeselare. Only those Chaffinch sittings which had been published in the journal were allowed by the Germans. Not until the winter of 1941 did the Federation start to revive. On Christmas Day, during a special meeting, some members of the previous chief council

Figure 28: The finches are kept in specially designed cages during the contest. This cage is from around 1900.

decided to lead the General Chaffinch Singing Federation. Gabriël Deschodt, the new chairman, contacted the Ministry of Economy to obtain legal status for the Federation. All chaffinch sitting associations were urged to pay their membership fee; those who neglected had to pay the double amount for announcements. Singing commissions were founded and the fincheneers were advised to release birds singing substandard songs. In March 1942, 174 associations were members of the Federation, and the number grew to 222 in 1943.

During the war years fincheneers tried to keep up activities. Despite difficult circumstances they practiced their hobby and continued constructing the internal organization of their own associations and the Federation. In September 1944 all journals which had been printed during the war were temporarily forbidden; *De Vinkenier* suffered as well. Only in 1945 could *De Vinkenier* be published again.

During the period after World War II until 1950 some remarkable events took place. The Federation continued developing and was transformed into a non-profit organization. Immediately after the war its name was changed into *Algemene Vinkeniersbond* (AVIBO – General Fincheneers Federation). Also some new statutes were passed and published in several editions of *De Vinkenier*. The Federation was headed by a chief council of seven members; daily management was taken care of by three, a chairman and two vice-chairmen. Once a year a general federation meeting – unconnected to the congress – would be held in Rumbeke, during which representatives of every member association had to be present. Likewise, after establishing districts or regions a general council of district leaders was held every year.

The Federation membership card, which had been in use for several years, became mandatory in the competition season of 1948. Every member had to have at least one Federation card. Before

Figure 29: A modern finch cage

1948 a federation card was only necessary if the fincheneer wanted to participate in the Tour of Flanders. Now the AVIBO also spread a printed rule book among its members and the set of regulations had to hang on the wall in the headquarters of every association, so it could be consulted when necessary. On 19 March 1950, thanks to the dedication of Roger van Oost, bookkeeper in service of Lieutenant Verstraete and also inspiring force within the AVIBO since 1947, the members of the chief council and the district leaders finally signed the charter of foundation of the non-profit organization, abbreviated as A. Vi. Bo. (short for *Algemene Vinkeniersbond,* General Fincheneers Federation). Only one month later the statutes of the organization were published in the *Het Staatsblad.* The founding process of the coordinating organ was in its final stage. Fifteen years had passed between the first call of Gaston Veys and the transformation of the factual association to a non-profit organization.

Everything that happened in the fincheneer world became precisely organized. Each month, the members of the chief council came together for a meeting, and once a year, their decisions were presented for approval at the assembly. All publications were spread through the Federation. Lieutenant Verstraete, honorary chairman, remained owner of the journal *De Vinkenier,* but daily management was in the hands of others and his involvement was limited. Not all fincheneers agreed with the method of organizing contests as stipulated in the rule book. Some associations, especially those from the region of Aalst, Ninove and Dendermonde, had slightly different opinions. They used different cages and also scored aberrant regional singing. In order to let them join the Federation as well, the chief council, upon request of Lieutenant Verstraete, allowed some exceptions to the general rule book concerning contests and singing regulations. Therefore a new separate division within the AVIBO was created, *Midden-België* (Middle Belgium). From 1947 onwards they held their own congress, led by H. Vander Poorten in Appelterre-Eichem. Every year *Midden-België* also participated in the assembly in Rumbeke.

At the end of the 1940s, media provoked large-scale protests against bird catching. The fincheneers were also criticized for example in the newspapers *Het Nieuws van de dag* (News of the Day) and *Het Volk* (The People). To meet the criticism, the bird organizations decided to work together. Representatives of the Belgian Cooperation of the Canary Fancy, the *Société Générale Pinsonnière* (Walloon Fincheneers Federation), the General Ornithological Federation, the Federation of Bird Catchers and the AVIBO gathered in Brussels. They founded a national union committee to defend the interests of bird lovers. Again in 1952 a bill frightened the fincheneers and called for immediate action. With the support of the Walloon Christian Democrat Philippart, members of parliament wanted to prohibit bird catching with nets. The bird catchers, with the support of Flemish fincheneers, immediately responded to the proposition with a petition. At the beginning of

1953 the bill of Phillipart was definitely rejected by the parliamentary committee for agriculture.

In collaboration with the Poperinghe politician Lahaye, the fincheneers made some important progress in 1954. Article 7 of the Royal Decree of 25 October 1929 was changed, which meant that in the Flemish part of Belgium Chaffinch sitting contests were allowed between 16 April and 31 August, one month longer than before. The law was enacted on 28 July 28 1954. Still, from 1962 the protest against bird catching revived again in media and parliament. In Tienen on 2 December the *Nationale Federatie van de Belgische Vogelvangers Verenigingen* (N.F.B.V.V., National Federation of Belgian Bird Catchers' Associations) organized a vast protest march in order to preserve bird catching. Their action could not prevent a new Royal Decree in 1964, which put heavy restrictions on bird catching. Caught birds had to live in cages of 20 x 20 x 20 cm, with a maximum of one bird per cage. The Royal Decree had also consequences for the fincheneers, as their contest cages did not meet these requirements. At end of 1964 again a protest march was organized in Roeselare. More than 30,000 people participated, with representatives of AVIBO, the N.F.B.V.V. and the Federation of Hunters. The protest had the expected result: The minimal measurements of the cages were revised.

In the beginning of the 1970s it became clear that the opponents of bird catching would win the battle. Stubbornly trying to maintain free bird catching, like the N.F.B.V.V. did, could in the long term be disadvantageous for the fincheneers' cause. On 10 June 1970 the Cabinet board approved the Benelux convention concerning hunting and bird protection, without preliminary advice or involvement of the bird lovers' federation. This is the reason why the AVIBO, which is mainly concerned with the representation of the fincheneers and only involved in bird trapping indirectly, and the AOB, the national federation of bird breeders, changed course in 1971. They opposed commercial bird trapping and trade in wild-caught birds. Both federations looked for a course which reconciled all parties involved and they devoted their efforts to obtain permission for limited bird catching during weekends with the sole purpose of supplying fincheneers and breeders. The idea was completely rejected by their former ally N.F.B.V.V., and the federations became antagonists.

In a comprehensive article published in *De Vinkenier* on 16 April 1971, Roger Van Oost promoted AVIBO's point of view regarding bird catching, and he described the cause of the breach with the N.F.B.V.V. The latter changed its name to *Belgische Ornithologische Federatie* (BOF, Belgian Ornithological Federation). Roger Van Oost explained that although they renounced the former name of bird catchers' federation, when AVIBO urged BOF to weaken their demands and to obey the call of the Minister of Agriculture, there was no response. The Minister had clearly stated that without any concession from the bird catchers, dialogue was no longer possible between the two parties. AVIBO feared that because of BOF and its irrefutable stance, bird catching would be abolished, and was therefore prepared to make some concessions, fighting mainly for bird catching during weekends. The breach with BOF was inevitable, but the average fincheneer, afraid of losing a precious pastime, supported the viewpoint of the Federation. Some die-hard bird catchers, however, caused anticipated havoc within the fincheneers' ranks.

The Minister of Agriculture Leo Tindemans made a final decision on 20 July 1972 and the Royal Decree was published in the *Het Staatsblad* on 1 August. Following the Benelux convention, trading, catching and transporting birds was now prohibited permanently. The bird catchers had lost and among fincheneers there was great consternation; their proposal of weekend catching had not even been considered. Yet the decree did provide that Chaffinches could be transported between 16 April and 31 August to participate in competitions, with a special transport certificate. Every bird keeper had to make an inventory of all birds that he or she possessed and the mayor had to examine those inventories and sign them.

Despite criticism by disappointed BOF members and member disappointment, the chief council of the Federation continued its efforts. At first it encouraged the breeding of birds and due to the efforts of the chief council, the AVIBO was able to obtain more favorable legislation in 1973, when people were allowed to catch some bird species again, but within a restricted period, under strict surveillance and prescribed methods. This limited bird catching was allowed to provide many bird lovers with the necessary "fresh blood" (young birds) for breeding. Every captured bird had to be ringed and listed in an inventory, which was filled out annually and handed over to the local authorities. This was not a perpetual right, and every year the Minister had to list the number of birds and the species allowed to be caught. The AVIBO divided 15,319 inventories among its members.

With this permission, the fincheneers managed to survive, but because of the additional administrative obligations, their pastime had not become any easier. The number of fincheneers kept growing and the problems concerning bird catching and collecting were the main concerns during the 1970s. The chief council of AVIBO had to put all its forces together to preserve the Chaffinch sport. In 1972, not only was capturing wild birds forbidden, but owning Chaffinches became more difficult, as every bird had to be ringed and inscribed in a register, with a green inventory for birds collected from the wild, and a red one for birds bred in captivity. The transportation certificate, which had existed since 1932, disappeared and was replaced by the second part of that register. Also in 1972, also a guideline was published by the European Commission to protect European bird species so that fincheneers not only had to deal with the national legislation, but with the international legislation as well.

On 20 April 1974 Lieutenant Verstraete died. Since he had been the owner of *De Vinkenier*, his family kept publishing the weekly journal, but without enthusiasm. The chief council succeeded in buying the rights of ownership, but *De Vinkenier* languished towards the end. With the 18 October number. the journal was only published with two printed pages. After the fortieth volume *De Vinkenier* ceased publication. Soon a new journal followed: the *Aviboblad* (Journal of AVIBO). A ministerial decree on 15 September 1980 simplified the inventory of caged birds for the Flemish district. The decree was supplemented and changed by a new Royal Decree of 9 September 1981 (applicable on 14 September 1981) concerning the protection of birds in the Flemish district. The green and red registers, in use since 1973, were replaced by new inventories and index cards. Every year on 31 January the fincheneers made an inventory of their European birds. A duplicate of that inventory was sent to the Services of Waters and Forests of their province. The index cards served as owner and transportation cards.

Every year AVIBO awaited with great suspicion the approval of the Minister for the necessary supplementary collecting. With one strike of the pen, the current recruitment from the wild by the fincheneers might be forbidden. On the proposal of Jan Verroken, former vice-chairman of the Belgian Chamber of Representatives, Marc Olivier drafted a decree in 1983. Its purpose was to safeguard the bird singing pastime in general, and the Chaffinch sport in particular, for the future. He also suggested restricting the collection of birds from the wild to a strictly necessary minimum, and regulating trapping methods. He also wished to involve the associations more than before in the policy-making, giving them additional responsibilities, and to limit the heavy administrative burden to an absolute minimum. After years of discussion, the responsible chamber commission withheld the decree following some agitation in the press and other forums. Both the global number of birds and the number of different species which were allowed to be caught were systematically restricted. In particular, in 1977 and 1982, the Minister drastically cut the collection quota. From 1990 onwards, Chaffinches were the only species allowed to be caught.

Recent changes

Several times the Chaffinch sport was criticised in the media, especially regarding the controversy surrounding the abolition of bird catching. Fincheneers were labeled as bogeymen. On 10 December 1977, the AVIBO organized a large-scale press conference for the first time, and the basic principles of Chaffinch sitting contests were explained. In this way the Federation hoped that the press would finally write something positive about the fincheneers. After the abolition in 1972, the fincheneers had to stay true to the legally permitted collection quotas. Since 1983 those quotas had been permitted without any problems by the responsible minister, but in 1993 this came to an end. Because of a ministerial decree on 30 July 1993 (published in *Het Staatsblad* 28 August) the fincheneers were disadvantaged again. The AVIBO received 16,900 rings, but half of them were meant for female Chaffinches (called poppets). Following a decisive protest action of the chief council, supported by sympathizing parliamentarians, Minister De Batselier partly revised his decision. He reached a compromise between the parties involved, which resulted in the signing of a protocol.

With this obligation to ring a considerable number of poppets, the Minister hoped to encourage the captive breeding of Chaffinches, albeit through some coercion. In 1997 the representatives from the fincheneers' associations spoke to the responsible minister again. Minister Kelchtermans, successor of De Batselier, renewed the protocol for five years. The obligation of catching poppets was omitted and each year the number decreased with 500 birds. In 1999 this new protocol, however, was abolished by the Council of State under the advice and incitement of bird protectionists. The new Minister, Ms. Dua, regulated the collection with a ministerial decree published every year. The bird protection federation kept on protesting to the Council of State. The minister listened to the advice of this organ and since 2003 there has been no more collecting from the wild in Flanders.

Without the coordinating organ AVIBO, the chaffinch sport would probably have been forgotten by now. Especially during the last couple of years, the chief council has shown its wisdom and consistency in a never-ending battle to preserve this old pastime. AVIBO has made a definite step towards breeding chaffinches instead of collecting from the wild. The number of captive-bred male birds and the number of breeders and owners of bred Chaffinches is clearly rising.At this moment AVIBO counts 47 districts: Aalst, Aalter, Alken, Antwerpen, Anzegem, Ardooie, Avelgem, Brakel, Brugge, Deerlijk, Deinze, Denderleeuw, Diksmuide, Eeklo, Gavere, Gent, Gistel, Haacht, Helkijn, Houthulst, Ieper, Ingelmunster, Izegem, Knokke, Komen, Kortrijk, Kruishoutem, Menen, Moeskroen, Ninove, Oostnieuwkerke, Oudenaarde, Poperinge, Proven, Roeselare, Ronse, Sint-Eloois-Vijve, Tielt, Torhout, Veurne, Vichte, Wakken, Waregem A, Waregem B, Zonnebeke, Zulte and Zwevezele. The regions where the Chaffinch sport is practiced the most are the two Flemish provinces of West and East Flanders. The 47 districts host 271 associations, a total of 12,000 fincheneers in 2008.

Walloon provinces

As in the Flemish provinces, in the Walloon provinces of Belgium Chaffinch sport is still very much alive. If West Flanders is the center in Flanders, the area around Verviers, Liège, the land of Herve, Namur, and Hainaut are the main places in the Walloon provinces. The method of "playing" in the Walloon provinces is entirely different from the one used in Flanders. The Chaffinches are put into cages of 15 cm in breadth, 20 cm in height and 20 cm in length. The barred front of the cage is open, which allows visitors of the contest to see the Chaffinches. Those cages are attached to a plank called *raie*, and are subsequently put up on poles of about 1.75 to 2 meters high. The distance between each hanger is only 40 cm. In front of the row of cages there is a row of benches, on which the scorers may sit during the concert.[259]

Contests in the Walloon provinces start at nine o'clock in the morning, three in the afternoon and six in the evening. They mainly take place in the afternoon and last about two hours. During the first hour the fincheneers put up their cages on the poles and the Chaffinches can get used to the surroundings, their audience, and opponents. The actual contest only starts in the second hour. The scorers take their places five minutes before the start of the second hour and begin marking the calls of the Chaffinches following the signal of the contest leader. The combination of listening to judged chaffinch calls and at the same time seeing them sing is a special experience. Contrary to Flanders, where only Flemish or indigenous singing is signed, variations are allowed in the Walloon provinces. Lambert Grevendal, a prominent fincheneer, once distinguished no less than 400 different kinds of Chaffinch calls, which he divided into fourteen different categories. Because of the closer distance between two cages, open cages and the longer duration of the contest, Walloon Chaffinches have fewer calls in one hour than in Flanders. The best Walloon singing birds perform about 500 to 600 calls.

Already around 1800 there were Chaffinch sitting contests in Liège and surrounding areas. A man called Tomassin, who wrote a description of the Department of Ourthe, the contemporary province of Liège, recounts how certain neighborhoods of Liège and Verviers and other villages in the area held Chaffinch singing contests on a regular basis. Approximately forty or more cages stood in one row surrounded by listeners and spectators. The fincheneers called their birds after famous generals or other celebrities. When one of the prize birds died, it was buried with the necessary decorum and with much attention.

Halfway into the nineteenth century the first true chaffinch singing associations were founded in the Walloon provinces. The still exant association *Les Oiseleur et Pinsonniers de Dolhain-Limbourg* (The bird keepers and fincheneers of Dolhain-Limbourg) was founded around 1846. Also Soumagne, Grand-Rechain, Petit-Rechain and Dison had associations organizing Chaffinch singing contests. About 1880 the city council of Mons organized a major Chaffinch sitting contest on the *Beestenmarkt* (Animal Market) on the Tuesday of the *kermess* (fair), and continued the tradition for about ten years. At the end of one such contest the winning Chaffinch was traded for a cow. In 1900 Liège bird friends founded the federation of bird amateurs and keepers (*La Fédération des tendeurs et amateur d'Oiseaux*). Chaffinch breeders as well as bird catchers, exhibitors and fincheneers were members of the federation. The idea was raised to establish a general fincheneers federation, a Walloon AVIBO or *La Société Générale Pinsonnière* or *Sogépin,* which would regulate all chaffinch singing contests. World War I thwarted the se plans and only in 1920 a Chaffinch sitting took place under the wings of Sogépin.

When competitions with blinded Chaffinches in the mid-1920s were forbidden, Chaffinch sport in the Walloon provinces seemed to die out. Yet a few prominent fincheneers, as in Flanders, succeeded in promoting the contests with sighted Chaffinches and the sport revived; the above mentioned Lambert Grevendal played an important part in this matter. In Liège, in 1937, there were 21 Chaffinch singing contests. The largest Chaffinch singing contest took place in Jalhay with around 375 Chaffinch sitters. Grevendal also tried to unite the existing associations in one federation and organized a congress in Vervier on 4 February 1940, but this time World War II stopped the initiative.

To the Walloon provinces Lieutenant Verstraete came as a Delivering Angel. In 1943, he left some space in his weekly journal for announcements and results of Walloon sittings. Grevendal regularly published contributions in French. In order to further encourage the Walloon bird lovers, Lieutenant Verstraete started a special competition called the Tour of the Walloon Provinces. Chaffinch *Anatole* from Verviers, owned by R. Pepinster, won the first Tour. In 1944, the back cover of the journal *De Vinkenier* was completely reserved for Walloon fincheneers. Lieutenant Verstraete baptised it *Le Pinsonnier* (fincheneer). In 1945, after the end of World War II, the *Sociéte Générale Pinsonnière* was re-established with the financial support of Lieutenant Verstraete and *Le Pinsonnier* became a separate weekly journal. By 1951 there were no less than 53 associations-members of *Sogépin* and they were spread over four regions: Verviers, Liège, Namur and Charleroi.

Frightened by the growing legal requirements and regulations, many Walloon fincheneers pulled out. At this moment around 3,000 fincheneers are members of *Sogépin* and the federation unites 33 associations. As in Flanders, Walloon fincheneers are dependent of the goodwill of the regional government. This body must grant permission each year to procure Chaffinches from nature. Probably, in the near future, in the Walloon provinces, the Chaffinch sport will also have to be practiced with home-bred Chaffinches.

France

In the nineteenth century Chaffinch sport was very popular in the north of France. The roots of this popularity can be traced back to the eighteenth century, when associations were founded in cities such as Dunkerque, Douai, Armentières and Bergues (Dutch: Sint-Winoksbergen). French fincheneers regularly participated in singing contests in West Flanders and the French method of playing was similar to the one in Flanders. In 181,1 in Armentières there was a big contest in which many guilds participated. The contest was announced in the press. Yprès took first prize before associations from Nieuwkerke (French: Neuve-Chapelle) and Armentières. A certain Valonges from Yprès was the individual winner, with 772 calls within one hour.[260]

The wars, the ban on blinding Chaffinches and abolition of bird catching at the beginning of the twentieth century caused the chaffinch sport to almost disappear in France, but there are still fincheneer competitions in the areas of Tourcoing, Halluin, Bousbecque, Roncq and Lincelle. Despite legislation which prohibits catching and keeping European bird species in France, and a government which never intervenes, a dozen fincheneers regularly organize chaffinch sitting contests, following the Flemish regulations of playing. Yet the sport is likely to disappear soon in France, as there are hardly any new fincheneers, the number of contests is rather limited and the activities are on the borderline of legal.[261]

In the 1950s Flemish fincheneers from Maritime Flanders and the area of Mouscron tried to revive the French Chaffinch sport. In 1953 they organized a couple of contests in Halluin and Neuville-en-Ferrain across the border. The border control point allowed Flemish fincheneers to cross the border and they could participate in the contests, but due to a lack of interest the initiatives died out quietly. Today there are some Chaffinch sitting contests in the border area near Belgian Mouscron and Rekkem. These contests are supported by Belgian fincheneers, but poor organization and lack of a clear rule book are not beneficial to the contests and probably they will gradually end.

The Netherlands

The origin of the Chaffinch sport in the Netherlands is unknown. In the Low Countries in the nineteenth century bird catching was very popular. There was even a period during the year called "the Chaffinch Holiday". Students and teachers from Leiden University and Amsterdam Grammar School had some days off in the beginning of October to catch Chaffinches. They were not caught for singing contests, but because they were a local culinary delicacy.[262] The bird law of Minister Dekkers in 1936 completely abolished bird catching and prohibited keeping of Chaffinches and the participation in Chaffinch singing competitions with protected birds. A new law from 1965 abated this law, and for periodsm the Dutch were allowed to breed and keep seven kinds of cage birds, Chaffinches included. Catching birds was unthinkable and strongly prohibited.

Despite the 1936 law, fincheneers of the General Limburg Fincheneers Federation (GLCSF) continued with Chaffinch singing contests, but only in Limburg, and more specifically in the area of Heerlen and Meerssen. The federation heavily opposed the ban on singing contests and tried to refute the main objections against Chaffinch competitions. According to their opponents, Chaffinches were put in too small cages and drugged with bird seed drenched in Dutch gin. The birds participating in singing contests were secretly caught and the contests ended up as gambling places.

Chairman Heidendal and secretary Van Heesch wrote the statutes and regulations of GLCSF. The organization united 59 associations and counted about a thousand members. It kept on arranging Chaffinch contests illegally. Lieutenant Verstraete, never shy for stunts, supported the Dutch fincheneers and in *De Vinkenier* published a couple of Dutch results. In 1946 he had the idea of financing a Tour of Holland similar to the Tour of Flanders, but because of a lack of Dutch subscriptions this plan was not executed. A short time after the Flemish fincheneers lost contact with their Dutch colleagues.

Before and after World War II, there were active fincheneers among the members of the Dutch Federation of Bird Lovers (DBLF), originally a federation busy with the keeping and breeding of European birds and later on also tropical birds and colored canaries. In 1939–1941 they arranged singing contests in Rijssen near Twente. In 1942 four fincheneer associations joined DBLF, Heerlen I, Heerlen II, Heerlerheide and Voerendaal. One year later a fifth association joined, Schaesberg. GLCSF did not like this and decided to unite with DBLF. In 1952 GLCSF with its 223 members operated as a separate division of DBLF. In 1976 J. Heinrich and J.H. Douven, frontmen of the Dutch fincheneers, tried to reinforce contacts with their Flemish friends and in 1985 members of AVIBO attended a contest of the association *Slag op Tijd* (Battle on Time) from Heerlen. Eighty-one

Dutch Chaffinches participated in the contest and the Chaffinch of Mr. Passen from Maastricht won. Today there are still weekly Chaffinch singing contests in Limburg. The method of playing in the Netherlands is considerably different from the Flemish and the Walloon methods. The birds are housed in barred cages covered with a white cloth. The cages are 25 x 25 x 25 cm. At ten o'clock, sometimes at nine o'clock, the cages are hung on poles two meters high. The hooks on the poles on which the cages are hung are exactly 50 cm apart from each other. During the first hour of the contest there is a kind of pre-selection. The Chaffinches listen to the other birds and those that are not singing are removed from the poles.

After one hour the actual contest starts. The amateur fincheneers give up their places to members of the jury, who score the Chaffinch calls. In front of every cage there are two members of the jury and they sign the calls for one minute. All calls, either Flemish or Walloon, are valid. After one minute the observers move and sign the next Chaffinch. Different groups of jury members judge the Chaffinches, so that every bird is listened to and signed at least six times an hour. The Chaffinch with the most songs in those six minutes wins the competition. The awards are similar to those in Flanders: Cups, flowers and a small amount of money.

Germany

The Chaffinch is a very common species in Germany and singing contests were held early. In the nineteenth century Chaffinch singing contests were arranged in many places. Especially in the Harz Mountains and the forest villages of Thuringia, chaffinch sport is still very popular today, and in the area around Aken, Wuppertal-Eberfeld, Essen, Ludwigshafen and Oberhausen fincheneers were and are active.[263] At the end of the eighteenth century, the renowned German ornithologist Johann Matthäus Bechstein described how smiths and pipe makers from Thuringia went to the Harz Mountains on foot to buy some excellent Chaffinches. Contrary to the Flemish, Walloon and French fincheneers, the German fincheneers did more than just arranging Chaffinch singing contests. They were especially productive when it came to studying the Chaffinch in general. The blinding of Chaffinches was unknown and not applied. In the nineteenth century numerous works appeared on how to keep and maintain Chaffinches. Many authors tried to define the different ways of singing and regional varieties and categories. When Lambert Grevendal did a similar thing for the Walloon provinces in 1933, he mainly based his work on previous German publications.

The first known Chaffinch singing contest – *Finkenmanöver* in German – was arranged in Thale in 1868. On Easter Monday fincheneers from Altenau, Benneckenstein, Blankenburg, Clausthal-Zellerfeld, Hohegeiß, St. Andreasberg, Wieda and Wildemann took the awards; almost all were from the Harz Mountains. In the 1930s especially, the Chaffinch singing contests became extremely popular and within the Imperial Federation of German Bird Tenders and Bird Breeders there was a research department for keeping Chaffinches that was very active. This research association united four national departments and counted more than a thousand members. In 1978 the Buchfinken guild Südniedersachsen (south part of Lower Saxony in northwestern Germany) united eight associations from the Harz Mountains and about 600 fincheneers.

In contrast to the Flemish fincheneers, the Germans only participate in a couple of contests each year. Almost all associations organize only one sitting, mostly around Whitsun, and there are only about ten Chaffinch singing contests each year. The German method of playing is quite unique. The birds are in barred wooden cages, length 26 cm, breadth 16 cm and height 22 cm. The cage is covered with a white cloth of 90 x 90 cm so that the bird cannot see outside and cannot be seen either. Many of the white cloths carry a carefully embroidered emblem. There are three methods of playing, called classes. Often all three methods of playing are used in one competition. The contests start very early in the morning, sometimes even at five o'clock.

The oldest method of playing, *Starkklasse* (strong class) or *Stark-Singen* (strong singing) tries to crown the "Strongest" Chaffinch, possessing the greatest endurance, as the winner. When the contest starts, all cages are put on the ground in a circle, 50 cm from each other. The owners move up to the left and score the bird of their neighbor for about five minutes, as in Flanders. All Chaffinches that have not sung during those five minutes are taken away from the circle. In the next round the circle becomes smaller and the cages get closer to each other. Again there are five minutes of scoring, and again the non-singing birds are taken away. They repeat this procedure five times, which diminishes the circle. In the last round all remaining Chaffinches are put on a table, one cage opposite another. The bird that gives the most calls in the final five minutes wins the contest. Fincheneers, who feel that their chaffinch is going to lose, can take away their bird and so end the battle. Earlier on people used to play until there were only two Chaffinches left. They were then put opposite each other and the "strongest" bird won the contest.

During the 1930s, a new method of playing was developed, *Kampfklasse* (competing class) or *Distanz-Singen* (distance singing), which seems related to the Flemish method. All cages are put in one row on an scaffold of about 1.6 meters; the distance between two cages is 50 cm. During half an hour all fincheneers sign the Chaffinch of the person sitting next to them; the Chaffinches remain in their places. The winner is the one singing most songs in that half an hour.

Also in the thirties a third method of playing originated, *Schönheits-Singing* (singing for beauty) or *Schönheitsklass* (beauty class). The Chaffinch singing the most beautiful call was crowned winner. The contest usually starts two hours before the Chaffinches of the *Kampfklasse* can show themselves, usually at five o'clock in the morning. The starting order of the participants is drawn by lot and the first participant bird is put on a table. Three judges (beauty judges) listen to the Chaffinch during five minutes. The different calls are translated into words and then judged following a complicated scoring scheme. The Chaffinch with most points wins the contest.

As elsewhere, in Germany, animal rights organizations are against the Chaffinch singing sport, especially the too small cages. The fincheneers refute this argument by answering that the birds spend most of their lives in spacious aviaries and only stay in such a cage for a couple of months. Many Chaffinches also live longer, some up to twenty years in captivity; in nature the oldest age of a Chaffinch is found to be thirteen years. The origin of these Chaffinches is a topic animal rights activists heavily criticize. Since the 1950s, bird catching has been completely forbidden in Germany and today's Chaffinch keepers claim that their birds are all, without exception, bred in a protected environment. Animal rights activists strongly doubt that and suspect that many chaffinches are still obtained through secret bird trapping. Yet, the fincheneers, not only bird lovers and bird specialists, also keep old traditions alive.

Chapter 6. Keeping Birds as Domestic Pest Control

Sabira Ståhlberg and Ingvar Svanberg

For thousands of years different kinds of animals have helped humans hunt other animals. Best known is the domesticated Wolf (*Canis lupus*), which developed into the dog some 15,000 years ago. Humans have also benefited greatly from the Horse (*Equus ferus*), and, in some parts of Asia, from the Cheetah (*Acinonyx jubatus*). Ferrets (*Mustela putorius*) and Eurasian and Smooth-coated Otters (*Lutra lutra* and *Lutrogale perspicillata*) are other examples of mammals that have been tamed in order to assist in hunt¬ing and fishing. Various birds of prey have also hunted or pursued game for humans in many parts of the world. Fishing assistants were e.g. the White-tailed Eagle (*Haliaeetus albicilla*) in Sweden, Osprey (*Pandion haliaetus*) in Hungary, Common Merganser (*Mergus merganser*) in various parts Europe and the Great Cormorant (*Phalacrocorax carbo*) and Japanese Cormorant (*Phalacrocorax capillatus*) in parts of Eurasia (China, Japan and England). Eagles, falcons and other raptors have since time immemorial been used for hunting mammals and birds in many parts of Eurasia. In old Russia the Grey Great Shrike (*Lanius excubitor*) was sometimes trained for catching small birds.[264]

Animals, especially birds, were also used as helpers to get rid of vermin in human dwellings. Pest control is a crucial issue since early in human history. Houses and huts were infested with all kinds of unwanted animals which damaged food, furniture and clothes, bit the inhabitants or infected them with illnesses, and in general disturbed domestic peace by their presence.

Mechanical destruction of vermin is maybe the most important and widely used method throughout history, but chemical (including poison) and biological methods are almost as widespread.

One way to keep the uninvited guests under control was to tame different animals. The Cat (*Felis catus*) is a well-known example, with a continuing global impact on mouse populations. In Europe Hedgehogs (*Erinaceus europaeus*) kept rodents and cockroaches under control; in southwestern Asia Marbled Polecats (*Vormela peregusna*) were used to exterminate rodents; in South Asia the Mongoose (*Herpestes* sp.) held snakes at bay; and in Scandi¬navia tame Stoats (*Mustela erminea*) were used to control rodents in barns. In western Siberia, explorer Peter Simon Pallas in the eighteenth century mentions that a certain kind of ant was encouraged to live in houses to clear out Bedbugs (*Cimex lectularius*).[265]

Controlling snakes and mice

Taming wild birds was a popular pest control method among peasants in several regions of the world. Free-ranging wild birds were often kept in the yard or flew about in homes. An often quoted example in ornithological literature is the Secretary Bird (*Sagittarius serpentarius*), which is reported to be kept on South African farms in order to control venomous snakes in the yard.[266] Domestic Ducks (*Anas platyrhynchos* forma *domesticus*) were according to Tao Huo (1487–1540) used in southern China during the Ming period for controlling amphibious crabs in rice paddies. In Mexico, tamed Roadrunners (*Geococcyx californianus*) were popular both within houses and outdoors, consuming mice, smaller snakes and other reptiles as well as insects.[267] On the Hungarian plain, peasants kept Common Kestrels (*Falco tinnunculus*) inside buildings to keep mice in check. In nineteenth-century Italy it was common to keep one or several specimens of the Little Owl (*Athene noctua*) in houses and gardens in order to capture rodents, slugs and insects.[268]

Also common was also the use of birds specifically as insect pest controllers. In agricultural areas, barns and fields are full of insects which destroy crops and consume produce. Some domestic birds such as Chickens (*Gallus gallus* forma *domestica*) keep worms and larvae under control in the yard. A few insect-eating wild birds which usually nest around humans, for instance swallows, wagtails and storks, have been welcomed to such an extent that they have entered folk beliefs and folk tales throughout Europe.

Using fly-catching birds in Sweden

Flies of different taxa, especially the Housefly (*Musca domestica*), are a general problem in agricultural environments. The proximity of cattle, faeces, food and all sorts of filth creates a favorable environment for them. Only a hundred years ago in northern Sweden the situation was described as very difficult: "In the small windows inside the house floated such huge clouds of flies that it would be impossible to form a fully accurate representation for someone, who has never seen anything like it; when food of any kind was served, it was instantly attacked by entire swarms of flies", wrote local historian Levi Johanson.[269]

Except for mechanical extermination of flies by home-made fly flaps or sticky, tar-smeared paper, Scandinavian peasants used various kinds of natural poison, for instance Fly Agaric (*Amanita muscaria*), Wolf's-bane (*Aconitum*) and Common Toadflax (*Linaria vulgaris*).[270] Another popular method was to catch wild insectivorous birds and release them in the house. This could be characterized as a simple form of biological pest control. Examples of wild birds as pest controllers abound in Sweden and northern Finland and there is also some information found in central Europe.

In a manuscript from the 1790s, Carl Linnaeus's disciple Pehr Osbeck describes in detail the biological diversity in the southern Swedish province Halland. Among many bird species, he mentions the *flygfängare*, 'fly-catcher'. From his reference to Linnaeus's scientific name, it is clear that Osbeck alludes to the species we know today as the Whinchat (*Saxicola rubetra*). As a devoted Linnaean, Osbeck did not miss the opportunity to highlight the economic benefit or usefulness of wild animals. A little note which explains the local popular name 'fly-catcher' also suggests that the peasants in Halland (or at least some) caught Whinchats and set them free within the house to keep flies away. He wrote: "It is sad to keep birds in a room without a cage, because of their filth, but if the bird may fly free, as in the case in many peasant homes here, it will soon eradicate all flies in the house."[271]

For the method to be efficient, humans need knowledge about the diet and behavior of both the prey and the catcher. The first step is to catch the pest controllers. Olof Christoffersson in Skåne

Figure 30: Chinese Quail

(southernmost Sweden) in the late nineteenth century tells how Goldcrests (*Regulus regulus*), settling on trees in gardens in the autumn, were captured by peasants with a stick dipped in tar. It was usually easy to catch them, but one had to be careful not to put tar on the wings, because it hindered the birds from flying. The tar was removed with swine lard and the birds were released indoors, where they soon gathered courage and started catching flies. As soon as they had cleaned the house from flies they were set free, but the birds could be kept inside for even for a couple of weeks. The peasants also captured Wrens (*Troglodytes troglodytes*) and Robins (*Erithacus rubecula*) for the same purpose, but these never became tame in the same way as Goldcrests, and escaped as soon as they had a chance.[272]

Figure 31: Secretary Birds were kept in South African farms to get rid of snakes

There are eyewitness reports about birds in homes from other parts of Europe as well. In sixteenth-century Switzerland, Redstarts (*Phoenicurus phoenicurus*) were kept in homes to catch flies. According to an ornithological manuscript from 1575 by Jodocus Oesenbry, the Redstart was held in winter inside the house and it "secretly cleaned the room of flies", thereby singing pleasantly as a result. Still in the nineteenth century, the Common Redpoll (*Carduelis flammea*) was kept in peasant homes in the vicinity of Schlotheim in Thuringia in central Germany for the same reason. A Danish handbook recommends fly-catchers as best suited for the purpose.[273] Also Alfred Brehm informs that if the Pied Flycatcher (*Ficedula hypoleuca*) is set free in the room, it thoroughly cleans the space of flies and other insects. Karl Naumann tells how a German peasant boy caught a Spotted Flycatcher (*Muscicapa striata*) female with her young. The bird effectively cleared the house and in order to feed it further, the boy carried his bird from cottage to cottage, thus also relieving the neighbors of their difficult guests.[274]

Cockroach eaters in Finnish farm-houses

Cockroaches compete with humans for living space, and most importantly, for food, clothes and different materials such as leather. Flies take up visible space and their attacks can be seen and diverted, but cockroaches are mostly nocturnal and wander about in places of the house, where they cannot be easily observed. Once infested, a house could be cleaned only with greatest difficulty before the invention of modern pesticides. Around the Arctic Circle, Forest Cockroaches (*Ectobius lapponicus*) were constant inhabitants of human settlements only a century ago. To combat them local people, both Swedes and Saami, tamed waders such as the Greenshank (*Tringa nebularia*), Spotted Redshank (*Tringa erythropus*) and Wood Sandpiper (*Tringa glareola*). In the eighteenth century, Carl Linnaeus himself observed these pests: "They live in Lapland and do great damage to food, especially to the fish when it is dried in the sun." He adds that the cockroaches could eat up all food stores in one single night.[275]

In summer 1906 northeast of Rovaniemi in Finnish Lapland, a tamed Greenshank was observed in a cottage. It had lived in the house for a couple of years and was sometimes lent to neighbors, who needed assistance in the fight against the abundant vermin. It was also common in the region to catch waders, both young and old, for cockroach control. "A great pleasure", wrote Ernst Orstadius who received this information from a friend, "it is to see how he [the bird] again and again, quick as a rat, runs over the floor to catch cockroaches as soon as they show themselves." Ruffs (*Philomachus pugnax*) were also tamed for the same purpose. They were caught in the spring with horsehair snares and easily kept at home, feeding on the cockroaches.[276]

Most wild birds caught for fly-catching were set free as soon as they had cleaned the house, but waders and bigger birds were kept for longer periods. They were easy to take care of and could live for several years in a home, like a domestic animal and constant pest controller. In a neighboring village of the cottage in the previous example, a forest watchman kept several birds: a Spotted Redshank, Ruff and a Wood Sandpiper in his cottage. One of them had been there for several years; the two others had been caught in the previous spring. Aside from cockroaches, which were their staple food, their diet also included also of other insects, bread and *piimä*, a kind of sour milk. The food had to be thrown on the floor, as the birds did not eat out of human hands. When for instance a ball of bread was thrown to them, they immediately pounced upon it. On the floor the family had put a tin pan with sand on the bottom and water, where the birds would drink after eating bread.

When the birds slept, they lay down on the floor, gathering their feet underneath the body. They could also discern between their owners and strangers and never left the cottage even when the door was open. All year round they kept to the house, freely roaming in the room, and they appeared to lead a good life. How these birds co-existed with predatory cats, dogs and other domestic animals we do not know, but the same eyewitness informs that in another cottage, a Common Sandpiper (*Actitis hypoleucos*) and a squirrel lived together on most friendly terms.[277]

In a completely different area and climate, Cuthbert Christy, at the end of the nineteenth century, reported from Santo Domingo that the Double-striped Thick-knee (*Burhinus bistriatus dominicensis)* sometimes was kept in houses in order to capture cockroaches which invaded the islanders' homes. In Chaco and Paraguay, the Guira Cuckoo (*Guira guira*) was "said to rid houses of all the creeping and running insects", according to an

Figure 32: The Muscovy Duck has become increasingly popular because it eats slugs and snails in gardens

ornithological study from 1926.[278] In several parts of Europe, Quail (*Coturnix coturnix*) were kept in houses to eat insects. There are also observations from Japan in the 1860s on keeping insectivorous birds as insect predators on farms.[279]

Other vermin controllers

Full-grown birds were not the only ones to keep vermin under control. In Joensuu, eastern Finland, the eggs of Curlew (*Numenius arquata*), hidden somewhere in the house, would according to popular belief chase away bugs.[280]

Another Finnish belief from the Archangel area in northeastern Karelia tells that when the Black-throated Loon (*Gavia arctica*) is under the roof, "bugs do not enter with the guest [the bird]".[281] At the turn of the twentieth century, British boys were encouraged to catch Jackdaws (*Corvus monedula*). This bird would become very tame and sit on a fist, and it could also be useful. If the boy shone a torch on a hut or garden-house, infested with many earwigs and other insects, the bird would promptly pick the insects off the wall with its powerful beak.[282]

In southern Tyrol, a peasant told a television crew, in 2003, that his great-grandfather caught Winter Tits (*Parus spp*). The wings were cut and then the birds were set free in the cottage, where they soon would eat fleas, which were very common at the time. In February, some more feathers were cut because they grew back, and in June the birds were set completely free. They had become very tame and one of the birds of his great-grandfather returned for four years to the same house and stayed the winter there.[283]

Starlings were important birds in the fields, as well as in the house. Lorens Wolter Rothof recommends in a book, from 1762, to domesticate starlings, which could fly freely around the house and feed on bugs.[284] In Siberia, locusts regularly demolished crops, gardens and even grass in the steppe. Swedish explorer Johan Peter Falck noted around 1770 locust invasions every three or four years, sometimes even every second year.

"The locusts would be even more", he commented, "if the Common Starling (*Sturnus vulgaris*) would not eat so many of them".[285] The starlings, added Falck, eat also other insects and is very useful for the fields. In southern Russia, people would hang old pots and bowls from the roof for the starlings to build their nests in, so that the birds would stay around the house with their families. In the Bukhara area (now in Uzbekistan), another starling locally called *ala targos* was supposed to be more beautiful and more dangerous for migratory locusts than the Common Starling. It moved with the locusts, and Falck was told by Bukharans that it consumed millions of locusts; one could see clearly within the locust clouds how the insects disappeared. According to popular belief, the locusts had their revenge: They gnawed at the feathers of *ala targos* and other starlings so that the birds could not fly any more.[286]

Keeping wild birds as domestic pest cont-rollers was a fairly widespread practice among the farmers in several parts of Europe, northern Asia, Africa and Latin America, but not only peasants used this method. The scientist, Carl Linnaeus, was familiar with the habit and applied it to his home. He was notoriously interested in all kinds of pets and used to visit the local bakery, from where he brought home House Crickets (*Acheta domesticus*); he found their signing calming. Apparently the crickets enjoyed his house and multiplied very quickly, and they would drop into the food when too numerous.

Then Linnaeus would either poison or or smoke them out or, in the autumn, capture the insectivorous Tree Creeper (*Certhia familiaris*) and keep it in the house until the quantity of insects diminished.[287]

Excepting these examples, the custom of

keeping birds in houses as free-flying pest controllers is unfortunately only sparsely documented. Handbooks for at least five centuries recommend the method and we can assume it was rather popular. People found the birds to be positive, efficient additions to their households.

Modern uses

In the past century, industrialization, urbanization and the increasing understanding of hygiene have created a vast array of other methods to keep flies and other insects at bay, and the practice of keeping wild birds for pest control has diminished or disappeared. In modern times, however, with growing interest in biological pest control as a sustainable alternative to artificial pesticides, the method of using birds is returning.

The Muscovy Duck (*Cairina moschata*) is reported to be capable of eating flies thirty times faster than for instance fly traps and fly paper strips can capture. They also eat snails and slugs. This species is therefore becoming increasingly popular in Scandinavia and elsewhere in northern Europe, because it is reputed to keep the invasive Spanish Slug (*Arion lusitanicus*) away from gardens. Turkeys and Guineafowls can also be used in gardens as pest controllers for the same purpose.[288]

Many botanical gardens in Europe keep Chinese Painted Quails (*Coturnix chinensis*) in their greenhouses. It is believed that the quails can help in pest management. These birds do not tear up plants in the same way as chickens (which otherwise can be good in clearing slugs from plants) and other larger birds do, but as yet no research on their usefulness for pest control exists. The botanical gardens we have interviewed in Sweden reports that quail presence in greenhouses is very much appreciated by the visitors; besides being entertaining, they also indicate that no pesticides are used within the premises. In the Netherlands, Glossy-starlings (*Lamprotornis* sp.) are kept for effective pest control in modern cattle buildings. Anton van der Linden in Wageningen has recently has researched the use of Dusky Fulvetta (*Schoeniparus brunnea*), a Chinese bird, in Dutch commercial greenhouses. These birds were very effective in reducing harmful bugs, but the promising results have not yet been developed into a common practice.[289]

Acknowledgement

Thanks to Christopher Gasser (Italy) and Anton van der Linden (The Netherlands) for providing us with additional references.

Figure 33: A young street trader in London selling cage birds. Engraving from the 18th century. Painting by Daniël van den Queborn, 1604

Chapter 7. Innovations in Early Modern Aviculture

From chaffinch to canary

Tom De Roo

Aviculture is not an invention of early modernity. Already, in Antiquity, there were reports of people keeping birds in cages and the practice continues throughout the Middle Ages up to the present. From the beginning of the early modern period, the predominantly indigenous birds kept as companion animals would endure exotic competition, first by the introduction of the Insular Canary (*Serinus canarius*) from northwestern Africa, and later with the import of several species of West African and Central and South American parrots. These parrots were probably unique to the higher echelons of society especially in the beginning.

Anecdotes in literary sources refer almost exclusively to a context of local and national elites, where dukes and kings admired their colorful attire and seemingly intelligent verbal prowess, and the birds themselves played an active role in many a humorous adventure. It is therefore not surprising that these spectacular exotics are considered to be interesting subjects of historical study.[290] Yet the early modern attitudes towards, and the keeping of, small singing birds have caught little attention, let alone that the introduction of the canary and its particular evolution have not been much described. What was this "common" aviculture? Which birds were being kept, and why? How did the "exotic" canary fit into the picture?

I wish to explore these aspects using sources ranging from the first works of scientific natural history in the sixteenth century until the appearance of leisure-oriented canary literature in the eighteenth century. The authors, language and spreading of the sources, as well as the geographical areas mentioned, serve to construct the boundaries of this article, with the focus therefore lying on Central Western Europe (the Netherlands, Great Britain, France and to a lesser extent Germany and the region of Tyrol in Austria). It will be argued that the canary rose from a scarcely mentioned and vaguely known exotic in the sixteenth century to a popular and commercialized pinnacle of avicultural pastime among the urban middle class and elite in the eighteenth century. Concurrently, "common" aviculture was fundamental in the rise of canary popularity and was at the same time subtly transformed and innovated through this evolution.

Early modern aviculture

Ever since late medieval times, the keeping of small songbirds in cages was a fairly common practice.[291] Members of the nobility kept birds in cages or aviaries, perhaps as part of their often renowned "menageries".[292] One would expect a much wider social diffusion, however. Not only was the capture of smaller indigenous birds with lures and nets legally unrestricted in many parts of Europe, but the relevant tools could easily be fashioned using fairly cheap and widely available basic materials.[293] We can assume that the vast majority of the captured songbirds eventually ended up on the dinner plate, not just because the hunt with nets had an innate high fatality rate, but also since many literary and visual sources refer to those indigenous birds as part of the human diet.[294] It is likely that only if, and when, a bird managed to survive the wiry violence, it might end up in a cage to continue its life as a companion animal.

Hunting with nets was not only a popular and easy way to catch birds, it was also very rewarding when targeted specifically at group-flying species like finches, which would have been fairly common.

The thrill of the hunt culminated in the pulling of a wire and subsequently seeing dozens of birds, lured by grains or an already captive congener, getting buried underneath several square feet of net.[295]

In the eighteenth century, a permanent "finch lane" (the installation of side-by-side nets that could collapse on each other) became part of many a country house, and could easily serve to catch fifty to 100 birds per try, up to 1,000 birds per day.[296] It will come as no surprise that this increasingly professionalized hunting method eventually became more than just another way to acquire edibles. Between 1600 and 1850, it transformed into a truly recreational and social activity, especially popular amongst the elite. The middle and lower classes would carry the practice well into the twentieth century.[297]

As already stated, the chance to acquire a fairly undamaged specimen was rather small. Hunting with nets had one big disadvantage that made it particularly unsuitable to catch birds intended to be companion animals. Methods of bird catching arose that were tuned to guarantee the survival of the target. In England, for instance, bird catchers who specialized in selling live animals were known to use hand-picked seeds and plants to lure birds unharmed into captivity.[298] This leads us to believe that these methods paralleled a rising market interest in companion animals, especially considering the fact that the sale of cage birds seemed to have happened in specific marketplaces, through traveling salesmen and eventually even in permanent shops.[299]

Aspects of a "perfect" companion bird

Indigenous birds are mentioned many times as companion animals in early modern works of natural history. The ready availability of most kinds of finches, due to their large numbers and easy capture, is the most likely explanation for their presumable dominant presence. Their varied coloring and often quite agreeable vocal capacities undoubtedly added to their value. Finches (the *Fringillidae* family generally comprising two tribes, namely the Fringillini and the *Carduelini*) are reportedly among the most easygoing cage birds. Greenfinches (*Carduelis chloris*), for instance, are known to become so trusting that they eat out of the owner's hand. Yet certain early modern natural history authors also attest to how they die after voluntary starvation or refuse to whistle for months out of depression.[300]

Goldfinches (*Carduelis carduelis*) seem to be the most well-known and appreciated members of the family. Their behavior of hauling up small reservoirs filled with food and liquids using an attached wire, as if they were getting water from a well, is admired in such a way that from as early as the sixteenth century cages are designed specifically to cater to this activity.[301] Its emblematic significance of handiness and craftiness is directly tributary to this action, [302]and they clearly offer more amusement in a cage than in a frying pan.[303]

Their singing is quite appreciated as well and can be subtly altered, according to taste, by placing the Goldfinch in the vicinity of other birds.[304] They can easily cross-breed with, for instance, canaries to form a mule, a very strong hybrid. Goldfinches were popular well into the eighteenth century; in 1780 Buffon stated that "fanciness requires neither exoticness nor rarity".[305] Archival sources, although vague in their description of animals, attest to the keeping of finches as well. Besides the numerous generic birds reportedly being kept in cages, the Linnet (*Acanthis cannabina*) is one of the few named species of cage-birds in early modern Antwerp probate inventories; the others are Canaries.[306]

Literary sources attest also to other families of birds. Larks (of the Alaudidae family) are mentioned both as savory parts of the average buffet and as appreciated singing birds. Not surprisingly, their ancient Germanic name *laiwrikon* is derived from a word for 'sounding'.[307] A lark is a studious and easily tameable bird, often kept in cages for its singing capacities, "yet at all times requiring a small patch of grass to sit on".[308] Turtle Doves (*Streptopelia turtur*) and Common Quail (*Coturnix coturnix*) are tasty menu items as well as interesting cage-birds, be it more for their attire and actions than their vocal capacity.[309] They are deemed essential in country house aviaries from the sixteenth century onwards.[310] Also mentioned are tamed Jackdaws (*Corvus monedula*), sweetly singing tits (members of the Paridae family) and a caged Red Crossbill (*Loxia curvirostra*), climbing up and down his wiry prison "like a parrot would do".[311]

Even House Sparrows (*Passer domesticus*) are often free to fly in houses; naturally, these particular individuals are expected to be tame enough to come when humans called out their given names. The rebellious Hawfinches (*Coccothraustes coccothraustes*), on the other hand, are horrible singers and cannot be taught to do anything worthwhile to boot (i.e. that is appealing to a typical early modern bird admirer). Starlings (*Sturnus vulgaris*) are equally off-key singers and babble so much that they are tiring to the ears, but at least they can be taught quite a few tricks, and will even mimic human speech under the right circumstances.[312] Lastly we have to mention the Nightingale (*Luscinia megarhynchos*), the true pinnacle of vocal ability, and coincidentally also quite easy to educate. It is, however, perceived as easily giving in to depression in captivity and being notoriously difficult to feed and keep.[313]

These very diverse descriptions can serve as a testimony to avicultural variety, but their worth is more fundamental. They inform us about the aspects of cage birds deemed important and appreciable, as these will have a particularly defining role in the coming popularity of the Canary. Those aspects are: Ready availability, musicality, appealing physical appearance (whatever the subjective taste may be), pleasant behavior and ability to survive, be reared and procreate in captivity.

Canary introduction

The Canary (*Serinus canaria)*, in early modernity also called *Passer canaribus* or Canary Sparrow was most probably "discovered" by the Spanish on its native namesake isles in the early fourteenth century. By the end of the fifteenth century it would make its appearance in a world already familiar with keeping small singing birds in cages. This naturally pale yellow-greenish bird did at first probably not cause much ado amongst the general bird-possessing public. Yet some literary sources state romantic stories about feverish domestication in secluded monasteries, infatuated statesmen like French king Louis XI, whose passion repeatedly is said to have caused him to board a ship to the Canary Isles, and a Spanish monopoly artificially kept alive by only trading female birds, which was even more miraculously broken in Italy, in the form of a ship carrying hundreds of birds of mixed sex wrecked on the coast.[314]

When romance raised its voice, presumably more objective sources, like the emerging natural history literature and the "natural" iconographic tradition, remained deafeningly silent. English naturalist William Turner appears to be the first to mention the canary in a work of natural history. Admittedly, his *Avium Praecipuarum* (1544) refers to the Canary in a somewhat side-way fashion. A note on the "Luteola" (Siskin (*Carduelis spinus*)) states that of this kind of birds "are those which

England calls Canary Birds".[315] In his *lemma* on Canary Grass (*Phalaris canariensis*) in the second part of his *Herball* (1562), Turner indirectly testifies to international trade in canaries:

"For they that brought canary birds out of Spain brought of the seed of the Phalaris to feed them with". It can be assumed that he is no longer talking about siskins here.[316] French naturalist Pierre Belon du Mans, however, does not refer to the Canary in his acclaimed masterpiece *L'histoire de la nature des oyseaux* (1555).[317] The Swiss botanist and all-round natural expert Conrad Gessner has long been considered the first to mention the bird. In his *lemma* on the "Citrinella" (Citril Finch (*Carduelis cintrinella*)) in the posthumously published copy of his *Historiae Animalium Liber III, qui est de Auium natura* (1585), Gessner designates the Canary as a bird that is "like sugar a native of and carried over by ships from the Canary Islands".[318]

In the following sentence he also mentions a golden bird, resembling the Citrinella, that is bred in captivity in Italy, which Tim Birkhead considers an early testimonial to domesticated yellow Canaries.[319] The Bologna professor Ulysse Aldrovandi, curator of his own museum of natural history and a driving force behind a whopping thirteen volumes on the natural (and somewhat supernatural) world, allocates one page of his *Ornithologiae: Liber Decimusterius* (1600) to the Canary or 'sugarbird', reinforcing Gessner's earlier association. He knows little more than that this bird is quite rare and presumably a good singer.[320]

The association of the Canary Bird with sugar is interesting, especially because it carries over into later works of natural history. It is however not at all surprising. Many ships carrying sugar, then the primary export product of the Canary Islands, set sail for the European mainland from the start of early modernity onward. If these ships carried Canaries as a secondary cargo, it is somewhat logical that the birds were referred to as Sugarbirds in their adoptive countries; this is quite analogous to the naming of the Bengalese Finch (*Lonchura domestica*), where Ingvar Svanberg presumes the associative nomenclature originated from English retailers, who received this bird delivered by ships arriving from India.[321] This non-biological association aside, it is telling that Aldrovandi writes but one page on the Canary, whereas he fills entire gatherings on larks, other species of finches and

Figure 34: An 18th-century illustration of two Canary Birds and an egg

exotic parrots.

Half a century later, Aldrovandi's already limited information is paraphrased by reputed Scottish-Polish scientist Iohannes Ionstons in his *De Avibus.*[322] These and other literary sources of the sixteenth and seventeenth centuries mention a diverse variety of indigenous species of birds as well as exotic parrots, stating biological information and recording anecdotes of these subjects in a context of animal companionship; similar references to the Canary are very rare, again telling of its presumed relative lack of popularity.

Although there are some illustrations in unpublished works and manuscripts, as far as commercial art and iconography is considered, there is a notable, yet unsurprising lack of Canaries, further affirming the disregard in literary sources.[323]

On top of the rare and rather unrealistic imagery in natural history sources, the only known commercial picture of what could be a Canary originating from this period, is a late seventeenth-century etching of a bird peddler in the streets of London, by Marcellus Laroon the Elder.[324] Renowned animal painters, like Jan Brueghel, Melchior d'Hondecoeter, Jan Fyt, Roelandt Savery etc. as far as we know, paid no attention whatsoever to the bird, while many indigenous species as well as parrots appeared in their works. The so-called bird concertos, for instance by Jan Kessel and Frans Snyders, picture a wide (if not the entire) variety of known birds, of which some are even unable to sing (like the eagle and the buzzard); the Canary is nowhere to be seen, while birds closely related to it, like siskins and goldfinches, are present.

The Canary is thus strikingly absent from both objective depositions of natural history as well as subjective (commercial) imagery during the sixteenth and seventeenth centuries. A logical explanation is that the Canary, as a bird of status, may have been an exclusively elitist cultural entity at best.[325] But why were parrots, equally exclusive birds, differently treated?

Many kinds of exotic parrots were mentioned extensively in literary sources of natural history and often present in works of art, so the elitist explanation does not suffice. Could it be that Canaries were simply not interesting to the general public? Were they not perceived during the sixteenth and most of the seventeenth century to be that different from readily available indigenous birds?

This could reinforce (and be reinforced by) a limited distribution of the canary. Eighteenth-century literary sources mention that ownership of Canaries only started growing by the middle of the seventeenth century, when the Swiss-Italian-Austrian border region of Tyrol evolved into a specialized commercial center for breeding and trading Canary Birds.

Figure 35: Young Louis of Nassau-Beverweerd (1602-1665) with a tame goldfinch. Painting by Daniël van den Queborn, 1604

Every year dozens of merchants would leave this region for destinations far and wide across the European continent, carrying several hundreds of birds on their backs.[326] The few reports of archival sources serve to affirm this; only roughly ten per cent of Dutch and Flemish probate inventories from the second half of the seventeenth and the eighteenth centuries contain Canaries, with an apparent concentration in the upper social strata.[327] As an eighteenth-century literary source states, Canaries were at the end of the seventeenth century still "sold at very high prices, and kept only for the amusement of the great – they have since been greatly multiplied, and their price is diminished in proportion". [328]This development may well be considered as a democratization of the Canary, since it seems indicative and decisive for its growing popularity in the eighteenth century.

Glory days

Natural history allocates more pages to the Canary during the eighteenth century than ever before. The widespread and highly acclaimed works of French naturalists, Georges-Louis Leclerc de Buffon and Jacques Christophe Valmont de Bomare, as well as Dutch popular scientist Martinus Houttuyn, not only copy already known facts and fictions, but contain much new information, which their authors often acquired first or second hand from breeders and fanciers.[329] For the first time Canaries appear on "mainstream" paintings: *La Serinette* by Siméon Chardin (1750–51, Musée du Louvre, Paris) offers one of the earliest truly artistic images of a Canary Bird. This painting also shows the leisure activity of keeping Canaries, since ordering them to sing certain melodies was not only an integral part of the hobby, but also primarily accomplished by using musical instruments like the *serinette* (small barrel organ).

The quite inconspicuous anonymous devotional painting *Two standing girls* (ca. 1756, Schreiboomgodshuis, Ghent), is exemplary of the rise of the Canary. This painting shows a Canary in the role of a "stool bird". This kind of toy, where a bird is connected via a wire to a wooden or metal stool, is shown on many children's portraits in the seventeenth and eighteenth centuries. It symbolizes obedience and willingness to learn, and religious conformity. In addition, most probably while being a real form of amusement, it was an ordinary toy, too, which had been elevated to a motif with underlying meaning in art.[330]

The bird species is irrelevant with regard to the iconographic meaning, and that is exactly why this *topos* can be so revealing. In the seventeenth century, stool birds are, without exception, indigenous species, mostly finches with an occasional tit or lark. Suddenly in the eighteenth century, the Canary joins them on the pint-sized throne. In essence this serves to illustrate that the Canary had become more commonplace, equal in cultural meaning and reception to the already well known and widely available indigenous birds.[331]

These cultural developments, revealing as they are, cannot compare to the final and strongest proof of the growing popularity of the Canary, the introduction and marketing of literature exclusively devoted to the keeping of caged Canaries as a leisure activity. Jean-Claude Hervieux de Chanteloup (1683–1745), Curator of the Canary aviaries of the Duchess of Berry, wrote the first standard work exclusively dealing with Canaries.[332] His thorough and well-versed *Traité curieux des Serins de Canarie* (1709) was translated and edited multiple times, and continued to be printed and sold throughout the century, influencing or inspiring almost every later work on the subject. In England, its translation competed with dozens of smaller and cheaper booklets and larger volumes on collected leisure activities and popular natural history.

In Germany it remained the standard for quite some time, although later adaptations by German authors often added gatherings on the Nightingale (*Luscinia megarhynchos*), a bird that seems to have remained intensely popular. In the Netherlands it was first translated in 1712 and remained the sole work on the subject, until Dutch self-proclaimed fancier Frans Van Wickede delivered the first edition of his book on Canaries in 1750, meant to be an improvement of Hervieux's original; it remained equally popular into the nineteenth century. Pirate prints appeared in England and the Netherlands. Together with the existing works on natural history and broader "domestic dictionaries", this specific literary field served to create a knowledge community backing the new Canary hobby, which can in fact be considered to be a transformation of the earlier practice of keeping small birds in cages. Designed mostly as how-to-manuals and semi-objective reports, a reading of the eighteenth-century literature offers an interesting view into the practices and attitudes concerning Canary-fancy.

As stated before, I would like to argue that the popularity of the Canary in the eighteenth century can be attributed to the fact that it was supposedly an improved and more readily available version of the traditional, indigenous cage-birds. Then it comes as no surprise that literary sources attest to Canaries being "easily bred and kept in cages and aviaries all across Europe", and offer practical advice on how to accomplish this with only the slightest effort and investment.[333] The song of a Canary is deemed absolutely wonderful, more or less equal to that of the Nightingale. Its musicality is often considered the primary attraction, and the prime reason mentioned by Anna-Maria Moretus-de Neuf, wealthy director of the Plantin-press in Antwerp, when she bought a few Canaries with the intention to have them sing.[334]

Their song could easily be improved through training, by using the *flageolet* or *serinette* to play melodies that the bird was supposed to learn. This training was an application of its natural tendency to emulate often-heard noises, whether it be the singing of other birds or even, like in the case of the pet Canary of the chief treasurer of Dijon, the particular sound of rolling coins. They can also be taught to come when their name is called out, and most often become very tame.

Buffon did not hesitate to emphasize the enjoyable companionship of the Canary, even comparing its behavior to that of "innocent children". The Canary epitomized the natural element that can be "cultured" into something that is beneficial to man in some way; keeping Canaries was pleasant, especially because they are so easily instructed.[335]

The ease with which Canaries can be trained betrayed their overly generous and caring nature towards their keepers and each other. Yet not all canaries were considered equally good. A differentiating categorization was projected upon male Canaries, not surprisingly since they are the only ones that sing and thus the only ones that seemed to matter.

Some authors reported four kinds, the first with a mediocre physique and not a really good sort of bird. The others were arranged according to personality; their difference was shown even in morality, according to Buffon. The second bird was a truly evil bird which mistreated and often killed his spouse. The third one was caring and friendly towards his mate, yet killed his offspring or destroyed the eggs before they hatched. The fourth, finally, was "always gentle and upbeat" and caring towards his mate, offspring and master alike, truly the perfect Canary.[336] Better yet, it was an irreproachable reflection of the eighteenth-century ideal family man.[337]

With the concept of paternity, we reach another element in the success of Canaries: their relative ease of procreation, upbringing and keep. Here they differ from traditional and often high-maintenance birds like the nightingale. Unlike the latter, canaries are easily fed with simple food like grains and seeds, which do not require any other preparation than mixing them into a preferred blend. Authors discussed the best mix, but differences were minute. Most included canary-, moon-, rape- and hemp-seed. Sometimes more sophisticated edibles were mentioned, such as hard-boiled eggs, bread dipped in milk and even certain kinds of cookies and biscuits.[338] In addition to food-related advice, how-to-authors provided information on common diseases and injuries, with special attention to the occurrence of the moulting period, which according to Hervieux did prove to be fatal to a multitude of birds, yet fortunately also kept Canaries from becoming as common as Siskins or Tits "which would considerably lower their value".

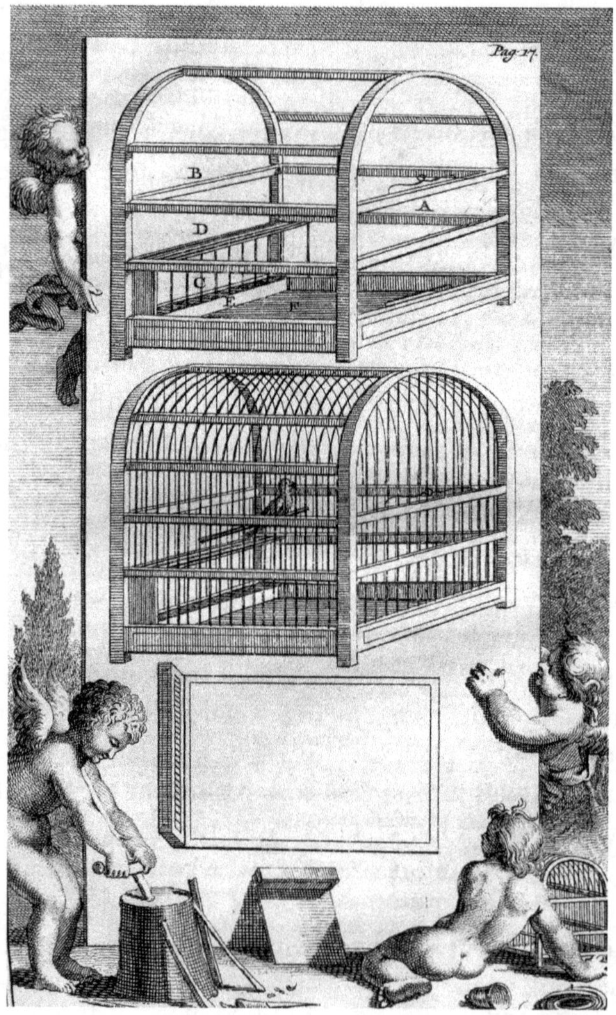

Figure 36: Small cages for Canary Birds. From Hervieux' book on Canary Birds, 1713

Authors also offered chapters on the intricacies of breeding (using special types of cages), the rearing of young birds by hand, how to complete inter-species breeding, etc. No matter which of these necessities drove one into action, the canary owner at all times had to treat the bird in a gentle and caring manner.[339]

Keeping and maintaining Canaries

Probably, in the final decades of the seventeenth century, Canaries could already be bought from traveling salesmen of Tyrol, Germany or Switzerland. They were known to travel across Europe to deliver their wares, paying particular attention to larger cities like Amsterdam, London, Paris and Constantinople. Their birds were well known and highly appreciated.[340] London newspapers e.g. advertised their arrival and the places where they sold their wares, usually inns and taverns.[341] In the middle of the eighteenth century many authors, however, advised to acquire canaries from fellow hobbyists instead, since foreign merchants were prone to deception, selling unhealthy or unworthy birds.[342] The salesman's primary goal was making money and they preyed upon the ignorant customer, who could be cheated by an assumed exoticness of the outlandish birds. These warnings testify at least that local breeding and a non-official commercial circuit already existed. Local breeding was reported to occur in an ever growing area, with Germany, France, the Netherlands and England being the most mentioned regions.[343]

The next step was choosing a bird, a task which appears easier than it is. First, there is the matter of diverse physical appearance, increasing in importance for the attraction. Hervieux mentioned 28 different variations, the Common Grouse being the cheapest at three and a half Dutch guilders and the black-and-bold-citron-colored the most expensive at 25 guilders. Varying feather patterns or slightly off-key coloration could push the price even further, depending on "how willing the buyer is to part with his money". Prices as high as 500 guilders were reported, yet they appeared to be exceptional. In the early eighteenth century, Hervieux attested to the Canary becoming more and more common. He even complained how the ever-expanding scale of hobby breeding was making it difficult to find pure-breeds of a certain variety.[344] Almost at the end of the century, Buffon for the most part still used Hervieux's categorization, probably with the addition of the so-called *Frisé* or 'tuft' Canary, which supposedly appeared in the 1730s and was mentioned in a pirate copy of the work of Dutch author Van Wickede.[345]

The existence of several varieties with a relatively wide price-range and a presumed equal rarity offered a method for distinction amongst fanciers. A similar form of conspicuous consumption was possible through the housing of the birds. Many authors attested to the wide variety of cages "that are constructed and sold on a daily basis, for each buyer to follow his heart and his wallet", and offered advice on how to construct them.[346] This "home improvement" was no less conspicuous, since custom-made cages could include silver wiring, mirrored floorings and painted walls. Still, despite possible elements of conspicuous consumption, the purpose of housing, whether custom-made or bought as standard design, was always made crystal-clear. It should be attempted to mimic as closely as possible the natural environment of the imprisoned canary, to do everything feasible to make the captive bird feel comfortable and at home, and in essence provide it with as large a sense of freedom as possible.[347] The eighteenth century sentiment was that this kind of captivity was actually beneficial; it is probably no coincidence that moralistic literature often let canaries explain how their captivity equaled a protection from the wild natural elements and insured comfort and stability.[348]

Commercialization

A German author testifies how an Ymst (Imst in Tyrol) corporation handled sales logistics by sending agents to local breeders in different parts of Switzerland and Germany to buy birds, after which the company's traveling salesmen distributed them around Europe.[349] The birds were undoubtedly bred for profit. There was a significant similarity to live capture of indigenous birds with the intent of selling them, the difference being that breeding was more stable and efficient, and in comparison to indigenous birds, the breeding and keeping of Canaries easy.

In the eighteenth century, further development of the commercialization of Canaries concurred with a broader movement of popularization of scientific knowledge for commercial purposes. [350]Professional sedentary and local hobby breeders sold birds in a sort of semi-commercial circuit, providing friends and relatives with cheap Canaries. Canaries were reported to be bred not only in Tyrol, but also in Germany, France, the Netherlands and England, increasing the availability. There was a growing market of paraphernalia like cages, cage-attributes and readily made food-mixtures,[351] which developed with the specialized literature. Besides famous authors like Hervieux, small-scale, often anonymous and mostly cheaper booklets were published. They were more similar to pamphlets and more commercial, providing a self-proclaimed "better way" of breeding, feeding, etc. than their predecessors. They found readers among those interested in keeping Canaries and looking to improve a pastime activity, yet wishing to pay less money for information.[352] This also shows an expansion of the interest base, probably even of a democratization of the keeping of canaries; the more serious literary sources described this "trickling down" in negative terms, stating that keeping Canaries used to be a rare occupation and only for the upper classes. At the end of the century the Canary was considered to be common, breeding occurring in cities and townships on the continent and in England, e.g. Ipswich and Colchester. The rise of local breeding did not replace the Tyrol import of birds, but complemented it and was further proof for the growth of Canary-keeping.[353]

Behind the facade of a genuine guide to keeping Canaries, some professional salesmen published their own how-to-books, as a form of advertising for their birds or paraphernalia. The anonymous author of *Exact and sure directions how to dress and keep canary-birds*, for instance, referred to himself as a salesman of bird-keeping attributes, and did not hesitate to promote his own "new fashion bird cages" and available goods, "all sorts of bird-seeds for all these birds, viz. canary and rape, maw-seeds, and turnip-seeds: also, hemp-seeds for other birds, pullies, lines, and counterweights, brushes and scrapers, glasses, and whatsoever belongs to these birds, and cages, to keep, and hang them up conveniently".[354] Bird salesman John Piercy used his *Natural history of Irish song birds* to promote his goods. His store, aptly named The Great Bird-Cage, was located at the intersection of Capel Street and Essex Bridge, a known retail axis in eighteenth-century Dublin, showing that this sedentary merchant was neither an amateur nor unsuccessful.[355] Still, visual art and prosaic literature continued to mention ambulant street criers advertising and selling birds.[356]

End of indigenous bird-keeping?

Spectacular as the growth of keeping Canaries may seem with regard to aviculture, indigenous birds were also very much appreciated and kept.

In the eighteenth century, various types of literature mention the intricacies of keeping and rearing indigenous birds, along similar wordings on Canaries.[357] At the beginning of the nineteenth century, an Amsterdam commercial leaflet, advertising birds for children, mentioned the once exotic Canaries alongside the actually exotic parrots and truly indigenous birds like the Chaffinch (*Fringilla coelebs*), Nightingale and Linnet.[358] The Canary could be treated as a superior version of the indigenous birds, with regard to appealing characteristics, especially being a lot easier to keep, which probably lowered the threshold for the average person to get a bird. This means that some people now favored Canaries over local birds, but it probably also meant that many people chose canaries instead of not keeping a bird at all; the literature often explicitly addresses these "new bird fanciers".[359] We would therefore assume that aviculture in general rose with the spreading of the Canary, meaning that the Canary did not replace indigenous birds per se, but rather served to complement them.

Early modern aviculture witnessed the introduction of an exotic novelty, the Canary. This bird found its niche in a context different from the parrot (undoubtedly the most spectacular avicultural newcomer), since it was not only biologically, but also culturally closer to the small songbirds already kept in cages by widely different groups in society. The traditional birds remained popular until the end of the early modern era and despite exotic origins, the newcomers did not replace the indigenous birds. This shows that not the species, but the general attitudes and motivations carry the most cultural weight. The birds were appreciated first and foremost because of certain characteristics: pleasing vocal capacities, an appealing physique and a conforming character being the most prominent.

An analysis of literary sources suggests that the Canary should be classified not as an invention, despite being a novelty, but rather as an innovation. It gained popularity gradually, not because it was very different from the traditional species, but rather because it was an improved version of them. The Canary was appreciated for very much the same aspects and elements, considered valuable in traditional species long before its arrival. It was considered an excellent singer, with a caring and agreeable character, offered a wide variety of domestic breeds for almost everyone's taste, and most importantly, it was easy to acquire and keep. When this last aspect became reality, through a steadily growing practice of breeding, the bird was easily included into the already existing practices of aviculture and its commercial development.

Figure 37: Various English breeds (Lizard, Norwich) of the Canary Bird in the 1870s

Chapter 8. Chamber and Songbirds in the North

Ingvar Svanberg

One day, most likely in 1907, an elderly, elegantly dressed gentleman entered Axel Cohn's Bird Shop in the center of Stockholm. He raised his top hat politely as a greeting, and asked the shop owner, if he would be allowed to study the behavior of the cage birds in the store. The shop-owner, who did not recognize the visitor, granted this modest request. His nephew Helge Cohn, who retold the incident for posterity, immediately knew the visitor as the famous novelist August Strindberg (1849–1912).

The writer carefully observed how the weaver birds built nests, as well as the peculiar behavior of the cardinals. He stayed for a while in the store and observed in detail the birds in their cages. Strindberg's interest in birds is well-documented and was founded early in his life. He was not a novice to cage birds, since in his childhood he became familiar with the breeding and care of Canaries (*Serinus canarius*)and other songbirds in cages, which he described in his novel *The Son of a Servant* (in Swedish *Tjänstekvinnans son* 1886–87).

Strindberg was raised in a solid middle-class home with both money and servants. There were lots of animals in his neighborhood; he grew up in a part of Stockholm that still possessed some rural character. Strindberg seems to have disliked dogs; a local butcher kept savage beasts during his childhood, which made a life-long impression on the writer.

Horses and cage birds amused him. In the mid-nineteenth century it was common to keep songbirds in well-off homes, especially as entertainment for the children. For those who could afford it, it was possible to obtain a Canary Bird. In Strindberg's childhood home there was no shortage of money, and the social capital available within the family let the son keep canary birds. During Strindberg's lifetime the conditions for keeping pet birds changed radically, from a rather limited means of pleasure for a privileged few, to companion animals for a more general public. Cohn's cage bird store, founded in 1895, in the Klara Church area in Stockholm's inner city, may be seen as a manifest example of this.

Ornamental birds in castles and manors

Before we deal with the changes following the

Figure 38: Tailor August Palm (later socialist activist) with a singing canary as company, 1890

urbanized industrial culture and modernity, generated towards the end of the nineteenth century, let us take a look at the history of keeping cage birds in Sweden.

Scandinavians have probably kept birds in captivity for various reasons since times immemorial. Swedish sources explicitly mention cage birds for the first time in the Late Middle Ages. Olaus Magnus devoted a separate chapter on the subject in his *Historia de Gentibus Septentrionalibus* (History of the Northern People) from 1555. He also depicted both parrots and songbirds in cages.[360]

During the reign of the Vasa Dynasty in the sixteenth century, the Swedish Royal Court kept parrots and so did the nobility. One Senegal Parrot (*Poicephalus senegalus*) is depicted in a painting from 1622. Paintings show the presence of species such as macaws, Amazon parrots and at least one Hawk-headed Parrot (*Deroptyus accipitrinus*). Scarlet Macaws (*Ara macao*) were kept as well, known as 'Indian ravens' in Swedish, probably because their ability to imitate human voices (like ravens did). In 1779, in manors and castles, species like Blue-and-yellow Macaw (*Ara ararauna*), Green-winged Macaw (*Ara chloropterus*), White-fronted Amazon (*Amazona albifrons*) and probably Sulphur-crested Cockatoo (*Cacatua galerita*) were kept. A Blue-crowned Hanging Parrot (*Loriculus galgulus*) was imported in 1755.[361]

For centuries it was common, as in other parts of Europe, especially among aristocrats and the upper middle class, to keep Chaffinches (*Fringilla coelebs*), Bullfinches (*Pyrrhula pyrrhula*), Linnets (*Acanthis cannabina*), Siskins (*Carduelis spinus*), Goldfinches (*Carduelis carduelis*), and even Canary Birds in cages for their singing qualities. The cultural history of songbirds and other cage birds in Sweden, as in the rest of Europe, reflects the Zeitgeist, i.e. the economic, cultural and political currents that dominated during various periods.

Poet-musician Carl Michael Bellman's (1740–1795) Epistle no. 36 describes an interior of an urban inn during the eighteenth century: "In the ceiling on a peg in the cage, already the innkeeper's Linnet chirps." Native songbirds dominated still during the first half of the nineteenth century. In 1835, zoologist Sven Nilsson praised the Bullfinch as a songbird. It was said to be "hardly shy, and becomes easily tame in captivity, and is soon taught to imitate several arias and tunes".[362]

Carl Linnaeus had an aviary in his so-called *vivarium*, which he had arranged in the Academic Garden in Uppsala. There he kept various exotic and native birds, such as several species of parrots: Scarlet Macaw, Yellow-crested Cockatoo (*Cacatua sulphurea*), Blue-fronted Amazon (*Amazona aestiva*), African Grey Parrot (*Psittacus erithacus*), and Blue-headed Parrot (*Pionus menstruus*), as well as Crested Curassow (*Crax alector*), Spoonbills (*Platalea* sp.), Southern Cassowary (*Casuarius casuarius*) and Ringed Turtle Doves (*Streptopelia risoria*).

Aviaries were also found at mansions in the countryside. From Charles De Geer's Leufsta estate in Uppland province, Daniel Tilas reports in September 1768 that "the owner had founded an aviary there for his amusement". On Mälby estate, the President of the Court of Appeal and ornithologist Gustaf von Carlson kept a large aviary with many native birds. When Carl Johan Linnerhielm visited Mälby manor and estate in 1795 he observed: "There, on a protruding point, an aviary was built for various wild birds." An aviary literally shimmering with life – obviously with native birds – was founded by patron and zoologist Baron Axel Gustaf Gyllenkrok in his winter home at Lund, depicted in a book from 1846.[363]

For those with space and money, there were also possibilities to obtain ornamental birds with which to gild parks and gardens. A few pheasants found their way to Sweden, a Golden Pheasant

Figure 39: Bird-catcher in Stockholm. Lithograph by Nils Andersson in 1863

(*Chrysolophus pictus*) was mentioned at the Royal Court as early as the 1740s. Indian Peafowl (*Pavo cristatus*) were kept in manor parks for several centuries.

Sven Nilsson wrote in 1835 that the peacocks did not only exist in the southern part of Sweden, but were also kept in the north "in towns and by aristocrats in the countryside. They are kept in chicken runs together with poultry, and they adorn gardens and parks". Guineafowl (*Numida meleagris*), known from the estate Åkerö of Carl Gustaf Tessin in 1754, as well as other exotic duck breeds, Muscovy Ducks (*Cairina moschata*) and Swan Geese (*Anser cygnoides* forma *domestica*), were kept. Sven Nilsson noted that Guineafowl was not commonly kept in Sweden in the 1830s, "only here and there, by estate owners and other nobilities".

Although occasional pheasant gardens existed at the Royal Court by the eighteenth century, only towards the middle of the next century pheasant breeding began to develop on a larger scale in Sweden. In 1846 during King Oscar I's reign, a pheasantry was founded on Djurgården in Stockholm and soon after Johan Carl Adelsvärd tried to breed pheasants at Adelsnäs in Östergötland. Another pheasantry was added in the 1850s at Näsbyholm estate in Skåne.[364]

In castle gardens there were usually ponds with Mute Swans (*Cygnus olor*). At Leufsta Manor in the 1760s, many "tame swans" were kept in the garden pool. Swan pools seem to have been quite common in large manors until recently. According to Sven Nilsson in the 1830s, the Mute Swan "is kept at manors in southern and middle Sweden".[365]

Play-birds and dove cots

One peculiar Swedish custom, of which there are several examples from the vicinity of Stockholm in the late eighteenth and early nineteenth centuries, was the keeping of Black Grouse cocks (*Tetrao tetrix*) in small cages as so-called play birds, in inns and also in many homes. According to a letter in 1761, Carl Linnaeus tried to trade a couple of Black Grouse for a "blue parrot" of unknown species, with the famous animal dealer company Blauw Jan in Amsterdam. Carl Michael Bellman described in a poem from 1772 an encounter with such a Black Grouse cock at Mummen's Inn in central Stockholm. The noble lady Märta Helena Reenstierna described in her diary in 1794 how during a garden party on her estate Årsta outside Stockholm she received, as a gift, a pair of Black Grouse in a beautifully ornamented cage. It seems to have been a widespread custom to keep the birds as pets.[366]

In many manors there was also a tradition of dove cots, and pigeons were kept at some vicarages. Carl Linnaeus's father had a dove cot at Stenbrohult vicarage, and during a visit to Björkskog parish in Västmanland province in 1746, Linnaeus met a clergyman who kept various pigeon breeds. "Domestic pigeons are to be found in cities and among noble homes in the countryside all over Scandinavia all the way up to the North Cape," wrote Sven Nilsson in 1835. Linguist Per Arvid Säve certified from nineteenth-century Gotland that among wealthy people, particularly their children, pigeon breeding was enjoyable work: "Because it is with the greatest eagerness that young animal lovers between school work build a dove cot in the attic with a few old boards."

Painter Ferdinand von Wright (1822–1906) established a dove cot at his rural studio on Orust in the 1850s. Pigeon breeding culture seems to have been well-established in Stockholm during the 1840s. Mr. Wahlström, who wrote a book about the subject in 1848 and mentioned several pigeon breeds, complained that in Stockholm "it is regarded as a legitimate pleasure for pigeon lovers to catch alien pigeons, thus quarreling with their neighbors, who also keep pigeons".[367]

Canaries arrive

Among those who could afford, the Canary was very much sought after as a cage bird. Canaries were probably domesticated in southwestern Europe in the early sixteenth century. It is not known when they first reached Sweden, but canaries are mentioned in Swedish-language literature as a songbird in the early seventeenth century.

Scientist Sigfrid Aron Forsius talked about *Canariefoglar* in a manuscript from 1611, and the superintendent for Visby diocese, Haqvin Spegel, was, in 1685, obviously impressed by the fine-singing *Canari Foglarna,* which "could bend the tongue so skilfully, that the voice suddenly increased, suddenly decreased again, that we with pleasure listen to how such a small animal could create such music".[368]

Canary import is sometimes mentioned in the sources. In 1648 a few Canaries were imported by Dutch merchants to Stockholm, and in 1726 merchant Volrath Tham imported 44 Canary Birds from London to Gothenburg. Itinerant bird traders from Tyrol also brought birds to Sweden. An advertisement in *Stockholmske Post-Tidningar* in the 1730s informed: "Joseph Kychel from Tyrol has arrived here with fine Canary Birds of a great many colors and he is living at wine cellar Tre Kongar [Three Kings]."

Linnaeus confirmed in 1748 that most song canaries were from the Tyrol. In the mid-eighteenth century it was not uncommon for Canary Birds to be kept at pharmacists' homes, vicarages and other middle-class houses all over the country. German-born pharmacist Franz Mikaël von Aken (1698–1760) published in 1749 a brief manual for Canary keeping and breeding based on his own extensive experience.

The increasing interest for Canaries led to a Swedish version of Joseph Charles Hervieux de Chanteloup's famous *Nouveu traité de Serins de Canarie* (1709 and later editions) with practical advice. The Swedish paraphrase *Underrättelse om Canarie-Foglar och deras ansning*, was published in two editions reflecting the demand, in 1770 and 1787. Erik Norén (1732–1808), an ex-major living in Vimmerby, was an enthusiastic breeder. He translated Johann Paul Friedrich's famous *Erfahrungen für Liebhaber der Canarienvögel* (1790), one of the most popular handbooks. Norén's translation was finished in 1806, but never published.[369]

Canaries were highly regarded; sentimentality often characterized the relationship with companion animals. Countess Beata Douglas had in the 1780s a Canary Bird which reached the considerable age of twenty years. When the birds passed away, they became the subject of poems for mourning, for instance by Bengt Bergius, Olof von Dalin and Carl Michael Bellman. "It hurts me to see you for faithful duty and effort, getting a blood-stained evening", wrote Bellman about the Canary Bird Putte, who was killed by a cat one evening in May 1768.

Märta Helena Reenstierna provided in the 1790s several interesting insights into the care of Canary Birds. She wrote about her protégés, with names such as Petit Roy and Chevailler D'éon, about their breeding and rearing and how they were singing to burning candles on Christmas Eve, but also how one of them met an unhappy destiny in the form of a cat.

In the early nineteenth century the demand for manuals on cage birds increased, but still most were translations from German. The first was August Immanuel Kellner's *Anwisning till Canarie-Fåglars Kännedom, fortplantning och skötsel* (1816), which soon was followed by others.[370]

In memoirs and correspondence we can catch glimpses of the Canary, but we never get an opportunity to have a closer look at it. Animal painter Magnus von Wright (1805–68) in Finland, the oldest of a talented trio of brothers, visited vicar

Carl Ulrik Ekström at Mörkö in Sweden in 1828. According to Wright's diary for 19 and 20 September he painted a Canary Bird from a stuffed specimen. The bird had belonged to Ekström's wife. The painting would have given us an opportunity to see what the Canaries in Sweden looked like, but unfortunately it is impossible to trace the painting.

Birds among the peasantry

Peafowl, swan pools and songbirds kept by the upper classes contrast with the situation among the majority. Before the large land divisions in the nineteenth century and with a majority population of peasants, who earned their living on locally available resources and their own produce, ornamental and songbirds in captivity belonged to what sociologist Börje Hanssen called a social stratum with an urban field of activity, that is the ways of consumption and living common with the aristocracy, higher upper classes, higher artisans and clergy.

Among the peasantry there was no interest or capacity for such an odd interest as cage birds. Domestic animals were kept for their economic value, although children were certainly trusted to care for young animals as companions. During the nineteenth century, a rather widespread habit and pleasure for boys in Skåne was bringing home Jackdaw (*Corvus monedula*) nestlings and taming them.

Of course peasants could also appreciate the wild birds singing and their presence in the vicinity; various kinds of bird behavior were interpreted as signs as to when work in fields and forest should begin, and rhymes could be read from bird songs. The presence of certain species, such as the Redstart (*Phoenicurus phoenicurus*), White Stork (*Ciconia ciconia*) and Barn Swallow (*Hirundo rustica*), also brought luck to the farm.[371]

Peasants captured small birds for various purposes, including food. Snow Buntings (*Plectrophenax nivalis*) were still being captured in large quantities in the early twentieth century in the interior of northern Sweden. They were viewed as a welcome addition to an otherwise meagre spring diet. From southern Skåne it is reported that the farm hands used to capture small birds and roast them baked in clay, when the baking ovens were heated in winter. The tender bird bodies spread a wonderful scent in the room.

Small table birds of various kinds, such as Fieldfares (*Turdus pilaris*), Song Thrushes (*Turdus philmelos*), Blackbirds (*Turdus merula*), Bohemian Waxwings (*Bombycilla garrulus*), and Yellowhammers (*Emberiza citrinella*), belonged to regular game brought by peasant hunters to markets in cities at the end of the nineteenth century. Opposition to the capture of birds developed during the second part of the century, but as cookbook writer Charles Emil Hagdahl noted in 1896, "should you feel pity with the whole world, you could hardly eat anything else".[372]

On Gotland, peasants placed nesting boxes for Starlings (*Sturnus vulgaris*) on the roofs in order to harvest eggs and nestlings, a much appreciated spring delicacy at a time of the year when other animal products were scarce. Forest officer Johan Otto Sylvan reported in the 1870s that between 2,000 and 3,000 pairs of starlings bred yearly in such nesting boxes on the island. Peasants also brought back nestlings of various wild birds in order to fatten them on the farm and slaughter when needed. Wild swans and geese could be fattened during the winter season, but also Black Guillemots (*Cepphus grille*) and Ruffs (*Philomachus pugnax*) were subject to home rearing. Nesting boxes for Goldeneyes (*Bucephala clangula*) and Goosanders (*Mergus merganser*) set up in order to regularly gather eggs were common along the coastal areas and watercourses in northern Sweden.

All this can be regarded as a kind of primitive aviculture for economic purposes.[373] Peasants also captured Goldcrests (*Regulus regulus),* Wrens (*Troglodytes troglodytes*) and Robins (*Erithacus rubecula*) for controlling insects and vermin, as a kind of biological pest-control.[374]

A Christmas prank

The role of captured small birds, usually House Sparrows (*Passer domesticus*), in a Christmas prank played in nineteenth-century southeastern Skåne, should be mentioned. It was called "running with the Christmas bird". In this case the birds were released indoors, but the purpose was different. House sparrows were captured at dusk when they had settled for the night. Signe Forsberg told the story from the eastern part of the province:

They were easy to get hold of, since they kept under the eaves of the low buildings. Someone who was more or less clever in rhymes wrote on a piece of paper and tied it either around the neck or the leg of the sparrow. When darkness came and all the people in the house were gathered around the large table, the door was slowly opened and a bird flew through the half-opened door. It flew straight to the candle, which flames were immediately extinguished by the wind of its wings. Inside the dark room people began to laugh, in great confusion, before someone was able to light the candle again and capture the frightened sparrow, which had been sitting hidden in a dark corner.[375]

It was important to catch the rogue, who had released the bird, and then send it onwards in the same way to a neighbor. This custom can be regarded as a kind of chain letter game.[376]

Bird catchers

Where did the bird lovers get their birds before industrialization? The birds in Sweden were generally, as we have seen, native species. Sometimes skilled aristocrats, with the help of small home-made cage traps, horse snares or birdlime (made of mistletoe berries or, probably more commonly, linseed oil), captured the bird themselves.

Figure 40: Mules between various finches. Carl J. Forsling, a porter at Uppsala University, was in the early 20th century a successful breeder of finch hybrids

Hunting was a well-established form of pleasure among the upper class in the nineteenth century, and many of these gentlemen spent considerable time on hunting trips, often in the company of peasant hunters. Many of them showed an interest in hunting in all its forms, although capturing small birds must have been a rather trivial matter for a devoted hunter, who had mastered various ways of trapping.

They had been taught these skills since early

childhood, especially in the company of peasant children, for whom trapping small birds was a form of entertainment. Several small handbooks on the subject were published, often translations from German, and they could be used as guidance in the art of capturing small songbirds.

In towns and cities it was possible to buy captured Bullfinches, Siskins and Goldfinches in the market, or from itinerant traders. There was no guild of bird traders in Sweden as in eighteenth-century France, but there was a group of professional bird catchers, at least in Stockholm where they are mentioned in sources from the early eighteenth century onwards.

Bird catchers could be seen among herring packers and baker boys, as noted by poet Carl Michael Bellman in his description of the crowd on Skeppsbron in Stockholm in his Epistle No. 33. During the eighteenth century many songbirds in cages were sold under more or less imaginative names, such as *svensk kanariefågel* 'Swedish canary bird' (European Greenfinch (*Carduelis chloris*)), *paradisfågel* 'Bird of paradise' (Snow Bunting) and *svensk papegoja* 'Swedish parrot' (Pine Grosbeak (*Pinicola enucleator*)), in order to increase their value on the market. *Siska* (Siskin) and *steglitsa* (Goldfinch) as Swedish bird names are trade names of German origin, which bear witness to the fact that these species were found in the bird trade early and were first and foremost connected with bird-keeping.[377]

A magazine report in 1860 by August Blanche noted the bird catcher group, which was despised by the urban population. Many of them belonged to the socially marginalized group of Swedish Travelers or Roma, but appreciated because they could supply buyers with songbirds. The essay, which gives a rather negative picture of the bird dealers, says that they bought or built the bird cages themselves and that they were found in the forests around the capital, where they lived a "complete Gypsy life".

They captured Siskins, Bullfinches, Waxwings and, for want of something better, Sparrows, which they then "bring to the city and offer for sale to prospective buyers". They were "skilled in the art of taming Canary Birds and teaching starlings a language which is mixed with Romani, which they are especially familiar with". According to the essay, one could see a group of two or three bird capturers on the northern end of Norrbro offering birds for sale almost daily.[378]

Peasant boys also captured birds in order to sell them in the market square or in front of the churches on Sundays. Zoologist August Wilhelm Malm referred to an encounter with a boy in November 1854 near Agnesberg outside Gothenburg. The boy used to catch Goldfinches with birdlime, and he had a certain decoy bird in a cage, which he placed close to the tree with lime.

Local historian Mattias Oscar Johannes Svedberg revealed that in Skön parish in Medelpad the boys used to capture Siskins with the help of a live female in a cage and with a surrounding net. By selling the songbirds they earned a small amount of money. From the coastal areas of the northern province of Västerbotten, zoologist Otto Holm described how peasant boys during the second half of the nineteenth century eagerly caught Siskins for cage birds and sold them outside churches. Geologist Arvid Gustav Högbom remembered how boys in the 1850s used to capture Siskins in Umeå in order to keep them in cages.[379]

Peasant children also kept songbirds in cages themselves during the nineteenth century. Historian and linguist Lars Levander reported from Älvdalen in northwestern Dalecarlia that the boys used to capture Bullfinches in order to have them as cage birds.

Per Arvid Säve gave an account from mid-nineteenth century Gotland, telling that the boys, with the help of lime-covered sticks, traps and bird nets, caught and traded with Bullfinches, Common Redpolls, (*Carduelis flammea*), Goldfinches, Siskins and other finches, especially in Visby. Olof Christoffersson remarked about his own childhood experiences in nineteenth-century southern Skåne, that with a net he captured Chaffinches, which when they had adapted to captivity, became tame and "twittered very beautifully"; Bramblings (*Fringilla montifringilla*), which had an angry temper and picked out the eyes of other cage birds, if they had the opportunity; Redpolls "which sang the most beautifully of all of them and adapted well to captivity"; and the Greenfinch, which was "a very rewarding and friendly bird which sang beautifully with a mild and pleasant tone".

He captured other birds as well: "Goldfinches, Linnets and Siskins could not be captured in nets, since they never came close to the farms, but kept themselves in the ditches and roadsides, where burdocks, thistles, chicory, greater knapweed and such plants grew. If you wanted to catch these birds you must make a cage trap and bait it with burdocks or other plant heads on the trip peg placed in the middle of the cage. If you were able to get a female as a decoy bird, it was easy to capture the males, because they were not afraid to sit on the seed heads on the peg and begin eating, where the upper part fell down to trap them."[380]

Cages could be made by the bird owners, local craftsmen or purchased from itinerant hawkers. The information available is scanty. The cage birds also required a supply of food. The commonest cage-bird food was seeds of flax and hemp, which the farmers were widely producing on their lands. Seeds of turnip were also commonly used as cage bird food.

Enterprising cage bird breeders could of course gather seeds and green parts from wild plants, which also became important food. The sources reported the use, among others, of seeds of plantain (*Plantago major*), burdock (*Arctium* sp.), ragwort (*Senecio* sp.), as well as leaves of dandelions (*Taraxacum* sp.) and chickweed (*Stellaria vulgaris*).[381]

Cage birds among the bourgeoisie

In comparison with central Europe, the cage bird hobby in Sweden was still very underdeveloped in the second half of the nineteenth century. While it became a wide and mature hobby on the British Isles and in German- and Romance-speaking areas, involving not only the upper classes, but also artisans and industrial workers, the love for cage birds remained very restricted in Sweden. The reasons for this are not clear, but an important cause is probably the delayed urbanization.

On the continent the hobby was indulged in by an urban layer of petty bourgeois, and on the British Isles by an established working class. In Sweden, most people still lived in the countryside and the surplus population emigrated abroad and not to nearby towns and cities, which had a limited amount of industries. Bird shows in Scotland, England or Germany gathered thousands of Canaries in the 1880s, while the species still was rare in Sweden and ownership restricted to a few well-to-do individuals. In Denmark the situation was different, which can be seen in local overviews.[382]

German zoologist Alfred Brehm, who briefly described the bird trade in Europe in the 1860s, concluded that in Sweden the trade was on the same level as in Greece and Turkey. Sweden mostly kept native seed-eaters. His correspondent, a German-born taxidermist at the Natural History Museum in Stockholm, Wilhelm Meves, testified that there was hardly any bird trade in the country. If one was looking for canaries, parrots or other exotic birds, it was necessary to either import them one's self to ports like Gothenburg and Malmö, or one was referred to German hawkers, who from time to time arrived in Stockholm with Canaries. The bird dealers from the Harz Mountains arrived in the capital with several hundred Canaries, Meves noted. They were soon sold out, although a great part of the goods turned out to be females, not the beautifully singing male birds the customers expected.

Of course some Canaries were locally bred, but this was on a limited scale among a small circle of friends. In this way also the young Strindberg received access to his Canaries.[383] If the international bird trade was still undeveloped, native birds were available on a regular basis. Bird handbooks from the late nineteenth century show that also many ornithologists kept them in cages for pleasure. Knowledge of the birds was often based on observations made in captivity.[384]

Sailor birds

Parrots remained invisible in Swedish sources during the entire nineteenth century. They were expensive and rarely found their way to Sweden, probably only with returning sailors and the odd private import.

Commander Erik Hägg remembered that in the late nineteenth century parrots were brought home by sailors in the neighborhood Björkholmen in Karlskrona. Mathilda Langlet wrote in 1884 that the reason why parrots are rare as cage birds is the difficulty in fostering a tame one, "so she does not become a nuisance rather than a pleasure". According to a pamphlet from 1903, the Amazon Parrot, probably referring to the Blue-fronted species, was the most common bird kept in Sweden, but the author also mentioned the Cockatiels (*Nymphicus hollandicus*) and Budgerigars (*Melopsittacus undulatus*).[385]

A specific setting associated with parrots is the cabin of old sailing ships. It was not uncommon for crew members to bring various animals as companions on board. From the steam corvette *Danzig*'s tour on the Mediterranean in 1853 there are reports of chained falcons and tame turtles. On the sailing ship *Herzogin Cecilie* from Åland the crew brought along a very much loved Canary Bird in the 1830s. When the bird one day escaped and fell exhausted into the sea, the first mate jumped, unconcerned about the shark risk, fully dressed into the sea and saved it.[386]

Modernization

As late as 1896, Danish V. Wiese reported that the bird trade in Sweden was not developed. He noted that the interest for tropical birds was still in its infancy and far behind Denmark. This was explained by the lack of relevant technical literature, but also the troublesome transport situation contributed. Some people interested in birds lived in the towns along the Öresund, according to Wiese. They could buy their birds directly from bird dealers in Copenhagen; he did not know of any dealer in exotic birds in Sweden. Problems with transportation hampering the development of bird-keeping culture were also stressed by Swedish authors at the turn of the century.[387]

However, during the last decades of the eighteenth century, it was possible to discern an increasing interest for cage bird-keeping among the bourgeoisie in Sweden. The new currents were reflected by the increase in published handbooks and guides, often published in several editions. Magazines also wrote articles about the care of Canary Birds. Care instructions were adapted to the situation; cages should never be hung from ceilings, wrote an author in 1882, especially not in a room where gas is burnt, "since the air thereby becomes hot and is deprived of all its oxygen".

In 1894 the first edition of Lorens Emanuel

Björkman's manual on Canary care and breeding appeared, the first book published on the subject written by a Swedish author. Björkman was born in Malmö in 1840, but he grew up in Schleswig. He first served as a school teacher and later as a journalist in Borås. A few years later, Danish immigrant and bird trader Axel Cohn wrote a few booklets on keeping exotic birds and Canary Birds respectively.[388]

The Society for the Promotion of Aviculture in Sweden organized exhibitions in Stockholm during the 1880s and 1890s. The first aviculture show took place in connection with a regular meeting held by the society at the so-called Moriska paviljongen (Alhambra) at Hasselbacken in Stockholm on 23–25 March 1885. The show included poultry, turkeys, geese, ducks, pigeons and the category "song and splendid birds". At the fair also other birds could be found, for example, "three specimens of Chaffinches, which attracted a special interest because of their unusual tameness". The 1886 exhibition had thirteen participants, but we have no information about which birds were shown. Three named exhibitioners, N.J. Österlund, Corny Murray och W. Holmström, all from Stockholm, received extra awards for their birds. At the fair the following year there were thirteen cages with "song and splendid birds", although no information about species is available.

At the fourth exhibition in the center of Stockholm in September 1889, there were beside poultry, turkeys and geese also a larger collection of songbirds, among other "Song Thrushes, Blackbirds, Garden Warblers (*Sylvia borin*), Robins (*Erithacus rubecula*), Roller and other Canary Birds, Chaffinches, Redpolls etc.", which "through their songs contributed to making the visit at the exhibition as pleasant as possible". Apart from the Canaries there were probably only native birds. The exhibitions of the society continued into the 1890s.[389]

From this small base of interested bird lovers in Sweden, eventually special societies develo-ped. In 1900 the Danish society *Foreningen for Fuglevenner* (Society for Bird Lovers) had eight members in Sweden, three in Gothenburg, four in Uddevalla and one in Malmö. Southern and western Sweden are still today the main areas for organized bird-keeping.

Among these pioneers we may mention Hjalmar Rodin (–1942), active in Gothenburg, who around the turn of the century published several articles illustrated with his pictures in the Danish journal *Tidsskrift for Fuglevenner* (*Magazine for Bird Lovers*).[390]

In the late nineteenth century bourgeoisie circles, songbirds were especially cherished among women. The bird was kept for its song, but also for company. Actor Anders de Wahl remembered how his paternal grandmother sat in the kitchen chatting with her Canary Bird Kalle, "while her fingers, gout-bent from scrubbing and washing, but always nimble, were sewing and repairing and stuffing". In her handbook for housewives (1884), Mathilda Langlet also wrote a chapter about the canary. They are "usually born and raised in a cage and therefore do not suffer very much from captivity", she commented, and recommended a spacious cage with a small bathing section and that they should be fed with seeds of turnip, hemp and canary grass, as well as green food.[391]

Parallel with the growth of cage bird-keeping, the interest in breeding Pigeons (*Columba livia* forma *domestica*) increased, although this hobby retained a lower profile. It was also a more masculine sport; since the 1880s the use of messenger pigeons became of military importance, and pigeon activity was established. Gradually also civil pigeon breeding developed.

The first association was founded and soon others followed. *Svenska Duvavelföreningen* (Swedish Pigeon Breeding Association) was founded in 1904. It is still active and was commemorated with a memorial stamp in 2004. The affinity between cage bird-keeping and pigeon-keeping is shown by the association *Sveriges duv och burfågelvänner* (Swedish Pigeon and Cage Bird Lovers), which was founded after a schism within the Swedish Pigeon Breeding Association in 1919. The period after World War I seems to have been unusually important for pigeon breeding in Sweden. Interestingly a native breed, the Swedish tumbler (*svensk tumlet*), originating from domestic pigeons kept in the countryside in the nineteenth century, was developed. Another local breed (*göteborgstumlare*) was developed in the vicinity of Gothenburg.[392]

Dovecots could be found in some neighborhoods of central Stockholm around the turn of the last century. In the early twentieth century, the city provided a certain area on the hill Bellevuehöjden for the local pigeon club, and their dovecots still exist. In Gothenburg the pigeon owners had their dovecots in the vicinity of Olskroken railway station. The Swedish Messenger Pigeon Association (*Svenska Brevduveförbundet*), an umbrella organization for all messenger pigeon clubs, was founded in 1917 by Oscar Gyllenhammar in Gothenburg. Pigeon breeding has remained a relatively anonymous activity in Sweden.[393]

Trade and import

Commercial breeding and training of songbirds developed during the nineteenth century in several parts of Germany. Training of song-birds, dictated by an imperative demand for song quality and extensive skills, was an old tradition, but its peak came now. For instance Bullfinches had traditionally been raised and trained in Hesse and Thuringia. Well-trained Bullfinches were sold at high prices on the market. Chaffinches were trained in the same way, not only in Germany and the Netherlands, but also in Belgium, where song contests with Chaffinches still take place.[394]

Through migration from Tyrol in the early nineteenth century, the knowledge of instructing good singers spread to the mining districts of the Harz Mountains in Lower Saxony. The high-quality singing Harzersänger, Harz Mountain Roller Canary, developed especially around the small mining town Sankt Andreasberg and surrounding villages. German zoologist Harald Othmar Lenz, who in the 1830s traveled in various German-speaking areas where canaries were bred, noted that the birds from the Harz Mountains were superior to all others. These Rollers were continually perfected and most desired. More than 400 households in the region were said to have been involved in the Roller business. Rollers soon became an item for export and an important source of income for the locals.

"Many earn as much as 70 to 80 dalers from it", according to Lenz, who added that "around 12,000 canary birds a year are exported from there". Females were sold by itinerant hawkers during the summer, while young or well-trained older males were sold from October onwards. In 1866 a retail shop was opened in St. Petersburg solely for Harz Mountain rollers; the golden age for breeding lasted from the latter half of the nineteenth century until World War I. Hundreds of thousands of trained rollers were produced and many went into the export market.[395]

The German traders reached Scandinavia, but Canaries were also imported from England. As communications improved and freight service became cheaper, the access to exotic birds was facilitated. Swedish bird lovers could import their cage birds from Denmark. Danish bird retailers, among them Jysk Fugle-Export in Randers and H.C. Hansen in Horsens on Jutland advertised in Swedish magazines.[396]

Around 1890 import companies had become

established in Sweden as well. The trade no longer involved only Canaries, but various exotic birds from Africa, Asia and America through agents in Germany and Denmark began to reach Sweden. An important meeting point for bird lovers was Axel Cohn's store, established in Stockholm in 1898. Cohn was a very skilled bird trader and he later became a pioneer in the aquarium hobby, which became popular in the 1920s.

Cohn's shop remained an important meeting place for everyone interested in cage birds and was still in existence in the 1970s. Knut Johansson, Seiler & Bolle and after the war, Nils Englund established stores. Like Cohn, the owners wrote articles about the care of cage birds in various contemporary magazines. Famous Swedish explorer Sten Bergman, who acquired his first Java Sparrows (*Padda oryzivora*) in a Stockholm bird store around 1905, confirmed that there were three main bird traders in Stockholm at the time of World War I. In the 1950s, Sten Bergman was the first to breed a King Bird of Paradise (*Cicinnurus regius*) in his aviaries outside the capital.[397]

In southern Sweden, Carl M. Clementson developed an important import trade company through his Malmö Fågelimport, which in 1906 advertised among others "English Type Canaries". A few years later we find a store in Malmö under the management of Agda Jaeppelt. Her store claimed to be "Sweden's oldest and largest special store" in the trade. Beside cage birds, small mammals and turtles were also sold in these stores.

The seed stores also played an important role in the promotion of keeping cage birds as companion animals. They provided not only the food for the birds, but also cages and necessary utensils. In some cases it was also possible to buy birds in the seed stores. An important retailer of birds in the 1890s in Stockholm was Tjäder's seed store. In a price list from the company, the Siskin especially is praised as "the joker of the bird collection", which with patience could become very tame. Swedish manufacturing of steel wire cages began in Malmö, where the company Ståltrådsverkstaden produced bird cages of various kinds.[398]

The Army of Bird Lovers

By the early twentieth century, exotic birds dominated the trade and competed with native species. At the same time resistance to keeping native species in captivity developed. In the countryside young boys continued to trap Siskins and Bullfinches for cage birds or to sell them. Opposition was possibly encouraged by late nineteenth-century children's literature, in which new values appeared. L.E. Björkman wrote a chapter about bird trapping in his book about cage birds, the first originally in Swedish, where he regretted that there was no law protecting birds against capture. Nevertheless, he gave some simple advice on how newly trapped birds should be taken care of. He dissociated himself from capturing adult birds and suggested that only very young birds should be used within aviculture. In 1907 a law protecting birds during the breeding season was passed.[399]

Songbirds, Canary Birds, and with an increasing availability, also more exotic birds began to fill Swedish homes in the early twentieth century. The social sphere expanded from a rather narrow bourgeoisie to wider circles. Bird lovers may be divided in three categories, noted Emil F. Clementson in the first issue of *Ornitologisk Tidskrift* in December 1905, namely "breeders and friends of Canary Birds", those "who value delightful bird song", and those who kept birds in order "to study their life and behavior and if possible to breed but preferring something more original than the canary bird". He also stated that the "army of bird lovers" probably was larger than for instance pigeon breeders and owners of bantam poultry, but "very little is heard from them".[400]

Hobby aviculture in Sweden survived the war years without any serious problems. Soon after World War I the Canary had reached the wider population. A report from Varberg mentioned that they were known as shoe-maker birds, as almost "all shoe-makers had canary birds. It was therefore nice to visit the shoemakers in order to listen to and see all the birds. There was no radio at that time, and the bird-song had to be the music", reported a woman born in 1913. The connection between shoe-makers and cage-birds was reported from various places, including Denmark.

Siskins and European Goldfinches continued to be popular as songbirds, but those in the commerce were often imported from the continent, although some local bird catchers were still in business. The Twit (*Carduelis flavirostris*) was according to Alarik Behm "one of those birds that almost every winter is offered for sale in the vicinity of Stockholm from November to April".[401]

Soon the interest for tropical cage birds was gaining ground. Stockholm physiotherapist Axel Adlersparre (1881–1956), who, in the 1910s and 1920s, become one of the country's most prominent cage bird experts and a successful breeder in his apartment in central Stockholm, wrote a richly illustrated article in 1914. He asserted that the Swedish bird hobby was increasing, although it kept a "more modest life" than on the continent. The supply of tropical birds increased during the last few years, although literature in Swedish was lacking.

Publishing articles in popular and specialized journals, Adlersparre worked towards a modern approach to bird-keeping. In an article from 1918 he wrote that "keeping birds of course involves certain elementary duties, which are often overlooked by both domestic and foreign bird lovers". The fundamental rule was to try to imitate as much as possible the living conditions of the birds in the wild, when one kept them in cages. Good bird cages, fresh air and satisfactory light conditions were important. Sanitation was an important subject of the time, also for cage bird-keeping, and scientific insights became increasingly important features in bird-keeping.[402]

Most imported birds still arrived via the wholesalers in Denmark and Germany. In the 1920s new handbooks on the subject were published. Types of food were modernized. Native birds could be fed with crushed hemp and turnip seeds, while canaries needed imported canary-seed. Exotic species demanded a more varied menu; greens were necessary for their health. Earlier recommendations about sugar disappeared. Critical voices were raised in some of the handbooks being published. Under the motto "do not keep birds in cages, and if you do, take care of them", a small booklet was published in 1920, written by a woman, for a change. This anonymous author advocated larger cages, better sanitary conditions and that the cage birds should be kept because of love for nature, and not "because it is the latest craze or something that will fit with other decorations".[403]

With the growing import, especially from Africa and Asia, but also from Australia and to some extent America, a lot of new birds appeared. The bird lovers divided their protégés into seed-eaters and softbills. A small parakeet from Australia became quickly popular. Budgerigars had been introduced to Europe in the 1840s and quickly attracted public attention. Towards the end of the decade they had already bred in France, Belgium and the Netherlands.

In the 1850s, mass export of birds from Australia started and continued intensely to the end of the century, when a total halt to exports of the species was introduced. However, the Budgerigar was easy to care for and breed, and soon new color variations appeared, a development which increased the interest for it. In the 1870s, the first Bengalese Finches (*Lonchura domestica*), a well-established domestic bird in Japan, arrived in Europe. Easy to care for and inexpensive, they soon became popular. The South-East Asian Java Sparrow also belonged to the popular species within the trade.[404]

Fairs were again organized after World War I, and in the 1920s they were arranged regularly in Sweden with both native and exotic birds. Judges usually came from Denmark in order to examine the birds. At an exhibition in Stockholm in 1923 a large number of exotic birds were shown. At the zoological garden at Skansen open-air museum in Stockholm, the Director, Alarik Behm, had built up a nice collection of birds, which also served as an inspiration for those interested in aviculture. During the 1910s and 1920s the zoo had many open-air aviaries and a bird house with several exotic and native species.[405]

Rollers and mules

In southern and western Sweden, interest in the German roller developed in the early twentieth century. A kind of local competence in the breeding and training of the Harz Mountain rollers was built, and the quality of the Swedish birds could match the German-trained ones. Breeding was concentrated to the Malmö-Lund and Gothenburg areas.

The now centenarian society *Kanaria* (their first statutes were passed in 1906) in Gothenburg is still active. The vigorous Harz Roller Breeding Association (*Harzerodlare-förbundet*) had local branches in Malmö, Lund, Trelleborg and Gothenburg.[406]

The Roller Canaries were originally imported from Germany and the Swedish breeding, training and competition activities followed the German pattern with a German terminology. The song was judged after the so-called German unity scale from 1922. In 1934 a Swedish translation of a handbook with the various song patterns of the rollers appeared.[407]

In the 1930s, the time was ready for specialized journals to deal with the subject. Anton Hillerman in Malmö published the magazine *Burfågelvännen* (Cage Bird Lover) in 1930, and between 1932 and 1937 Gösta Niehoff, also in Malmö, published *Kanaria*. Annual shows guaranteed the high quality of the birds and proud breeders could sell their award-winning Rollers of high class.[408]

A strange feature of European Canary breeding was the enthusiasm for breeding hybrids between Canaries and other finches. This kind of hybrid breeding has been known from the end of the sixteenth and early seventeenth centuries, when the Nuremberg doctor Lazarus Röting began to depict Canary Birds. In a preserved painting from 1610 there is an illustration of a cross-breed between a Goldfinch male and a Canary female, a combination which later became the most common among breeders.[409]

This method of breeding canary hybrids has been especially popular in Great Britain, although it has also existed on the continent. Cross-breeds, so-called *mules*, during the nineteenth century received a separate category at exhibitions, a tradition continuing until today. Breeding hybrids was known as a practice in Sweden in the eighteenth century. During Linnaeus's provincial excursion in June 1746, he visited the German-born pharmacist Franz Mikaël von Aken, who gave him details about hybrids between Canaries and Goldfinches. From his university lectures in 1748 it is obvious that Linnaeus was very fascinated by the topic.[410]

Despite the fact that most hybrids are not fertile, this kind of breeding has continued to attract breeders. "The author had once taken eggs from a Linnet's nest and let them be hatched by a Canary. It succeeded well and the Linnets were not only tame, but also mated and gave hybrids with the Canary birds", wrote August Emil Holmgren. A large number of finch species have been reported giving hybrids with the Canary Bird. In the 1920s, mules between Goldfinches and Canaries belonged to the normal supply among the bird dealers in Stockholm and they were eulogized for being good singers.[411] In the early twentieth century, the porter at Uppsala University, Carl J. Forsling, was a very successful breeder of hybrids between various native finches.

The new understanding of heredity, especially following the rediscovery of Gregor Mendel's findings around 1900, caused zoologists to take a closer look at the offspring. Forsling's hybrids therefore were subjected to detailed studies by zoologist Einar Lönnberg, who described a hybrid between a European Goldfinch and a Bullfinch, a rare combination which bears witness to Forsling's skill as a breeder.[412]

Around the time of World War I, German geneticist Hans Duncker began a collaboration with bird breeder Karl Reich in a conscious attempt to transfer the genes for the red color from the South American Red Siskin (*Carduelis cucullatus*) to the Canary.

This early example of gene manipulation, long before the sheep, Dolly, would even be considered, was eventually crowned with success and had consequences for the supply of canary birds in Sweden.[413] In 1917, Axel Adlersparre bred Australian Gouldian Finches (*Erythrura gouldiae*), a colorful, expensive and sensitive bird. He made advanced observations in heredity, directly influenced by Gregor Mendel's theories. Adlersparre published a detailed and sophisticated study of the heredity of cross-breeds between the red-headed and the black-headed Gouldian Finches, which were still regarded as separate species.

Adlersparre was obviously an early breeder of the species. The first reproduction on the European continent is documented in 1915, about the same time that the species was also bred in Sweden. Adlersparre's successful breeding was much commented on in international literature.[414]

In 1920 young Curt af Enehjelm (1903–1982) arrived in Denmark from Österbotten in Finland, and almost immediately established himself as a successful breeder and exhibitor. He soon became an important figure throughout Scandinavia, and from the 1930s was also an author of an increasing number of avicultural handbooks published and sold in all northern countries.[415]

Foreign eyewitness

When World War I ended, the interest in cage bird-keeping was well established in Sweden. Several factors were responsible for this, among them improved communications, increasing influences from the European continent and an urbanized way of living, which provided free time to develop a hobby.

Gone were the sentimental relations of the eighteenth century for companion animals; instead the relations were characterized by scientific objectivity and a greater social distribution. The status of the companion animals had changed. Both children and adults were amused by birds in cages, especially exotic species arriving in increasing numbers. In the same way as Strindberg in his later days was fascinated by the selection of weaver birds and estrilds in Mr Cohn's store, many of his contemporaries were attracted by what the bird dealers could provide. Long before nature films became an everyday occurrence, the cage bird store was evidence of a magnificence from distant parts of the world. The trade in exotic birds provided an opportunity for escapism and a concrete way of getting in touch with the tropics. Writer Eva Alexanderson, born in 1911, related in a self-biographic novel about the Stockholm girl Mona who visits the same bird store as Strindberg a few years earlier:

> It was the world's most wonderful store. It was dirty and smelled, but after she had been

standing there for a while she did not notice the odour. Cages in long rows were filled with brilliant colors and life. It was like the light from a broken, feathered rainbow, small downy soft bodies in all colors. They were eyewitnesses brought from far-away foreign countries with sun and life. She had learnt the names of them: olive green Chinese Robins, Lazuli Buntings, Zebra Finches, Saffron Finches; and there were also the Pink Cockatoos and the fire-red and emerald-green parrots and cream-colored Java Sparrows with red bills. All of them had black shining inquiring eyes that looked at her, apart from the parrots who glared cold and indifferent. Soft, downy bodies which she longed to touch with her fingertips and put her cheeks against in order to feel their warmth and softness.[416]

In the girl's otherwise grey and gloomy life, Mr Cohn's store was a kaleidoscope of colors and experiences with a breeze from another, happier outside world.

Chapter 9. Two Asian Domesticated Species

The Bengalese Finch (*Lonchura domestica*) and Java Sparrow (*Padda oryzivora*)

Ingvar Svanberg

Japan possesses a long tradition of aviculture and songbirds have probably been kept in small cages since ancient times. Most birds in captivity were, as Ken Kawata has pointed out, native species, such as Japanese Bush-warbler (*Cettia diphone*), Blue-and-white Flycatcher (*Cyanoptila cyanomelana*), Japanese Robin (*Erithacus akahige*), Japanese White-eyes (*Zosterops japonicus*), Skylarks (*Alauda arvensis*) and Siberian Meadow Buntings (*Emberizia cioides*). Some birds, especially V.aried Tits (*Paris varius*), trained to perform tricks, were a popular entertainment in the streets. Imports from other parts of Asia and from Europe became possible when Dutch traders reached Japan. Caged birds were given as gifts to the shogun and officials in the government, who were directly involved in trade. Many of the imported exotic birds (lories, cockatoos, mynahs, cassowaries, pheasants, tragopans) given as gifts came from South-East Asia, but also from other parts of the world. Canaries, domesticated in Europe, were also brought to Japan and eventually bred there. The Dutch were more significantly involved in importing birds to Japan than the nearby Chinese.[417]

Although overseas traders brought several cage birds to Japan earlier, only in the mid-nineteenth century did Japanese birds reach aviculturalists in Europe. The Bengalese Finch (*Lonchura domestica*) and Java Sparrow (*Padda oryzivora*) are two truly domesticated species with long traditions in Asian aviculture. Today they are popular among breeders, because of the variety of colorings that exists among them. They are also appreciated as companion birds. Together with the Zebra Finch (*Taeniopygia guttata*) they are also very easy to keep and thus recommended for beginners. The Zebra Finch is a relative newcomer, introduced into aviculture in the late nineteenth century (yet already kept, and probably bred, in France in the late eighteenth century), but it has quickly became domesticated and many color mutations have occurred.[418] The other two species have supposedly been bred for centuries in Japan. They were already domesticated birds when they reached Europe through trade in the nineteenth century. Java Sparrows, known as *rijstvogel* in Dutch, *Reisvogel* in German, *oiseaux de rix* in French and sometimes 'Java rice bird' in English, captured in the wild, were brought to Europe from South-East Asia even earlier, and were among the earliest imported exotic finches to be kept as cage birds. "There are few vessels from Java and the Cape of Good Hope that do not bring numbers of these birds", wrote Johann Matthaeus Bechstein in the early nineteenth century. "When the first specimens may have reached Europe is impossible to tell, because every writer speaks of the Rice-bird as a well-known cage-bird, and very large numbers have been exported from Java ever since Europeans visited the island", noted August F. Wiener around 1880.[419]

The bird was known in England in the early eighteenth century. George Edwards observed in 1743 one Java Sparrow at Sir Hans Sloan's home. Java sparrows are mentioned as one of the more popular seed-eating species imported to France in the 1750s. French ornithologist Louis Jean Pierre Vieillot kept, in the late eighteenth century, a Java sparrow that survived for seven years in a cage. Because of its handsome appearance and easy care it was very popular. "To be a pet and plaything of a lady in her boudoir seems to me the proper place of the Java Sparrow", wrote

Wiener and continued: "But the lady must not mind being pinched sometimes in her fingers by her pet." Wild-captured birds were also bred in the early 1870s in St. Fiden, Switzerland, by a Dr Carl Stölker.[420] The domesticated variety, usually white or pied, was bred in Asia. It reached Europe from Japan about the same time as the Bengalese Finch in the mid-nineteenth century.[421]

Origins of the Bengalese Finch

It is easy to agree with the fine characterization given in a popular English handbook that the Bengalese Finch (*Lonchura domestica*) is probably "one of the most charming foreigners that one can keep". Many love this small bird and I remember in the early 1960s, while still a child, when I bought my first cage birds. We had only one small pet store with a very limited selection in my home town. The owner suggested that I start with Bengalese Finches. They were attractive birds, easy to care for and they bred without trouble. My Bengalese became the beginning of many years of hobby aviculture.

Figure 41: Grey and white Java Sparrows in 1880.The Java Sparrows are flanked by a White-headed and a Black-headed Munia respectively

The origins of the bird has long been debated. "This delightful little fertile hybrid was produced by the Japanese", wrote Arthur G. Butler in his famous book on foreign finches in 1899. "The Bengalese is not a wild but a domesticated bird", noted Alec Brooksbank of the renowned Keston Foreign Bird Farm in his popular book on foreign birds for garden aviaries in 1950, adding that it is a "most unusual thing, a fertile hybrid". He speculated that Indian Silverbill(*Lonchura malabarica*), "Striated Finches (*Lonchura acuticauda*)", and "Sharp-tailed Finches (*Lonchura striata*)" had all taken part in its production.[422] Butler and Brooksbank were not alone in their views; it seems to have been the common opinion until rather recently. "Its true origin is obscure but it is generally supposed to be a fertile hybrid", thought D.H.S. Risdon, manager of Dudley Zoological Society. Actually, many handbooks still regarded the Bengalese Finch as a hybrid originating from China, which once was imported to Japan and from there to Europe.[423]

The Bengalese Finch is indeed very popular among bird fanciers. It is today kept all over the world. In 1998, around 25,000 Bengalese Finches were produced annually by members of a German bird association. The global figures are of course much higher. Many color varieties (including white, chocolate, chestnut, fawn, grey, pearl, pied, dilute, tricolor, and clearwing) exist; also various crested forms are known. It is a very prolific breeder, easy to maintain and its biology in captivity has been studied in detail. Newer research deals with ethology, song pattern, heredity and neurobiology. The Bengalese Finch is from many points of view a well-studied bird, but still its origin remains a mystery.[424]

In Europe

Japan remained for a long time a secluded society. From 1603 during the Tokugawa Shogunate the country adopted a policy of isolation and kept limited relations with the outside world, including China. When it was forced to open up to the surrounding world after the Convention of Kanagawa in 1854, many foreigners visited and Japanese trade products arrived in Europe and North America. Among the many wonderful things, the foreign traders also brought home two white domestic birds, the Bengalese Finch and the Java Sparrow.

The earliest importers of Bengalese Finches to Europe were zoological gardens. According to August F. Wiener, two specimens of the "White Finches" were purchased by the Zoological Society in London in October 1860, becoming the first known of the species in Europe, but soon the birds became common goods. Mr. Wiener purchased the first white Bengalese Finches from a London dealer in 1869 or 1870, "the first bird of the kind I had seen". C.E. Dyson reported in 1878: "I had a pair of white birds [...], kindly sent to me by Mr. Hawkins on their first introduction into this country."[425]

Another early aviculturalist, with some years of experience of the Bengalese Finch, was Charles William Gedney, who claimed he had imported the birds himself from Japan. In Antwerp around the same time, Jacques Vekemans was an enthusiastic pioneer in breeding and keeping cage birds, including Budgerigars (*Melopsittacus undulatus*). He was director at the local zoological gardens from 1865 until 1888 and he received some Bengalese Finches in a shipment of African estrilds in the 1860s. In 1871, a number of white and variegated Bengalese were offered on the market by the zoo in Antwerp. Also zoological gardens in the Netherlands had the new species in the 1880s. A few years later, the Bengalese Finch was bred annually in thousands in Europe.[426]

The first birds in Europe were probably completely white, but soon also variegated "Bengalis" were available. Three color varieties were found: pure white, fawn pied with white background and chocolate brown pied with white background. They soon became hugely popular in Europe and North America, not only because they were easy to breed and care for, but also because they could be used as fostering birds for more difficult breeding species, such as the Lady Gouldian Finch (*Erythrura gouldiae*) and other species, especially Grass-finches.[427]

Naming

The Japanese name for the Bengalese Finch is *jūshimatsu.* The name itself appears to originate in Chinese, meaning 'in group'. English sellers for some reason associated the bird with Bengal in South Asia; not aware about the longer trip the birds had made, the bird retailers took them off ships arriving from India. The bird has no connection whatsoever with Bengal. Another explanation is that in the early days "Bengali" was used indiscriminately for many exotic finches appearing in the bird trade. According to August F. Wiener,

Figure 42: White Java Sparrow, Uddevalla, Sweden, died in 1904

the English dealers called it *White Bengalese*, and in French a similar name was used, *Bengali blancs*. Wiener prefered the name *pied mannikin*. In the early twentieth century, it was called Japan Munia by the Indian bird dealers in Calcutta.

In North America it was sometimes called *Bengalese*, but in the 1880s commonly named *Japanese Nun*, according to George H. Holden. Today it is called *Society finch* in the United States, a rather recent and probably post-World War II name. Alternative names are *Bengalese mannikin, Bengalese finch* or just *Bengalee* in English. Oddly enough, hardly any dictionaries document the names of this bird taxon. *Webster's New International Dictionary* has *Bengalee* as an entry for "any of several small tropical songbirds commonly kept as cage birds".[428]

While English-speaking aviculturalists never recognized its Japanese origins in the name, their German colleagues did. Famous German aviculturalist Karl Russ, who bought his first specimens in 1871, baptised this newcomer obscurely *Japanesisches Mövchen*. The reason behind the name (diminutive of *Möwe* 'Gull') is not that the small and lively pure white birds reminded him of little gulls, but more likely that it was created as a kind of opposite to *Chinesesisches Mövchen,* a pigeon (*Columba livia* forma *domestica*) breed imported from France. The pigeon breed name was probably invented by French pigeon dealer Mr. Destriveaux, who in 1865 sent *Chinesesische Mövchen* to a breeder in Germany, as also mentioned in the first issues of *Gefiederte Welt* in 1872. The pigeon, known as *Chinese owl* in English, has no connection with China, but probably originates in Spain and can be traced back to around 1850. The various kinds of pigeons belonging to the *Mövchen* group are said to originally have been similar to Herring Gulls (*Larus argentatus*), German *Silbermöve*. Possibly the original Bengalese Finches imported from Japan in shape and coloration brought to mind those pigeons.[429]

Karl Russ' German name for the Bengalese Finch, today spelled *Japanische Möwchen*, has been translated into several languages, *Japanse meeuw* in Dutch and *mewka japońska* in Polish. Scandinavian bird retailers had their birds from German traders and followed their naming pattern, *Mågefinke* in Danish and *måsfink* in Swedish. The geographical background as indicated in the German determiner was seldom mentioned. The Swedish name *japansk bronsfink*, suggested by L.E. Björkman in the beginning of last century, never was accepted in the cage bird business. In French it is commonly called *Moineau du Japon* or *Domino*. Other French names used within aviculture are *Muscade Blanche*, *Societé*, *Pinson société*, and *Bengali*.[430]

Theories about background

Many handbooks and websites still maintain that the origin of the Bengalese Finch is obscure, because it does not occur in the wild, and it is a purely domestic bird, arrived in Europe from Japan in the 1860s. Domestication is here defined as a process of hereditary reorganization of wild taxa into forms more accommodating to the interests of people. Obviously very few facts for a historiography are available, and anecdotes and non-corroborated stories are flourishing.

Some writers believed that the Bengalese Finch was a result of cross-breeding done in China centuries ago; others have tried to trace it back to one of the many Mannikin species found in East Asia. The finch is actually easy to cross with related *Lonchura* species, which results in fertile hybrids. According to Karl Russ, the Bengalese Finch was a domestic form of "Sharp-tailed Finch (*Lonchura acuticauda*)"; Mr. Wiener viewed it as a domestic variety of "White-rumped Munia (*Lonchura striata*)". Both authors were convinced that the Bengalese Finch originated from one species. These two *Lonchura* "species" are now considered as being subspecies of *Lonchura striata*.[431]

Joseph Abrahams, an experienced breeder of the Bengalese Finch at the end of the previous century, also known to have bred the first yellow budgerigar, launched a third theory. Abrahams based his notion on morphological features, and was convinced that the Bengalese Finch was a cross between the White-rumped Munia and the Indian Silverbill. British author and finch specialist Arthur Gardiner Butler, who received his first Bengalese Finches from the renowned banker and bird fancier Walter Rothschild, agreed, because of Abrahams' impressive experience with the species in captivity. Butler introduced the theory about hybridization into cage bird literature and it was accepted among many aviculturalists. The view that the finch is a "cultural product" of many *Lonchura*-species is still maintained in several handbooks. Hybridization during the domestication process has been suggested for many species, while recent genetic studies have shown that this is not so common.[432]

Japanese versus Chinese

Despite the generally accepted view that the Bengalese Finch was first domesticated in China, almost no information is available to prove it. Songbirds (and birds kept by fortune-tellers) have been popular for centuries in China. We also know that the Chinese traded with cage birds, including Yellow-crested Cockatoos (*Cacatua sulphurea*), imported on Europan merchant ships returning from Canton in the mid-eighteenth century.[433]

While Chinese always have taken great pleasure in keeping birds, they have preferred songbirds, such as White-eyes (*Zosterops* sp), Bulbuls (*Pucnonotus* sp.), and Laughingthrushes (*Leiothrichidae*), which were captured in the wild and not bred in captivity. It is therefore strange that they should have been interested in such insignificant singers as the ancestors of the Bengalese Finch. Many writers, without quoting any sources, maintain that the Bengalese Finch can be traced back to the Tang Dynasty (from the seventh century) or even earlier, thus making it the oldest domestic cage bird known.

So far no Chinese sources that can prove this background have been revealed. Certainly, the Chinese domesticated in early times different animals as pets, such as the goldfish (*Carassius auratus*), and for centuries, songbirds sang in cages, but we have no information about ancestors of the finch in Chinese culture. The information can be traced to one Japanese source, which stated that the Bengalese Finch was imported from China.[434]

Historical literature on Japanese aviculture is generally scarce. Japanese encyclopedias give no information, and hardly any books, or non,e on this issue, appear in Japan. A recent overview about exotic birds introduced to Japan through the bird trade also gives no details.

The basic source remains Prince Nobusuke Takatsukasa's essay about the history of Japanese aviculture, published in 1922. He wrote several books on cage birds in Japanese, such as *Kaidori* 'Aviculture' (1917, several editions) and *Kaidori shûsei* 'Collected works on cage birds' (1930). Takatsukasa claimed, referring to an old unidentified Japanese cage bird book, that the first Bengalese Finches were imported to Japan from China in the early eighteenth century, despite the isolationist policy at that time. Private Chinese merchants could, like the Dutch, keep up some commerce through the Dejima Island in Nagasaki harbor, and here their ships brought live birds.

Although Japan remained fairly closed during the eighteenth century, we have evidence that also Europeans tried to deal in exotic birds with Japanese buyers; Carl Peter Thunberg mentions an attempt to smuggle a parrot into the tightly controlled country. According to two newer publications, there are records of *jūshimatsu* arriving from China in 1717, 1762 and 1816. In Japan the finch became subject to further domestication and several color variations developed. There is a record of a Dutch ship bringing Bengalese Finches back from Japan as early as in 1833, but this is not confirmed by any European source.[435]

As already mentioned, the first Bengalese Finches arriving from Japan were white. Possibly the breeders gave priority to white birds, the color of purity, and the pied or variegated ones were seen as bi-products. The Japanese were, in contrary to the Chinese, skilled breeders who developed an expertise that defies our modern opinion about how to breed birds successfully. The Bengalese Finches were bred by the Japanese in very small cages but in large numbers.[436]

White Java Sparrow

The Japanese also bred the white Java Sparrow (*Padda oryzivora;* Linnaeus, 1758) in great numbers. The wild variety of this species, grey with a white-cheeked black head, was imported in large numbers to Europe from Java, Bali and Bawean already in the eighteenth century. The species was already known when the domesticated bird arrived probably about the same time as the Bengalese Finch. The Java Sparrow has also been introduced to many parts of South-East and East Asia, as well as other regions of the world. Due to habitat loss, hunting and the use of pesticides the native population in Bali and Java is now almost gone. Feral populations exist in various parts of Indonesia, Maluku Islands, Sri Lanka, Singapore, Taiwan, Zanzibar, Hawaii and Puerto Rico. Wild-captured Java Sparrow remained one of the most common species in the trade until the mid-twentieth century.[437]

Wild-captured birds were brought by Dutch merchants to Japan in the late eighteenth century, but Japanese breeders had possibly earlier received Java Sparrows from China. The birds eventually bred in Japan and it is likely that the Japanese developed the white domesticated variety; it was popular for a long time and Nagoya is mentioned as an early breeding center. The Japanese bred the white Java in large numbers. It was kept "very much like a breeding machine". It was a favorite among Japanese artists and is therefore commonly depicted in early nineteenth-century woodblock prints. The early twentieth-century author Natsume Sōseki wrote his famous book Bunchō (1908) about his pet, a beloved Java Sparrow, which in the end dies of neglect.[438]

The white Java Sparrow, which also produces grey offspring, must be regarded as a true domesticated bird. Its history is still not very well known. "The clear white color is another evidence of the superiority of the scientific breeding of the bird-fanciers" in East Asia, noted American author George H. Holden. He told also about the breeding: "The Japanese make a business of breeding the White Java Sparrows; and each pair has a small cage, in which is placed a perch and nest. By careful and stimulating feeding, they are made to lay freely; and, as soon as a brood is hatched, the old birds are placed in another cage; where they continue hatching and breeding. The young are brought up by hand, being fed on a scalded vegetable decoction mixed with millet-seed".[439] When finally imported to Europe in the 1860s, the domesticated variety became popular in some circles. C.E. Dyson wrote in 1878: "At the Crystal Palace show in 1868, a pair of Java Sparrows were exhibited by Mr. Hawkins, entirely white, with the black head and throat and rose-colored beak of the grey Java Sparrow, and with the plumage equally soft and downy, and I was told that they were most beautiful birds. The hen had laid four eggs. They came, I believe, from Japan, and were the first brought to England. Since then many pure white specimens have been exhibited."[440] The white domesticated birds were easy to breed, while the wild-captured ones were difficult. Arthur Butler wrote:

> Although it is possible to purchase newly imported grey birds at a ridiculously low price, it is of little use to attempt to breed with these excepting in a netted-in garden where natural nesting sites abound [...] With white birds, on the other hand, breeding could always be accomplished because (probably for centuries) they have been bred in boxes. I failed to breed Java Sparrows until I crossed a wild cock bird with a white hen. The offspring from these birds varied not a little, some being pure greys like their father, others pied grey and white in all degrees. Since that time I have bred many every year, and have selected pairs from my stock with which other aviculturalists have bred successfully. The white blood, and the fact that the young are born and brought up in a nest-box, seem at once to ensure a similar form of nidification in these home-bred birds. [441]

Pereira mentioned that the Java Sparrow was not a great favorite as a cage bird among the Chinese in Hong Kong in the 1930s, at least not in its original colors: "In Japan, where they have been bred in captivity for the past thirty years or more, the Japanese have produced a white variety and it is this bird that is much in demand as a pet—not for its beauty, nor the wild bird surpasses it in this point, but simply because it is out of the ordinary. Good white birds without the least sign of grey or black are fifty times as expensive as their commoner brothers, if a careful examination is made, most of the white birds that are offered for sale in the shops of Hong Kong will be seen to have some form of blemish or other". [442]

He also mentioned that the Java Sparrows were very popular with Chinese fortune-tellers: "These gentry, who profess to tell everyone's fortunes except their own, are adept at training them to pick cards on which are various Chinese characters and hieroglyphics indicating the fortunes of the ignorant victim attempting to probe into the mysteries of the future. So that it will not fly away, the fortune-telling Java sparrow has invariably all his primaries pulled out and each time after drawing a card is rewarded with a single grain of paddy stowed in a match-box. The more the customers the more grains for him, otherwise the bird goes hungry until evening—as the wily fortune-teller runs no risk, knowing well that a bird with a full crop will draw no card". [443]

The wild variety kept its popularity because of its beauty. As long as the wild-captured grey birds kept a low price, the domesticated birds continued to be rather rare. In North America the grey bird is sometimes referred to as the *Java temple bird*. Even in the 1970s and 1980s, the wild-captured Java Sparrow was one of the most commonly kept cage-birds in Europe and North America.

In some places like California, it was illegal to own one, because of a perceived threat to agriculture. The species has been released in many areas and become an invasive threat to other birds. Today imports from South-East Asia have ceased and most grey birds and other color variations among bird fanciers, like agate, cinnamon, fawn, ivory, opal, isabel, pastel, pied and silver, all originate from the Japanese domesticated birds.[444]

Where was it domesticated?

The Japanese did not only domesticate the Bengalese Finch and the Java Sparrow, but also Japanese Quail (*Coturnix japonica*). It was bred as a song-bird and especially appreciated by the samurai. Japan also has a long tradition of developing strange breeds of chickens, known for their plumage and unusual bright colors, such as the long-tailed *onagadori*.[445]

Aviculturalists in Japan were skilled in domestication and breeding and continue to be. In the beginning of the twentieth century they bred Lady Gouldian Finches in huge numbers, as mentioned above, the Bengalese Finch and Java Sparrow are supposedly other products of these skilled breeders. China's role in the domestication of the species can of course not be ruled out without further investigations.[446] It is not known if any genetic tests have been done to determine the exact origin of the Bengalese Finch, but Japanese scholars who recently have studied various aspects of its biology, are of the opinion that it became domesticated in Japan in the mid-eighteenth century.[447]

After a century of speculation, new research shows that *Lonchura domestica* is a domestic strain of the White-rumped Munia (*Lonchura striata*). One researcher compared a large number of skins of *L. striata* with Bengalese specimens and concluded that the domestic bird must originate from one of the geographical subspecies of *L. striata*. She suggested the subspecies *Lonchura striata squamicollis* (= *L. striata swinhoei*) in southeastern China as the ancestor of the Bengalese Finch and there is no reason to expect a hybrid background.

Robin Restall has confirmed that the Bengalese Finch is a domesticated form of the White-rumped Munia, "almost certainly" the Chinese subspecies (*L. striata swinhoei*).[448] Yet the process of domestication of the *Lonchura domestica* has taken place primarily in Japan, and the Japanese breeders selected white birds for aesthetic and cultural reasons, as with the Java sparrow, which was domesticated and bred in large numbers in Japan. The birds were probably kept as pets, not, for instance, as ritual release birds within the Buddhist tradition.[449]

During the domestication process, the finch species has been modified in many ways, especially as regards color and behavior. Through centuries of selection in Asia, and later in Europe and North America, it has like the dog, horse, Barbary Dove (*Streptopelia risoria*) and Canary, become a cultural product, entirely dependent on humans and adapted to living together with people.

Due to their pleasant character and easy care they continue to be the ideal species for any beginner interested in aviculture and cage birds. Japanese breeders still contribute with new breeds, e.g. the frilled Chiyoda, Bonten, Chuunagon, and Dainagon. Since the Bengalese Finch is a recognisable entity, it could still be named *Lonchura domestica*.

In the past few decades a new bird has turned up in North America, called *European society finch* or *Continental Bengalese*, which in fact is a crossbreed between the Bengalese Finch and various species of Munias and Mannikins. These new birds, which consequently are not "real" Bengalese Finches, are the result of efforts among European aviculturalists some 40–50 years ago.

These hybrid Society finches are sometimes said to be poor parents and often produce infertile eggs. Randall concluded that modern breeders are attempting to turn the Bengalese "into the hybrid it never was". Some breeders have managed to reproduce them for several decades now without any further hybridization. The approach to canaries is different; we do not regard red canaries as hybrids, although breeders in the past used the Red Siskin (*Carduelis cucullata*) for interbreeding with domesticated canaries in order to produce the varieties with red in the plumage.[450]

Today, especially, the Bengalese Finches produced in the Netherlands are of such high quality that they are exported to Japan, where the birds continue to be popular. The circle is closed. The white Java Sparrow arrived about the same time to Europe as the finch, and it is also very popular in Japan. Both are attractive contributions to contemporary aviculture.

Acknowledgements

Thanks to Bengalese Finch breeder Fred Pander (The Netherlands) and to Dr Roland Van Bocxstaele (Antwerp Zoo, Belgium).

Chapter 10. Talking Birds

The Ancient Tradition of Teaching Corvids to Imitate Human Voices

Pedro Fernandes and Ingvar Svanberg

"Talking" ravens can still be seen in a few places in Europe. In the mid-1990s, Ingvar Svanberg saw a tame Raven (*Corvus corax*) in the village of Vestmanna on the Faroe Islands, which could imitate some phrases, especially curses. It was kept in a simple outdoor cage. In general the custom of teaching native birds this ability is, due to legal restrictions and changing attitudes to local fauna, quickly disappearing. Older memoirs, however, are full of childhood stories on keeping tame corvids.

Jackdaws (*Corvus monedula*) were a favorite for many schoolboys in Great Britain. In Sweden, people still tell anecdotes about pet jackdaws: retired old fishermen in Limhamn outside Malmö used for instance to keep tame jackdaws that whistled if a pretty woman walked by. The practice has not survived, but the habit of taming corvids to "talk" still continues in some remote areas on the Iberian Peninsula. It is prohibited there, too, but apparently accepted in certain villages. This ancient practice, exercised in Europe for more than 2,000 years, is kept alive by a handful of villagers in Portugal and elsewhere; it is a kind of traditional aviculture, already described by Pliny and other authors in Antiquity.

Birds are of course not talking, but imitating human speech. The ability of some bird species, especially corvids, starlings and parrots, to imitate human speech has been an appreciated capability for ages. Today, birds such as the Grey Parrot (*Psittacus erithacus*), and Moluccan Cockatoo (*Cacatua moluccensis*), Budgerigar (*Melopsittacus undulatus*). and Cockatiel (*Nymphicus hollandicus*) are often kept in order to teach them to whistle, sing and imitate phrases. The Common Hill Mynah (*Gracula religiosa*) became very popular for this quality and was imported to Europe and the United States in large numbers in the 1970s and 1980s, but the thriving trade in wild-caught Hill Mynahs was brought to an end in the late 1990s. Now few specimens are found, because the species is protected and is a slow breeder. However, teaching birds (now mostly parrots) to talk is still a very popular way of keeping tame pet birds in modern society.

Corvids

Historically, the knowledge of how to teach birds to "talk" seems to have been common both in Eurasia and the Neotropics in pre-modern times. In Europe various species of the crow family were known as good talkers or imitators, for instance the Jackdaw, Eurasian Jay (*Garrulus glandarius*), Magpie (*Pica pica*) and Raven. Corvids

Figure 43: Ravens have always been popular as pet birds and are easily taught to imitate human voices

are usually amusing birds in captivity and easily tamed. They have always been popular in Europe. The Common Crow (*Corvus corone*) was easy to tame, but not often kept, because of its reputation for thieving and being very mischievous. The custom of taming and teaching crows, jackdaws, ravens and other corvids to imitate human voices is not limited to the Mediterranean region, but is known in northern Eurasia and North America as well. Blue Jay (*Cyanocitta cristata*) can also be taught to "talk", and Hyatt Verrill wrote that in his youth "I owned a pet blue jay that spoke a number of words and several sentences".[451]

The habit of keeping ravens for the purpose of "talking" was well-known in Roman times. Bones of ravens are among the commonest finds in Roman camps and the ravens were probably kept as pets by the soldiers. The story of the cobbler's tame raven, as retold by Pliny, was also popular. The bird attached itself to a cobbler's shop and learned to imitate human voices. Each morning it would fly "off to the Platform that faces the forum and salute Tiberius and then Germanicus and Drusus Caesar by name, and next the Roman public passing by". The bird became a favorite and eventually a kind of public character. When the raven was killed by a rival of the cobbler, "this caused such a disturbance among the public that the man was first driven out of the district and later actually made away with". Pliny even claims that this talking raven received a public funeral in Rome in 36 CE.[452]

Talking corvids are also known in the Middle Ages (for instance, Geoffrey Chaucer refers to them in the *Maunciple's Tale* written about 1390), and during the Renaissance. In the nineteenth century many handbooks on keeping cage birds recommend jackdaws and ravens because of this. "Young Ravens are easily tamed if taken from the nest and fed upon bread and milk, meat, worms, cockchafers etc.", wrote Dyson and continued: "When full grown they will eat anything; and when young they will learn to repeat any words by hearing them again and again, especially if they are rewarded by a nice titbit after they say their lesson, their memory and power of imitation being quite wonderful." [453]

Further north the raven has also a long reputation of being a good talker and therefore it was popular as a pet. Reverend Lucas Debes wrote in his book about the Faroe Islands in 1673 that he had raised a raven that every morning shouted for "Rasmus". That was the name of the man-servant of the house. Carl Linnaeus, who remembered Ravens from his youth in Småland, confirms that they can easily learn to talk, if you get them when they are very young.[454]

Several notes in literature give examples of how easy it was to bring home a Raven or other species of the crow family and to teach them to imitate human voices. The Raven is a large and elegant bird, which follows its caretaker as a dog. Fale Burman (1758 –1809) tells us, in the 1700s, about reverend Enok Huss in Jämtland, who tamed a Raven youngster. The bird knew and greeted his master, ate a lot, stole silver items, and hid the stolen goods and the meat which it could not finish, in the dunghill. Ellis S. Harris retold in the early 1880s an interesting story about a Gypsy who kept a pair of old Ravens that bred yearly in captivity. He had had offspring for a decade from the pair and sold the young ones.[456]

Jackdaws have also been widely kept in northern Europe, especially for their ability to whistle. "This beautiful bird [...] is docile and teachable and learns to utter words, though in a very unconnected manner", wrote Henry Gardener Adams in 1854. "A person who has never owned a pet crow, raven, or jay does not realize how much humor a bird may possess or how entertaining a pet bird can be", wrote Hyatt Verrill in his famous book on pet animals from 1915. When the former Soviet leader Nikita Khrushchev in the mid-1960s was forced to retire, according to contemporary newspaper reports, he became occupied trying to teach a Jackdaw to talk.[457]

Cutting the tongue

A widespread practice was to cut the tongue of the birds in order to improve their ability to talk. German cage bird expert Johann Matthäus Bechstein wrote in the late eighteenth century about the Raven: "Some people are accustomed to cut what is called the nerve of the tongue, supposing that it would make them better able to articulate sounds". He added, however: "It seems most probable that this cruel practice is of little use, and, like many others, only a vulgar prejudice, for I have heard ravens speak perfectly well without having the tongue touched".[458]

Despite the facts, this cruel custom lived on. Wallace wrote in 1887 about how to train magpies. If taken from the nest at an early stage and hand-reared, they could be taught to "talk well, and almost fluently". However, the author claimed that "it is necessary in most cases [...] to sever the ligature that binds the tongue with a pair of small, sharp scissors, but the operation should be performed either by an expert or a surgeon". He anyway concluded that some magpies have been known to repeat words without this operation having been performed "but they never articulate so well, or speak so freely, as birds which have had their tongue liberated". More recent authors condemned the custom. Verrill found the operation "as useless as it is cruel". [459]

Contemporary custom in Portugal

Some years ago during a visit to a Portuguese village, the inhabitants were observed to still keep corvids and teach them to talk. Ethnographic annotations of the impressions give a detailed description on how the birds were cared for. In one house the visiting researcher found a tame jay. It was named Galoulas, which was explained to mean in a regional or archaic Portuguese dialect 'Gobbler' (Glutton?). The name was passed from the person who caught it and referred to the quantities of food it took when it was still a chick. The bird was in good condition, its plumage complete, and there was no sign of clipping of wings or tongue. It was kept on its own in a cage, big enough to hop about, but not to fly.

One of the most interesting aspects of the cage was the fact that the owners had a little sack on the wire, where the bird could stock up extra food, allowing the bird to keep its behavioral trait. When the owners were questioned about why they had added this unusual cage feature, they said that they saw the previous jay trying to hide food in a little hole on the wall, and so decided to provide it with a little treasure sack. The cage was located at the entrance of the house, a little patio where the family walked through to get inside, which provided ample opportunities of contact for the jay. If the jay was ignored, it would sometimes get agitated and vocalize loudly or whistle until attention was paid to it. The bird was apparently a beloved pet, considered smart and witty, and said to understand when they were talking about it.

The jay did not speak much, though, for it was not taught from an early age, though it was bought while still a chick. The bird could say *Olá* ('hello' in Portuguese), and it whistled and mimicked perfectly the phone, the laundry machine and the artisan cheese production, which has a dripping sound. The phone mimic was so perfect that it had fooled the family on more than one occasion. Sometimes it seemed to produce a melancholic call when sad, and when excited, it raised its crest and squawked. The jay liked to play with a laundry clip and peck on fingers close to the cage. Sometimes it was taunted in a loving way, a frequent local habit with pets; whether or not it likes that is debatable.

This Jay was raised from a flightless chick on

flour-like chicken feed with water, which it still ate, though much less than it used to. As an adult it ate fruits, potatoes and table scraps. Sometimes as a treat it got dried almonds, which it liked to stash in the little bags. It also had grasshoppers, but not too big; it was afraid of the large ones.

An expert bird trainer

The jay originally came from a lady who lived in the village, also visited and interviewed. She had a Magpie named Rita, and three Jays all named Chico (short for Francisco). All of them talked to some extent and none had had its tongue clipped. The Magpie was kept by itself and the three jays together, though they used to peck at each other. Rita was said to get agitated and angry when called *Marracha* (a regional Portuguese light insult meaning 'slut/pig'). The lady and her husband provided details about keeping corvids and how they caught them. They were aware that it was illegal, but this did not seem to bother them, as the authorities thought lightly about this old tradition. During a fire the National Guard went in to extinguish it, and when leaving saw a cage for a jay, advising them to keep it away from the door, in case someone should see it and report it to the authorities (the National Guard should do the collection of illegal birds).

The couple did not appear to make any money out of the birds, though the husband mentioned that the Magpie was so valuable that it would not be sold even for fifty Euros (which left the interviewer in doubt if it was a bidding invitation, or a true appreciation of the pet). It was also mentioned that both jays and magpies were costly to raise. The wife was the bigger enthusiast and she collected the birds from the nests, distributing them to others. She had been doing it since she was young, just out of her love for birds.

Lately jays and magpies were hard to come by, because a fire a couple of years ago had swept through the region and burnt down favorite breeding grounds. Jays prefer bushier habitats (cork-oaks, stone and maritime pine trees, chestnut trees, avoiding eucalyptus areas), and will leave their nests even if they have already laid, if they suspect they are being watched. They need to be well advanced in their hatching for a chick collection to be successful. The lady mentioned that if a jay is collecting nest materials and spots someone watching it, it will immediately abandon that site.

These villagers are probably the last to maintain the tradition. New legislation does not allow Europeans to capture wild birds any longer, and will probably ensure the ending of the custom. Most corvids are easy to breed in captivity, but so far these local bird keepers have not been interested. Urban people are also not interested in keeping native birds in captivity; for those who are interested in "talking" birds, the colorful and prestigious parrots are more attractive.

Chapter 11. Aviaries in the Wilderness to Arks in the Metropolis

Some Notes on the Evolution of American Aviculture

Josef Lindholm, III

In 1682, the 39th year of the reign of Louis XIV, a letter was received by his cousin, Prince Louis II de Bourbon, Le Grande Condé, veteran of the Thirty Year's War, and widely considered one of the great military figures of all times. Its author was a clergyman resident in Paris, Bernard Lenet, L'abbé de la Victoire. In part, it translates thusly:

> The bird-seller having notified me that he had an extraordinary bird freshly arrived from the Indies, I went to see it, and I found it as rare in its figure as in its plumage. It is a little larger than a big lark; it is all fire-colored with a crest on its head, and a large reddish bill too, well made, and steady on its feet, very cheerful, very lively and eating well.

It has, says the bird-seller, a warbling song like a nightingale's, but I did not hear it, although I did hear an occasional whistle which was very pretty and gracious; and if one were assured that it would live a long time, it should not be passed up; it is true that it seems healthy and eats well, of a certain seed that is called *alpiste* [canary grass], which it likes very much. If Y.M.S.H. [Your Most Serene Highness] would like it, I will make the trip, pay for it and send it to you.[460]

As it happened, the purchase of this North American Cardinal (*Cardinalis cardinalis*) became a protracted affair. In a subsequent letter to the Prince, the Abbé wrote that he had managed to convince the seller – a member of the Guild of *Oiseleurs* – to reduce the price to ten louis (the better part of a year's wages for an artisan, or the price of "five writing desks"). The Abbé promptly purchased this bird, since an agent of His Majesty the King appeared ready to offer twelve louis! Thus, a *Coq de Virginie* was added to the extensive aviaries of His Most Serene Highness, upon which he had lavished attention since 1663, when he had to retire to his Chantilly estate in temporary disgrace.[461] Despite its treasured status, this was not the first Cardinal in France. Appropriately, the earliest import appears to have been of two "redbirds" presented to King Henri IV in 1607, by Samuel de Champlain, following his third expedition to the New World.[462]

Little more than a century later, another revered military figure received the following letter: "The delay was vexatious, yet I am richly paid in being able to preserve so much beauty. Before this time I had thought those birds which I had seen in the Chinease [sic] paintings were only works of fancy, but now I find them to be only aukerd [sic] portraits. I am sorry that their lives cannot be preserved. I did not find this body very lean. The musls [sic] of the thighs were strong which with the smallness of the wings makes me think that they run fast & fly but little. When you have the misfortune of losing the others [sic], if the weather should be warm be pleased to order the bowels to be taken out and some pepper put into the body. But no salt which would injure the feathers."[463]

The author of this letter was Charles Wilson Peale (1741–1827), one of the several brilliant polymaths who graced the United States of America in its infancy. His letter was in response to one written eleven days earlier, on 16 February 1787, by General George Washington, not quite two years before he became America's first President. Part of Washington's letter to Peale reads as follows: "You will receive by the Stage the body of my Gold Pheasant, packed up in wool, agreeable to your directions. He made his Exit yesterday [hence Peale's vexation], which enables me to comply with your request much sooner than I wished to do. I am afraid the others will follow him but too soon, as they all appear to be drooping."[464]

On Monday 27 November, 1786, George Washington recorded in his diary the arrival of a present from his dear friend and comrade-in-arms, the Marquis de Lafayette. Lafayette, whose own interest in game birds is commemorated by Lesson's description of the Sri Lanka Junglefowl as *Gallus lafayettei* in 1831, sent to Mount Vernon, by courier, seven Golden Pheasants (*Chrysolophus pictus*) from the Royal Aviary in Paris, along with a "partridge" and "three Maltese donkeys".[465]

In 1784, Charles Wilson Peale, whose fame then rested as a painter (and who had already executed three military portraits of Washington), established a museum in Philadelphia. Here (as he informed Washington in 1786) he intended to display "everything that is curious in this Country". The Peale Museum, which closed in 1849, would achieve renown for the first mounted Mastodon skeleton, excavated in 1802.

Part of the eventually large collection of animals preserved by Peale was displayed in habitat groups, which drew worldwide attention. And, as will be further discussed, it housed living exhibits as well.[466] After 1850, many of Peale's specimens were lost, or suffered damage and loss of data into the early twentieth century. Among the lost objects was one of the "two groups of mounted birds, arranged in two glass cases, presented to the Peale Museum by General George Washington". However, the contents of the of the other glass case, a male and female Golden Pheasant, are now, since 1914, treasures in the Harvard Museum of Natural History.[467]

I have chosen to contrast the correspondence between L'abbé and Le Grande Condé with that between Peale and Washington as an introduction to the diversity of historical aviculture in America. In the first letter, we see America as a source of exotic cage birds. In the second, we see an early example of collecting exotic species, as well as the beginnings of American research aviculture, of which there would subsequently be a distinguished record.

America as an avicultural resource

The commercial exploitation of North American Cardinals began early. In 1698, Gemelli Carreri documented the trade, in Havana harbor, of "a sort of scarlet birds to keep in cages, for their color called cardinals". They were brought to Cuba from Florida, by Native Americans who sold them ship-side to merchants in galleons. During one such transaction, these merchants "laid out eight thousand pieces of eight upon those foolish birds".

The price of individual birds varied from six to ten pieces of eight, so the number of birds in this consignment was in the neighborhood of a thousand. According to Carreri, one could live for a week in Havana on ten pieces of eight.[468] By 1749, Peter Kalm, a pupil of Carl Linnaeus, observed that "red birds" were "sent abundantly to London, in cages", "on account of their agreeable song", which "very nearly resembles that of our European Nightingale". He personally trapped a number, using a male as a decoy, noting "they have such strength in their bill that when you hold your hand to them they pinch it so hard as to cause the blood to issue forth". He kept his decoy for five months: "It ate both corn and buckwheat, for I gave it nothing else".[469]

While Peter Kalm also described the Cardinal's song as "exceedingly sweet", his contemporary, Antoine-Simon, Le Page du Pratz, who lived with the Natchez and published his *Histoire de las Louisiane* in 1758, considered it too loud to be enjoyed indoors. This was obviously not the opinion of Lord Botetourt, Royal Governer of Virginia who kept 28 "red birds" in individual cages at his estate in Williamsburg in 1769. In 1788, traveler Nicolas Louis Bourgoise noted Cardinals "are easy to maintain, and sing in their cage; as a result, all the ships take away a large number of them".

That same year, a "lemonade seller" in Paris offered for sale "10 Cardinals with scarlet plumage, having a 1-inch crest on the head, and the beak crowned with black; they have a very pleasant song and quickly learn airs from a bird-organ". 1788 was also the year Georges-Louis Leclerc, the Comte de Buffon died. His ten volumes *Histoire Naturelle des Oiseaux*, published from 1770 through 1783, was considered the authoritative reference of its time. In this, he "surmised [...] that when travelers described a cardinal with a pleasant song they were referring to the South American bird [the tanager (*Ramphocelus bresilius*)], because the North American species was a type of grosbeak and so would not sing [!]".[470]

Other North American species became available in Europe at the same time, in particular, Painted Buntings (*Passerina ciris*) and Indigo Buntings (*Passerina cyanea*), known in the trade respectively as *Papes* and Blue *Papes*. Indigo Buntings appear to have been a specialty of Ange-Auguste Chateau, Bird Seller to the King. While the journalist Helen Ossa mentions that the Cardinal was bred in Holland from the middle of the eighteenth century, I have not located further details. On the other hand, the English ornithologist Marmaduke Tunstall (1743–1790) "had a pair which built a nest in an orange tree placed in an aviary, and eggs were laid, but whilst the hen was sitting, a high wind blew down the nest and the eggs were broken; young birds were, however, found in them". The chronicler of this event, John Latham, whose *General History of Birds* was published in ten volumes, from 1821 to 1824, goes on to mention "another instance on record, but I have not heard whether the young birds were brought to perfection". Latham also mentions that Tunstall kept American Goldfinches (*Cardeulis tristis*). While he did not breed the "American Yellow Finch", Tunstall recorded that "both sexes lost their yellow color in winter and became pale olive, but recovered their beautiful yellow plumage in spring".[471]

By the 1820s, Latham was able to record that "the Dutch contrive to breed" the Painted Bunting, but "we believe the attempt has rarely succeeded in England". He continued: "I was assured by the late Mr. Tunstall that it has happened more than once, and that two pairs made nests and laid eggs in orange trees in a menagerie of a friend

Figure 44: Young Theodore Roosevelt, Jr (1887–1944) in the White House conservatory in 1902, with Ely Yale, a Hyacinth Macaw (*Anodorhynchus hyacinthinus*)

of his at Holderness, but the young were not hatched. The above gentleman has kept many, but seems to think that they gain their full plumage sooner than the third year".[472]

John James Audubon discovered an active trade in Painted Buntings in New Orleans. In his *Ornithological Biography*, he described in detail in how males were trapped, commencing in April, with a stuffed male used as a decoy: "Few vessels leave the port of New Orleans during the summer months, without taking some Painted Finches, and through this means they are transported probably to all parts of Europe. I have seen them offered for sale in London and Paris, with the trifling difference of value on each individual, which converted the sixpence paid for it at New Orleans to three guineas in London".[473] Audubon noted that the Buntings "feed almost immediately after being caught; and if able to support the loss of liberty for a few days may be kept for several years. I have known some instances of them being kept in confinement for upwards of ten years." He found the Indigo Bunting, which the Louisiana Creoles called the *Petit Papebleu,* "arrives in the Southern States from the direction of Mexico, along with its relative the Painted Finch, and is caught in trap-cages, but with more difficulty than the latter bird".

Audubon attempted imports of his own. In March 1830, he bought "about 350" Passenger Pigeons (*Ectopistes migratorius*)" in "the market of New York, at four cents apiece" (Audubon 1831–1839). He documents that "most of these I carried alive to England, and distributed them amongst several noblemen, presenting some at the same time to the Zoological Society". The Gardens of the Zoological Society of London (established in 1828) commenced breeding this species in 1832. That same year, one of the "noblemen" was also successful.[474]

Lord Stanley, thirteenth Earl of Derby, was a charter subscriber to Audubon's *Birds of America*. When Audubon was first presented to the Earl at Knowsley Hall near Liverpool, he expressed misgivings, being "troubled at the idea of meeting an English gentleman, called, moreover, a lord!" These were immediately dissipated when Lord Stanley spread Audubon's original watercolors on the floor and examined them on his knees. This should not have been surprising, as the Earl (1775–1851) was the premier aviculturist of his day, with 619 taxa of birds received at Knowsley by his death; 73 of these were propagated there. At the Earl's death seventy Passenger Pigeons were present.

The Earl was President of the Zoological Society of London from 1831 to his death. In 1834, he sent a Passenger Pigeon to London Zoo, two more in 1839, then four in 1843. In 1847 he sent six in June and five in September. Eight went to the zoo in 1849, and a final ten in 1850. A number were also liberated on the Knowsley grounds.[475]

Audubon observed: "A curious change of habits has taken place in England in those Pigeons which I presented to the Earl of Derby in 1830, that nobleman having assured me that ever since they began breeding in his aviaries, they have laid only one egg. My noble friend has raised a great number of these birds, and has distributed them freely. It is not therefore very surprising that some which have escaped from confinement have been shot; but that this species should naturally have a claim to be admitted into the British Fauna appear to me very doubtful".[476] A less fruitful import ation by Audubon involved the Blue Jay (*Cyanocitta cristata*):

> While at Louisville, in Kentucky, in the winter of 1830, I purchased twenty-five of these birds, at the rate of 61 cents each [compare this to his price for Passenger Pigeons!], which I shipped to New Orleans, and afterwards to

Liverpool, with the view of turning them out in the English woods. They were caught in common traps, baited with maize, and were brought to me one after another as soon as secured. In placing them in the large cage which I had ordered for the purpose of sending abroad, I was surprised to see how cowardly each new caught bird was when introduced to his brethren, who, on being in the cage a day or two, were as gay and frolicsome as if at liberty in the woods.

The new comer, on the contrary, would run into a corner, place his head almost in a perpendicular position, and remain silent and sulky, with an appearance of stupidity quite foreign to his nature. He would suffer all the rest to walk over him and trample him down, without ever changing his position. If corn was presented to him, or even placed close to his bill, he would not so much as look at it. If touched with the hand, he would cower, lie down on his side, and remain motionless.

The next day, however, things were altered: he was again a jay, taking up corn, placing it between his feet, hammering it with his bill, splitting the grain, picking out the kernel, and dropping the divided husks. When the cage was filled, it was amusing to listen to their hammering; all mounted on their perch side by side, each pecking at a grain of maize, like so many blacksmiths paid by the piece. They drank a great deal, ate broken pacan [sic] nuts, grapes, dried fruits of all sorts, and especially fresh beef, of which they were extremely fond, roosted very peaceably close together, and were very pleasing pets…

They bore the passage to Europe pretty well, and most of them reached Liverpool in good health; but a few days after their arrival, a disease occasioned by insects adhering to every part of the body, made such progress that some died every day. Many remedies were tried in vain, and only one individual reached London. The insects had so multiplied on it, that I immersed it in an infusion of tobacco, which, however, killed it in a few hours.[477]

As it happens, the Blue Jay was propagated by the French aviculturist Andelle in 1876. By then a number of North American species had been hatched in collections in Continental Europe and Britain. The following are the earliest records of which the avicultural historian Emilius Hopkinson was aware:

Indigo Bunting, Painted Bunting and Rose-breasted Grosbeak (*Pheucticus ludovicianus*) by the famous Karl Russ in Berlin before 1870; Eastern Towhee (*Pipilo erythrophthalmus*), by German aviculturist Landauer, before 1870; Common Grackle (*Quiscalus quiscula*) by German aviculturist Julius Wentko in 1884; Northern Mockingbird (*Mimus polyglottos*) by "Chiapella at Bordeaux in France before 1865"; Eastern Bluebird (*Sialia sialis*) at London Zoo in 1860.[478]

Hopkinson was less precise regarding the Cardinal. He was apparently not, at the time he published his *Records of Birds Bred in Captivity*, aware of Marmaduke Tunstall's nineteenth century near success, though he cited it later (Hopkinson, 1940). Instead, he proposed that "Viellot early in the nineteenth century was probably the earliest to attempt and to probably succeed in its breeding".

Ornithologist Louis Jean Pierre Viellot (1748–1831) was also a pioneering aviculturist, and certainly succeeded in propagating Cut-throat Finches (*Amadina fasciata*), the Red-billed Firefinch (*Lagonosticta senegala*) and the Zebra Finch (most likely the Timor subspecies *Poephila g. guttata*).[479] Hopkinson was also unaware of the first certain British success, achieved by the Earl of Derby in 1834, when three chicks were raised at Knowsley.

This has been proposed as the "possible first captive breeding", though occurring three years after Viellot's death, would obviously not be the case if Viellot had indeed been successful.[480]

Propagation in the Carolinas

Cardinals appear to have been captive-bred in the United States before 1834. Audubon's plate 182 in *The Birds of America* is of the Common Ground-Dove (*Columbina passerina*). This was engraved in 1833.[481] In the corresponding text in his *Ornithological Biography*, Audubon noted: "They readily become domesticated, and indeed so very gentle are they, that I have seen a pair which, having been caught at the time when their young were quite small, and placed in an aviary, at once covered the little ones, and continued to nourish them until full-grown.

They afterwards raised a second brood in the same nest, and shewed great spirit in keeping the jays and starlings from their charge. In this aviary, which belonged to Dr. Wilson of Charleston, several other species bred, among which were the Carolina Dove, the Cardinal-bird, the Blue Grosbeak, the White-throated Sparrow, the Towhe [sic] Bunting, the Common Partridge [Bobwhite], and the Wood Duck. The Ground Doves were fed on rice and other small grain".

In his profile of the Cardinal (plate. 159, also engraved in 1833), Audubon went on to note that this species is "easily raised when taken from the nest, and breed when kept in aviaries. My friend, Dr. SAMUEL WILSON of Charleston, has had them breeding with him, having placed straw-baskets for the purpose, in which the female deposited her eggs, without improving the nest any more than by placing a few grass blades, perhaps pilfered from some of her neighbors".[482]

Dr. Wilson and Audubon had been introduced through the Rev. Dr. John Bachman, a Lutheran Minister, whom he met in Charleston in 1831. Bachman, who eventually became father-in-law to both of Audubon's sons, and wrote the text for the *Quadrupeds of America*, was also an aviculturist. Regarding the Painted Bunting, he wrote:

"I have kept these birds for many years in aviaries. The males and females of a year old were of a uniform color, but I have invariably found them to assume their perfect plumage in the second year. This bird could be easily domesticated and multiplied in Europe, in the manner of the canary. I have had them to raise three broods of young in the year in confinement. The plumage, however, in this state, was never so brilliant as when wild".

Audubon noted that Bachman's "very interesting notice regarding its change of plumage" was "greatly at variance" with data published by Alexander Wilson. Bachman achieved interesting early results with a softbill, the Brown Thrasher (*Toxostoma rufum*), which he referred as the Ferruginous Thrush. Audubon noted:

> They breed well in aviaries, and are quite tractable in a closer state of confinement. The young are raised in the same manner, and with the same food, as those of the Mocking-bird. In cages it sings well, and has much of the movements of the latter bird, being full of activity, petulant, and occasionally apt to peck in resentment at the hand which happens to approach it. The young begin their musical studies in autumn, repeating passages with as much zeal as ever did Paganini. By the following spring their full powers of song are developed. My friend BACHMAN, who has raised many of these birds, has favored me with the following particulars respecting them: -- 'Though good-humoured towards the person who feeds them, they are always savage towards all other kinds of birds. I placed three sparrows in the cage of a Thrush

one evening, and found them killed, as well as nearly stripped of their feathers, the next morning. So perfectly gentle did this bird become, that when I opened its cage, it would follow me about the yard and garden. The instant it saw me take a spade or a hoe, it would follow at my heels, and, as I turned up the earth, would pick up every insect or worm thus exposed to its view. I kept it for three years, and its affection for me cost it its life. It usually slept on the back of my chair, in my study, and one night the door being accidentally left open, it was killed by a cat'.[483]

Presidential birds

Thomas Jefferson had a happier experience with "Dick", one of several pet Mockingbirds he owned over the years. When the Third President was busy, this thrasher relative resided in a "cage hung among the roses and geraniums [then a rarity] in a window recess at the White House", but when Jefferson had time to himself, Dick had the freedom of the room, where he would sing while his master played the violin, and take food from his lips. When he completed his second term in 1809 and retired to Monticello, Jefferson noted in his diary: "My birds arrived here safely, and are the delight of every hour." Dick was included among this retinue, and for years afterwards, accompanied his owner to bed, hopping up the stairs one step at a time.[484] "The delight of every hour" is apparently not how the first President of the United States might describe the bird which shared his abode during at least part of his two terms. The property of Martha Washington, one would guess from its description as a "pretty green pet", and the fact that Nellie Custis (Martha Washington's grand-daughter) was able to teach it to sing *Pauvre Medelon* in ten days, that this parrot was some sort of Amazon (*Amazona sp.*).

George Washington never got to live in the White House; his presidency was spent in Philadelphia. In the commotion of returning from there to Mount Vernon in 1797, he wrote: "On the one side I am called upon to remember the parrot, and on the other to remember the dog. For my own part I should not pine much if both were forgotten."

While digressing on the subject of Presidential Parrots, mention should be made of the Blue-and-Gold Macaw (*Ara arauana*) that Dolly Madison rescued (along with the Declaration of Indepen-dence and a painting of George Washington) when the British burned the White House in 1814. Also "Eli, the most gorgeous Macaw, with a bill that I think could bite through boilerplate, who crawls all over Ted, and whom I view with dark suspicion", listed among his children's many pets by Theodore Roosevelt, in a 9 June 1902 letter to Joel Chandler Harris (the author of the "Uncle Remus" stories), less than a year after Roosevelt became the 26th president.

A much published photograph makes it clear that "Eli Yale" was a Hyacinth Macaw (*Anodorhynchus hyacinthinus*), certainly a lordly bird for the household of a President widely and justly acknowledged as a first class amateur ornithologist.[485]

Industrialization of the American trade

Aviculture of the sort practiced by the doctors Bachman and Wilson was the exception. In general, native species kept by Americans before the twentieth century were companions, songbirds or ornaments. Cages for songbirds were considered important items of property among eighteenth-century American families.

Typical of the excesses of the "Gilded Age" following the American Civil War, large-scale commercial exploitation of North American native songbirds became a matter of concern to conservationists. In 1882, the *New York Times* reported: "The first-class aviary of the young ladies of fashion today contains fifty or more birds, usually a pair of each kind."

By the end of the nineteenth century, Louisiana was the center of the US bird trade, 90 per cent of the total collected originating there. In the age of steam, North American species were well represented in British and continental European aviculture.[486] The great zoo professional Lee S. Crandall (1887–1969) wrote many years later of the professional trappers whom he knew in his youth:

"Really involved techniques were developed for the capture, at various seasons, of Bobolinks, Blue Robins [Eastern Bluebirds], Rose-breasted Grosbeaks, Indigo Buntings and Purple Finches (*Carpodacus purpureus*)... I still have the feeling of amazement at the intricate methods that had been evolved, suited exactly to the habits of each species. The coming of the Audubon laws not only put a stop to this practice but choked the beginnings of the [American] avicultural impulse in its infancy. The old bird trappers have long since disappeared, and with them have gone their hard-won lore. The keepers of birds, forbidden to cage familiar native species have at last turned their attention to songsters of other lands."[487]

In 1886 the State of New York passed a law mandating that "no person in any of the counties of this state shall kill, wound, trap, nor snare, catch with birdlime or any similar substance... any songbird or linnet, bluebird, yellow hammer, yellowbird, thrush, woodpecker, catbird, pee-wee, swallow, martin, bluejay, oriole, killdeer, snowbird, grosbeak... nor shall any person purchase or expose for sale any song or wild bird or any part thereof after the same has been killed". This legislation was passed to end the wholesale use of dead birds as hat ornaments described in dreadful detail by Helen Ossa.

However, in 1890, the New York Supreme Court approved the extension of its interpretation to include living cage birds, when it upheld the conviction of "a woman who had been fined \$5 for each of the seventeen birds she had in cages in her shop". Other governments followed suit. In 1900, the owners of twelve Chicago bird shops were each fined \$100, following the seizure of "200 to 300 caged birds, chiefly mockingbirds and cardinals". Cincinnati, Cleveland, and Washington D.C. all prohibited the possession of native birds by 1902.[488]

Louisiana remained, as previously noted, the commercial center of the native bird trade, and the major source of exports to Europe. By the close of the nineteenth century there was serious concern that both Indigo and Painted Buntings would be trapped to extinction there. In 1904, the bird trade was ended in Louisiana. This was especially significant in that this prohibition was a Model Law, which as defined by lawyer and aviculturist Jerry Jennings, concerned about other legislation as well, defined thus:

"When the Model State Law is in its final form, it would be forwarded to the state legislature of each... state with recommendation that this law be adopted. Of course, each state will have to individually consider this model law and a bill must be introduced into each legislature and hearings must be held before any portion of the Model State Law is adopted by that state".

In this case things progressed, and in 1906 it became illegal everywhere in the US to keep native songbirds without a permit. That year 322,297 cage birds were imported to the US (of them 274,914 were Canaries (*Serinus canarius*)).

Meanwhile, Mexico became the commercial source for North American Passerines in Europe, continuing with brief interruptions until the European Union ended commercial import of birds in 2007. As of 2009, around forty Cardinals were distributed among at least thirteen European Zoos, and Painted Buntings continue to be present in European aviculture as well.[489]

The prohibition of trade in US passerines by 1906 did not entirely end the practice. The presence of the House Finch (*Carpodacus mexicanus*) throughout the contiguous US is partially due to cage-bird dealers attempting to circumvent the interruption of imports due to the outbreak of war in Europe in 1939.

According to ornithologist Oliver Austin, "Hollywood Finches" from California appeared in New York City shops in 1940. When agents of the Fish and Wildlife Service investigated, these birds were released *en masse* and shortly thereafter observed as breeding birds on Long Island. By the early 1960s, House Finches had spread to Connecticut and New Jersey, and before 2000, this formerly western bird was resident in all the lower 48 States, as well as portions of adjoining Canada.[490]

For decades, permits to collect and possess native passerines were granted to individual private aviculturists. By 1909, C.C. Worthington, who had extensive aviaries at Shawnee on Delaware, Pennsylvania, was credited with the first breeding of White-breasted Nuthatch (*Sitta carolinensis*) and Eastern Song Sparrow (*Melospiza m. melodia*). Anthony Kuser, at Bernardsville, New Jersey, also the first US breeder before 1909 of, among other species, Chilean Pintail (*Anas georgica spinicauda*), Cabot's Tragopan (*Tragopan caboti*), Swinhoe's Pheasant (*Lophura swinhoii*), and Demoiselle Crane (*Anthropoides virgo*), is credited with the first US breeding of the Northern Catbird (*Dumetella carolinensis*). (In England, Rev. Charles Farrar had achieved success with the Catbird in 1902.)

Lee Crandall observed that "American aviculturists who are privileged to keep indigenous birds are compelled to depend upon their own resources for securing specimens. Too stringent protection laws do not favor the development of expert bird catchers, without whose aid the formation or maintenance of a large collection of native bird sis a practical impossibility". One particularly resourceful specialist in native birds was Dr. Eric Kinsey, in Manor, Marin County near San Francisco. Active from the 1920s into the 1960s, he assembled a remarkable series of California passerines, hummingbirds, woodpeckers and other species. He was the first American to breed any species of piciform, the Acorn Woodpecker (*Melanerpes formicivorus*), commencing in 1942 and to the second generation. To my knowledge, he also achieved the first captive breeding of any of the plethora of Wood Warblers (Paurlinae), when the Western Yellow-breasted Chat (*Icteria virens longicauda*) was propagated in 1945 after years of effort. In the 1930s, he not only provided the Chicago Zoological Park (Brookfield Zoo) with Anna's and Allen's Hummingbirds (*Calypte anna* and *Selasphorus sasin*), but made three shipments of the same species to the Marquis Yamashina, in Tokyo, the first hummingbirds in Asia. When Jean Delacour made an extensive visit to America in 1936 and 1937, Dr. Kinsey took him on expedition in the California Sierra, and with others arranged permits for the export of 21 taxa of passerines as well as other birds, which were eventually distributed among the respective collections of Delacour, Alfred Ezra and the Zoological Society of London. Within months of their arrival at his Foxwarren Park in Surrey, Sir Alfred Ezra, President of the Avicultural Society, had bred three of these species for the first time in captivity, the Yellow-billed Magpie (*Pica nutalli*), and the Mountain and Western Bluebirds (*Sialia currucoides* and *Sialia mexicana*). These successes could be considered a fitting conclusion to one of the last legal major exports of passerine birds from the United States.[491]

New England

Between 1830, when Audubon took Passenger Pigeons and Blue Jays across the Atlantic, and 1848, when James Mann, "Taxidermist, Dealer in Birds, Natural Curiosities, etc.", published *The American Bird-Keeper's Manual* in Boston, the world had become a remarkably different place. In 1838, the launching of the Great Western inaugurated regularly scheduled Trans-Atlantic steamship crossings. In 1830, the first American Railroad in Quincy, Massachusetts, was only four years old, while eighteen years later railways were exploding up and down the Eastern Seaboard. Native New Englander Samuel Morse's development of a working telegraph several years earlier had profound effects. While the Northwest Trade, which had greatly enriched Boston, was already past its prime, due to the near-extinction of Sea Otters whose furs the Yankees collected and traded to China, oil whaling was at its prime, centered in the harbors of New England. At the same time, New England was thoroughly caught up in the Industrial Revolution. Mann's slender book is a remarkable documentation of aviculture in a rather young, very vibrant and highly literate culture.[492]

Born in the north of England, Mann gave every indication of having become a proud New Englander. He considered the "Yankee Fare" of yellow corn meal and milk to be an excellent basic softbill food, when supplemented with scraped beef and insects. The continuing importance of Boston's sea trade was obvious from his advice for starting a meal-worm culture: "In case you cannot procure meal worms, and they are sometimes very scarce, buy, or beg, from a vessel returned from a long voyage, some stale ship biscuit, soak it in cold water, squeeze it dry, and put it in a covered vegetable dish in a closet, or other warm place. In a few weeks you will obtain an ample supply of meal worms."

Mann discussed forty species of native passerines and eighteen European ones (including the canary). Two species of doves are presented, the European Turtle Dove and the Barbary Dove (*Streptopelia turtur* and *Streptopelia risoria*). As Mann did not consider Macaws and Cockatoos appropriate household pets, he recommended only four species of parrots: two Amazons, the African Gray (*Psittacus erithacus*), and the Carolina Parakeet (*Conuropsis carolinensis*), a bird Mann had never seen in a cage, but which he understood to be popular in Europe.

Of particular interest to the avicultural historian are the twelve "South American, Asiatic, and African Birds". Somewhat surprisingly, three of the four South American birds are softbills: The Troupial (*Icterus icterus*), Red-rumped and Yellow-rumped Caciques (*Cacicus haemorrhous* and *Cacicus cela*). The fourth is the Red-crested Cardinal (*Paroaria coronata*). Though all three identified African birds are seed-eaters, none are estrildines. They are the "Broad-shafted" (Paradise) and "Red-billed" (Pin-tailed) "Widahs" (*Vidua paradisea* and *Vidua macroura*), and the "Crimson-collared Widah" (*Euplectes ardens*).

Estrildines are represented by Java Sparrows (*Padda oryizivora*) and Strawberry Finches ("Amandava, or Avodavine Finch") (*Amandava amandava*), which considering Boston's China trade, is not surprising. A mystery is the "Senegal Finch, or Spice Bird".

Though Mann (who did not use Linnaean names in this book) states it is "Found in various parts of Asia", his description does not conform to the expected Spice Finch (*Lonchura punctulata*): "This pretty little songster is still smaller than the Amandava, and its note, although not quite so loud, is much more harmonious. The female also sings, and if there is a pair in the cage, they will sit closely together and sing alternatively, the male first, and then the female… The male has the bill red; sides and fore part of the head and chin, black; the top of the head reddish, with black spots; the rest of the plumage is of a light reddish brown, spotted with black and brown. The female has the upper parts of a light brown; the lower parts, reddish gray."

While this description is initially baffling, in light of Derek Goodwin's observation that the Gold-breasted or Zebra Waxbill (*Amandava subflava*) is "very prone to induced melanism under adverse captive conditions", it seems that Mann's "Senegal Finch" came from Africa after all.

Of the two remaining Asian species, the presence of the Hill Mynah ("Mino Bird") (*Gracula religiosa*) is not surprising, again considering Boston's Chinese connections. The Black-hooded Oriole (*Oriolus xanthornus*) is startling, as it has never been a particularly common or popular bird in the Asian bird trade. Mann stated that it "is an Asiatic species, and is sometimes brought to this country". He further specified the range as "Madras".[493]

James Mann's emphasis was on maintaining birds indoors in cages, as pets, songsters or ornaments. The one exception is the Canary, to the propagation of which he devoted a major portion of this book. Canary breeding appears to have been in practice in America by the 1820s, and by the1840s its popularity had grown to the point that even John Tyler, Tenth President of the United States (1841–1845), attempted to do so. Following his second marriage, not long after he assumed the presidency, Tyler and his new bride Julia decided to pair up their pet Johnny Ty: "Julia wrote to her family in New York asking them to select a wife for the bird." After a steamboat journey to Washington DC, by way of Richmond, Virginia, the new canary was introduced to Johnny Ty, who promptly went out of condition, sulking on the floor of its cage until its death a week later. It was only then that the Tylers became aware they had been shipped another male.[494]

Mann's discussion of canary breeding, obviously from extensive personal experience, includes interesting pre-Mendelian observations on the inheritance of colors, and as well as behavior notes. As a canary-breeding manual, however, this English immigrant's book was shortly eclipsed by one by an immigrant from the Continent. In the 1840s, Charles Reiche and his brother Henry, "late of Braunschweig, Germany" opened a shop in New York City, offering German and French Canaries and other European songbirds (including softbills), as well as North American natives. Up to 1853, when Charles Reiche published the first edition of his *The Bird Fancier's Companion*, the brothers had imported more than 20,000 canaries to New York.

In 1853 alone, the figure was 10,000 and by 1867, when the third edition of *The Bird Fancier's Companion* appeared, they were bringing 20,000 canaries to America each year (that year they took credit for two thirds of the cage birds imported to the country). The tenth edition appeared in 1871, the year the brothers opened a Boston store. By that time, James Mann's establishment may have been only a memory, as the Reiche's Boston branch styled itself as a purveyor of "all kinds of Rare Birds, Reptiles, Sea Shells, Stuffed Birds, and Curiosities", with a "Department of Taxidermy" where a "good assortment of Stuffed Birds" was "constantly on hand".[495]

Mann's *The American Bird-Keeper's Manual* is very readable and replete with fascinating anecdotes. The reader encounters a Northern Shrike (*Lanius excubitor*), which Mann recommends as "very amusing birds in a cage", which entered a house in New England by breaking a window "and succeeded in killing the poor Canary Bird", then "effected his escape through the breach he had made, on the entry of an inmate of the house, and without his prey after all, which he was in the act of pulling through the wires of the cage when interrupted".

A *Troopial* which had been kept by a family in Boston for six years was "completely domesticated as any member of it". Mann described this bird further: "He will visit every apartment in the house, will answer to his name, and come when he is called. He knows the voice, and even the footsteps of the inmates of the family. He will play or fight with them, (which he seems to enjoy very much,) and loves to be fondled, like a dog.

He seems to think also, that he has a right to protect his premises, and drive off all intruders. On leaving his cage one day he went up stairs, and on finding an artisan engaged repairing the lock of a room door, he commenced an attack upon him with great energy, and actually succeeded in drawing blood, before the man recovered from his surprise. He will sit on the hand of his mistress, and accompany her anywhere. When I saw him last his appearance indicated that he was taken good care of, and all his little wants supplied, for he was in good health and beautiful plumage. In addition to this other attractions he whistles melodiously".

Regarding the Indigo Bunting, Mann wrote: "It has [a] characteristic peculiar to itself. It does not seem to be the least annoyed when exposed to the rays of a summer sun. During the heat of the day, in summer, most of our birds retire to the shadow and stop singing. But I have seen this little bird sit for an hour or two, in an exposed situation, in the meridian heat of a July sun, piping its sweet little notes without intermission. It is a very clean and docile bird in a cage, and has still another recommendation, that the longer you keep it the prettier it grows. I had one in my possession which had been kept for four years in a cage, and it was the most beautiful Indigo Bird I ever saw in my life; it was like a piece of bright polished indigo, and it was also of a superior songster".

Of the Painted Bunting, Mann noted: "They are very docile, and both male and female, when in full feather, are lovely cage birds. The male has a sweet note, it bears some resemblance to the Indigo, but is weaker and more concise. The planters on the Mississippi breed them in their aviaries, and I think they might be raised here; but they require more care in winter than any of our Finches, as they are very susceptible of cold. They have been crossed with the Canary here, and make a beautiful variety". This is one of the very few dubious statements in the book.

Mann considered the Java Sparrow to be "the neatest and cleanest bird that ever was in a cage. I have had thirty of these birds in a bamboo cage, which I kept for some time. During the time I had them I never saw them wash, (although I have seen them wash since, but they wash but seldom,) and I never saw a spot on one of them; they were always neat and clean". He described a Hill Mynah "...at the Exchange Coffee House, in Boston, last summer, who uttered the following sentences,

Figure 45: The "Great Flying Cage" at the National Zoological Park, Washingon D.C., constructed in 1902. It was the site of the first US captive breedings of the Double-crested Cormorant, American White Pelican, Black-crowned Night Heron, American Egret, and Sacred Ibis

which no doubt it had learned on its passage to this country: 'Good morning. What are you going to do, Jerome? How's the wind, Jerome? Jerome, what's the clock?' [...] There was also one of these birds in New York, recently, which was celebrated for uttering sentences distinctly. I kept one myself about a year ago, for a short time; while with me he would whistle and imitate other sound, but did not pronounce words; since that time, I understand he has got the gift of speech also like the others ...

> These birds are great eaters; I fed mine on meal and milk, as a staple dish. If I approached the cage with anything in my hand that he could eat, (he was not particular) meat, cooked or raw, bread or potatoes, he would be down at the cage door putting his head through the wires, and uttering a whining cry; and would always get some of out of my hand, before I could open his cage door, or put it into the cage. They require plenty of gravel, and water to wash often. It may be worthy of remark here, that I have always found those birds, which eat a good deal of food, all fond of bathing, and wash every day at least, if they have an opportunity.[496]

Exhibits

Needless to say, James Mann heartily recommended the Cardinal: "The Cardinal Grossbeak, with his bright black and red plumage, and fine pointed crest, is really a splendid bird. His fame for beauty and song hath reached across the Atlantic, and in Europe he is much prized, and named the Virginia Nightingale, from the resemblance his notes bear to the notes of that bird, and which is said by judges, to be but little inferior [...]

They are clean in a cage, feeding principally upon seeds, which makes them, altogether, very desirable cage birds, and great favorites with those who possess them [...] They are a hardy and easily kept bird, and will live a long time in a cage. There was a preserved specimen in Peale's Museum some years ago, which, it was said, lived twenty-one years in the cage."[497]

While we cannot be certain where this remarkable Cardinal resided in life, evidence suggests it lived and died in the same place where it was eventually exhibited as a mounted specimen. Peale's Museum existed from 1784 to 1849, for most of this time in the Pennsylvania State House, now revered as Independence Hall. The bird collection, to which George Washington's 1787 Golden Pheasants were an early addition, had by 1802 grown to 1,800 species of birds displayed in rows of cabinets, most prepared by Charles Wilson Peale, an innovator in taxidermy. They shared this Philadelphia establishment with a grand series of preserved mammals, fishes, insects, molluscs and other animals, ethnographic artefacts and the first mounted skeleton of a mastodon, unearthed and prepared in 1801 (now in Hessiches Landesmuseum in Darmstadt).[498]

Living animals were featured as well. Outside the museum entrance, an 1810 visitor found bears, parrots and a monkey, with a nearby Bald Eagle (*Haliaeetus leucocephalus*) in a "high cage" with a sign proclaiming "feed me daily for a hundred years". This bird vocalized every time it saw Charles Wilson Peale. In 1805, a Monsieur Soissons donated, along with a hippo skull and African tribal artefacts, a collection of living birds: "Scarlet Ibis, Crown Heron [sic], Widow Birds, Senegal Finch." From the evidence of traveling menageries, it is likely that "heron" meant a crane. In the eighteenth century, the term *Senegal Finch* covered such birds as Combassous(*Vidua chalybeata*), (*Uraeginthus bengalus*), and Fire Finches (*Lagonosticta senegala*). At any rate, this is perhaps the earliest record of African Finches in the US

The pioneering ornithologist and illustrator

of American birds, Alexander Wilson recorded the arrival of Ruby-throated Hummingbirds (*Archilochus colubris*): "Mr. Charles Wilson Peale, proprietor of the museum, tells me that he had two young humming birds, which he raised from the nest. They used to fly about the room; and frequently perch on Mrs. Peale's shoulder to be fed. When the sun shone strongly in the chamber, he has observed them darting after the motes that floated in the light, as flycatchers would after flies".[499]

The arrival of what was to be the most famous bird to live at the Museum was announced in a letter of 21 October 1805 to Peale:

> Dear Sir
>
> The day before yesterday I sent…
>
> 1. a large box containing skins, skeletons & horns
>
> 2. small box containing the Polygraph
>
> 3. D:o. With minerals for the Phil. Society to be presented in Captain Lewis' name. A cage with a living magpie.
>
> These were delivered to Capt. Elwood as you will see by the inclosed [sic] receipt & the freight paid. He promised he would sail yesterday & I hope you will receive them in good order. The undressed skins arrived here full of worms. I fear you will be puzzled to put them into form. Accept friendly salutations.
>
> Th. Jefferson.[500]

The Magpie (the survivor of an original group of four), a live Prairie Dog (sent to Peale several days earlier), a collection of Native American artefacts (kept by Jefferson at Monticello), the minerals, and an extensive collection of skins and skeletons (which arrived crawling with "great numbers of vermin") arrived at the White House on 12 August 1805. They had traveled by way of New Orleans and Baltimore, from Fort Mandan (in present day North Dakota), the last outpost of civilization on the Missouri River, from whence they were sent by Lewis and Clark on 7 April 1805 before they plunged into The Unknown.

The President was in Monticello when this collection was delivered to the White House, where it was received by Etienne Lemaire, Jefferson's French butler, who wrote: "This is to inform you that I have just received by Baltimore a barrel and 4 boxes, and a kind of cage in which there is a little animal very much resembling the squirrel, and in the other a bird resembling the magpie of Europe". Jefferson replied on 17 August:

"Be so good as to have particular care taken of the squirrel & pie [sic] which came with the things from Baltimore that I may see them alive at my return." Lemaire wrote again on 20 August: "The magpie and the kind of squirrel are very well; they are in the room where Monsieur receives his callers…". That would have been the uncompleted East Room.[501]

Charles Wilson Pale wrote to President Jefferson on 22 October 1805: "I am surprised to see the Magpie so correctly like that of Europe, for I have always found some difference in the Birds which has been described [sic] in belonging to both continants [sic]. It is interesting to get the living one in good condition, for a better comparison and also to give it a place near one I have from Great Britain handsomely mounted".

On 3 November Peale again wrote to Jefferson: "The Magpye [sic] is in good health […] It may be well to inquire whether this breed of the Magpye has not origionally [sic] been from Europe". As it happens, the American Black-billed Magpie (*Pica pica hudsonia*) was not given a formal scientific description until 1823, following the British ornithologist Joseph Sabine's travels in the Arctic. It has "rather longer wings and tail and usually a very distinct white, or white and grayish, band across the rump". At any rate, no one in the scientific community had been aware of magpies in Western North America before the Lewis and Clark Expedition.[502]

Alexander Wilson used Jefferson's present as a model for the picture in his *American Ornithology* (1808–14): "The drawing was taken from a very beautiful specimen, sent from the Mandan nation, on the Missouri, to Mr. Jefferson, and by that gentleman presented to Mr. Peale of this city, in whose museum it lived for several months, and where I had an opportunity of examining it. On carefully comparing it with the European Magpie in the same collection, no material difference could be perceived". [503]

Among the earliest exhibits of living birds in the United States, Peale's Museum was not alone. In 1794, a "King of the Vultures, from South America" (*Sarcoramphus papa*) was advertised in an illustrated broadsheet for the "American Museum & Wax-Work at the Exchange" at Pearl Street and the Battery in New York City.

"This remarkable bird" was described as "about as large as a turkey, the head and neck is beautiful and resembles the head of an Indian when painted and decorated in an elegant style". Other living animals exhibited in this collection in 1794 included "The American Eagle, very large and gay", a "Porcupine from the East Indies, the only one ever seen in America", "The Ant Bear, from the coast of Patagonia, a beautiful animal, and the only one ever brought to this country", and "The American Grey Squirrel, in a machine, in which he grinds pepper for his living". Eminent circus historian Stuart Thayer notes the first record of an Ostrich (*Struthio camelus*) exhibited in North America in 1794, and the first Cassowary (*Casuarius* sp.) in 1800.[504]

In his *American Bird-Keeper's* Manual, Mann discouraged the keeping of macaws and cockatoos in private households, finding them "so extremely vociferous and noisy, that they are much more at home in the menagerie than in the parlor". America was more than a decade away from its first municipal zoo, but privately owned exhibits, both stationary and nomadic, were a familiar attraction.

Such collections are largely forgotten, but Stuart Thayer has devoted much attention to them. In a history of the traveling menageries that traversed Eastern North America from 1813 through 1834 Thayer lists among others a variety of birds: Ostriches, a Cassowary, "Emuses" (*Dromaius novaehollandiae*) (whose 1826 exhibitor claimed to have come from the Tower of London), pelicans, adjutants (*Leptoptilos sp.*), ibises, "Turkey Buzzards" (*Cathartes aura*), "American Eagles", Golden Eagles (*Aquila chrysaetos*), a "Gray Eagle", a "Black Eagle", "vultures", a "Red Bird of Paradise" (which, in 1828, was most likely a Golden Pheasant), Peafowl (*Pavo cristatus*), "Wild Turkey" (also referred to confusingly as a "Powhee", a corruption of "Pauji", a widespread South American name for Curassows), Crowned Cranes (*Balearica* sp.) (advertised as "crown birds", "crowned herons", and "Belaric cranes") a "Mexican Crane" (or "Grooganeria"), gulls, cockatoos, macaws, unspecified "parrots", and something called "Jackal Birds". Especially interesting were the two Secretary Birds (*Sagittarius serpentarius*) brought to Boston in 1834, as part of an African collection assembled by Zebedee Macomber (who, on behalf of a partnership dubbed the "Boston Zoological Association", made three such shipments in the 1830s. Among other animals in this collection were an Eland (*Taurotragus oryx*), five gnus, seven unspecified zebras, and four specimens of the now extinct Quagga (*Equus quagga quagga*), as well as fifteen "White Pelicans" (presumably *Pelecanus onocrotalus*) "two ibis", and other unnamed birds.[505]

In 1835, the owners of all sixteen existing traveling menageries in the US met with investors in Somers, New York, to form a capital stock company. Grandly titled the Zoological Institute, its mission statement proclaimed it would "more generally diffuse and promote the knowledge of natural history and gratify rational curiosity". That year, thirteen traveling collections were sent out under its auspices, but only seven in 1836.

"The Panic of 1837", when every bank refused payment in specie (silver and gold), resulted in five years of depression, the worst in America until the big stock market crash a hundred years later. The Zoological Institute did not survive the Panic.[506]

While the Zoological Society of Philadelphia was incorporated in 1859, its Zoological Garden was not opened until 1874. The first municipal zoo in America was New York's Central Park Menagerie, which began as an accumulation of donated animals in 1861. By 1873, when the first Director's Report was published, the collection was substantial and included an impressive series of birds.[507]

I am indebted to zoo historian Marvin Jones for providing me with his notes from the reports of William A. Conklin, Director of the Central Park Menagerie. In 1873, a visitor would have found two Harpy Eagles (*Harpia harpyja*), an Andean Condor (*Vultur gryphus*), a Saddle-billed Stork (*Ephippiorhynchus senegalensis*) and five Carolina Parakeets (acquired 7 August 1872). A male Whooper Swan (*Cygnus cygnus cygnus*) and a female Trumpeter Swan (*Cygnus cygnus buccinator*) produced a hybrid, which lived eight months in 1877.

In 1878 new acquisitions were the type specimen of *Amazona lactifrons* (whatever a "Milky-fronted Amazon" was) and the very rare *Megapodius pritchardii,* the Incubator Bird found only on the Tongan Island of Niuafo'ou (not seen again in American aviculture until San Diego Zoo acquired two females in 1991). Two Bare-throated Bellbirds (*Procnias nudiicollis*) – the first cotingas I know of in an American zoo – arrived in 1879.

Visitors in 1888 could still see one Harpy Eagle, as well as a pair of Whooping Cranes (*Grus americana*). The "Cape Crowned Crane" (*Balearica r. regulorum*) was new to the collection in 1889. In the same year, a Passenger Pigeon which had arrived in 1876 died, but another could still be seen there. In June 1891, less than a decade before this species was extinct in the wild, "several birds" were received, apparently replacing two specimens, whose carcasses were received by the American Museum of Natural History in January 1891, prepared as two of the less than twenty known skeletons in American collections.[508]

While the Passenger Pigeons received by the Central Park Menagerie in 1891 may possibly have been the last wild specimens acquired by an institution, another American zoo attained lasting fame as the home of the last known living specimen, as well as the last living Carolina Parakeet. Opened 18 September 1875, Cincinnati Zoo was established by German immigrants, who had previously distinguished themselves by an attempt to establish a sizeable fraction of the European Avifauna in the Ohio Valley.[509]

Experiments in introduction

Led by Andrew Erkenbrecher (1821–1885), who came to America at the age of fifteen and made his Cincinnati fortune in laundry starch, the Society for the Acclimatization of Birds spent $9,000 for 4,000 European birds imported from 1872 through 1874.

The major portion of this collection was liberated from "the garret of a towering old mansion standing in a part of the city known as Burnet Woods" on a single May morning in 1873, and included Corn Crakes (*Crex crex*), Skylarks (*Alauda arvensis*), Wagtails (*Motacilla*), Great Tits (*Parus major*), Blue Tits (*Parus caeruleus*), European Water Ouzels (*Cinclus cinclus*), Hedge Sparrows (*Prunella modularis*), Blackbirds (*Turdus merula*),

Mistle Thrushes (*Turdus viscivorus*), Song Thrushes (*Turdus philomelos*), "Hungarian" Thrushes (?), Nightingales (*Luscinia megarhynchus*), European Robins (*Erithacus rubecula*), Starlings (*Sturnus vulgaris*), House Sparrows (*Passer domesticus*), Linnets (*Carduelis cannabina*), Goldfinches (*Carduelis carduelis*), Bullfinches (*Pyrrhula pyrrhula*) and Crossbills (*Loxia sp.*).

It is often stated that this experiment was prompted by a disastrous plague of caterpillars in 1872. However, in marked contrast to Audubon, who had lived in there in the 1820s, and had a high regard for the local birdlife, Andrew Erkenbrecher found Cincinnati "in comparison with European Cities [...] so bare of feathered songsters", and expected "the ennobling influence of the song of birds will be felt by the inhabitants". At any rate, for the majority of the species involved, nothing in particular appears to have resulted from these efforts.[510]

Whether or not any of the sparrows or starlings included in this collection made a genetic contribution to the current population is not clear. The decade following the end of the American Civil War in 1865 saw a great interest in the establishment of House Sparrows in North America. The species was not mentioned at all by James Mann, who discussed eighteen species of European passerines in his *American Bird-Keeper's Manual* in 1848. The first appears to be in 1850, when sixteen birds were imported from England by the Brooklyn Institute (now an esteemed Art Museum). Kept through the winter at the Institute, they were liberated in 1851, but "did not thrive". In 1852, the Brooklyn Institute allocated $200 for another attempt. Nicholas Pike, a member of its board of directors and chairman of the introduction committee, on his way to assume the US consul-generalship of Portugal, stopped in Liverpool, submitting an order for "a large lot of Sparrows and song birds", which were sent to America on the steamship *Europa*. The ship's officer responsible for these birds was instructed to release sixty sparrows as the *Europa* entered New York Harbor through The Narrows between Brooklyn and Staten Island. The remainder spent the winter in captivity. In the spring of 1853 they were liberated in Brooklyn's famously scenic (and exclusive) Greenwood Cemetery, where with "a man... hired to watch them" they prospered.[511] According to some sources on the Greenwood Cemetery, it was also the site of one of earliest nestings of the now-widespread Monk Parrot (*Myiopsitta monachus*), reputedly resulting from a dropped crate of Argentine birds at the nearby JFK airport in the late 1960s. They still nest in the central spire of the much admired Gothic Revival 5th Avenue Gate.

In his classic book on the deleterious effects of introduced animals, George Laycock examined at length the establishment of the House (or "English") Sparrow in North America. He largely attributed these investments of money, time and energy to "cherished memories of the sparrows known to [immigrants] in their youth", though there also appeared to be some promotion based on a supposed value for caterpillar control.

At any rate he presents rather amazing statistics. In the 1870s, New York bird dealers were selling house sparrows for the then princely sum of a dollar each. In 1869, the City of Philadelphia bought and released a thousand. In New York City a dozen were released in Madison Square in 1860, followed by an unspecified number in Central Park in 1864, and 200 all at once in Union Park in 1866. Private persons everywhere were buying House Sparrows, acclimatizing them in aviaries and taking pride in establishing them in their neighborhoods.

It did not take long, however, for the general perception of this species to evolve towards the one typical in the United States today. Fourteen

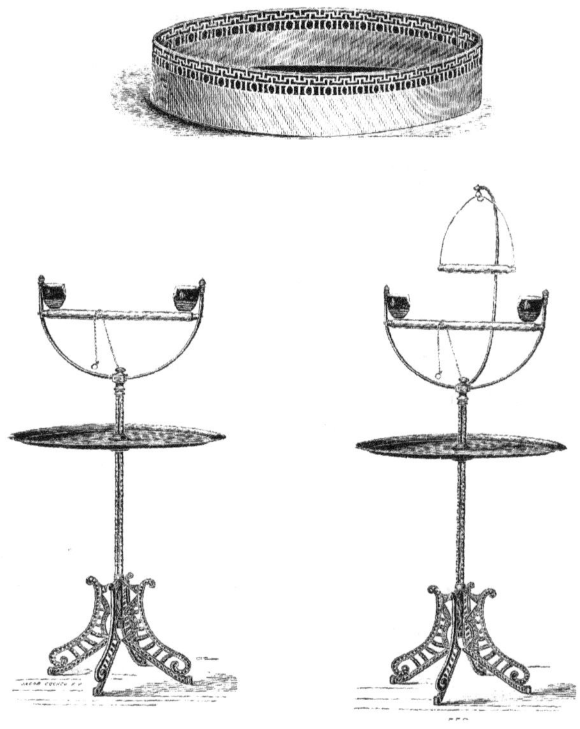

Figure 46: Parrot stands from the late 19th century

years after the mass liberation in Philadelphia, the state of Pennsylvania declared it "lawful at any season of the year to kill or in any way destroy the small bird known as the English Sparrow". In 1887 the State of Michigan offered a bounty of a penny a piece, if dead sparrows were delivered in lots of 25 or more. That same year they were being sold in Albany, the New York state capitol, for a hundred for a dollar as food. They were considered almost as good as Bobolinks. Avicultural historian Arthur Douglas told me that in Medieval Europe bird houses, often in the form of clay pots hung on the walls of buildings, served the practical purpose of breaking dietary monotony by insuring a constant supply of plump young sparrows.[512]

Carnivorous and changed attitudes towards English Sparrows by the 1880s did not prevent equally enthusiastic attempts to establish the European Starling in the 1890s. Since Roman times, starlings were renowned as talking pets. James Mann, who had ignored the House Sparrow in his 1848 book, highly recommended this species as "a lively and active bird, always gay and very familiar" which would "soon become acquainted with all the inmates of the house", adding that they could be "taught to whistle a tune, or speak a sentence very plainly, the male was well as the female". Mann also mentioned that on "the Continent of Europe, in some places, they put up boxes for these birds to breed in, as we do for the Purple Martin".

In *Henry IV, Part* I, the title character, contemplating ways of being especially annoying, declares: "I'll have a starling shall be taught to speak nothing but 'Mortimer'." This made the starling a target of New York pharmaceuticals maven Eugene Scheifflin, who felt America would not be complete without feral populations of every bird mentioned in the plays of Shakespeare. One often reads that Scheifflin set loose thirty pairs of English Starlings in Central Park in 1890 and liberated another twenty pairs in 1891. However, in his 1889 Annual Report, W.A. Conklin, Director of the Central Park Menagerie, reported that Scheifllin released 72 birds (which promptly migrated to Fort Schuyler).

At any rate, the first nesting was discovered in 1890 at Central Park West, not too surprisingly beneath the eves of the northeastern wing of the American Museum of Natural History, a leading center for ornithological research. Although independent releases of starlings were made in such places as Massachusetts and New Jersey in 1844, Cincinnati in 1873, Portland, Oregon in 1889 and 1892, and Pennsylvania and Massachusetts again in 1897, it is widely believed that Scheifflin's Central Park releases were the source of the entire North American population.[513]

About this time, it appeared that the European Goldfinch had become an established element of the American avifauna. From an initial release in Hoboken, New Jersey in 1878, a feral population spread to Central Park, soon reaching Long Island where it became well established.

It was abundant around Massapequa in the 1940s, but was essentially wiped out by the spread of Long Island bedroom communities by the mid-1950s. True to his Shakespearian mission, Eugene Scheifflin likewise conducted experiments with Skylarks, Chaffinches (*Fringilla coelebs*) and Nightingales, none of them successful.[514]

The aviculture of extinct birds

Andrew Erkenbrecher and his associates made a far more lasting contribution by reorganizing the Society for the Acclimatization of Birds as the Zoological Society of Cincinnati in 1873, initially "incorporated as a joint stock company for profit, patterned after similar European institutions".[515] Several months prior to this transition, the Acclimatization Society's Secretary Armin Tenner (who had procured the afore-mentioned European birds and escorted them across the Atlantic) was directed by Erkenbrecher to solicit the advice of no less a person than Dr. Alfred Brehm (1829–1884).

This ornithologist was not only the author of the enormously popular *Tierleben*, whose first edition was published from 1864 through 1869, but was the founding director of the (now defunct) Zoologischer Garten Hamburg (1863–1867), and the Berlin Aquarium Unter den Linden (1869–1874). Closed in 1910, the Aquarium featured grand indoor aviaries housing a major collection of birds.

Not long after the 11 July 1873 incorporation, in his capacity as the new Zoological Society's Corresponding Secretary, German-born Tenner was in his native country, negotiating Brehm's appointment as Director of the Cincinnati Zoo. This came to naught as a result of the Panic in 1873, commencing with the Vienna Stock Exchange in May and culminating in a plethora of US bank collapses in September, resulting in another American depression that only ended in 1879.[516]

Despite the difficult times, the site of Cincinnati Zoo was chosen and work began in 1874. Armin Tenner went back to Germany, researching the latest in zoo technology and successfully recruiting as Superintendent Dr. Herman Dorner, Scientific Secretary of Zoologischer Garten Hamburg. Dr. Dorner commenced his duties at Cincinnati in April 1875, just before construction of the zoo's primary buildings began. These included a series of outdoor aviaries connected by seven small stone buildings, each with on ornate roof of Japanese style.[517] The zoo was opened on 18 September 1875 and by the end of the year, the bird collection included 494 specimens of 84 species. While it is mentioned 200 of these birds were "pigeons", the collections held rarities from the start. An Abyssinian Ground Hornbill (*Bucorvus abyssinicus*), present at the opening, died in 1910; a record Hornbill longevity for many years.[518] Possibly it was part of a collection Dr. Dorner purchased from Carl Hagenbeck to form the zoo's nucleus.[519] Other birds obtained that year were a Bald Eagle from the Central Park Menagerie and a Golden Eagle from the Cincinnati Parks Board. Along with ten breeds of pure-bred dogs, Andrew Erkenbrecher donated a substantial portion of his personal bird collection, as well as several very ugly aviaries. One constructed in 1868 remained at the zoo well into the twentieth century, full of Barred Owls (*Strix varia*) as a memorial to Erkenbrecher.

Another founding benefactor was fellow German immigrant Adolph Strauch (1822–1883), who in his youth studied landscaping at Vienna's Schönbrunn Palace, the site of the oldest continuously existing zoo in the world. In America he became a famous and innovative landscape designer and Cincinnati Superintendent of Parks. While not a member of the Zoological Society's Board, he took an early interest in the zoo and gave much advice during the selection and preparation of the site. It was forgotten for decades that Strauch had successfully bred the Trumpeter Swan in the nineteenth century, producing more than fifty, some exported to Europe and England.[520] The authoritative *First Breeding of Wild Waterfowl in North America,* compiled by Greenwell and Sturgeon in 1988, credits Dr. W.H. Whealton of Chincoteague Island, Virginia in 1915.

At the Zoo's opening, Strauch donated "pairs of Owls, Mute Swans (*Cygnus olor*), and Canada Geese (*Branta canadensis*)". He appears to have eventually donated Trumpeter Swans as well. At any rate, Cincinnati Zoo bred them in 1880, another long-lost fact. When Philadelphia Zoo achieved success in 1965, it was widely believed this was the first such occurrence in a US zoo.[521] A Whooping Crane was added to the collection in the 1884 and in the 1890s, Blue Jays were bred in an aviary. This was eventually proclaimed a North American first breeding by the Bronx Zoo's Lee Crandall, who also credited Cincinnati with the first American breeding of the Strawberry Finch sometime before 1917.[522]

The Cincinnati Zoo's most famous early inhabitants were the last living Carolina Parakeet and Passenger Pigeon. Unfortunately the history of both these species at Cincinnati is not clearly documented, and data is contradictory. This is in part due to some disorder in the administration during the first decade. Dr. Dorner, who had arrived from Hamburg in 1875, was discharged in 1876.

An official history of Cincinnati Zoo states this was "for reasons unknown", but he in fact went to work for none other than Charles and Henry Reiche, who after more than thirty years in the business had turned their New York bird shop into an international animal importing empire, supplying large mammals to circuses. In partnership with the circus impresario William Cameron Coup, they opened the Great New York Aquarium in 1876. Dr. Dorner became its manager and shortly after wrote a ninety-page guidebook. Armin Tenner, who had arranged Dorner's Cincinnati appointment, took over the post himself, only to resign several weeks later. For seven years until 1885, the Superintendent was Captain Frank J. Thompson, who had resigned from Philadelphia Zoo in 1876.[523]

Thompson, Philadelphia's first Superintendent, had assembled an enormous collection of animals in Australia and nearby places, including a "wonderful assortment of colorful birds from the southwest Pacific area", forming a major portion of the collection at that zoo's opening on 1 July, 1874. Thompson eventually also became first superintendent of Buffalo Zoo (whose nucleus was deer presented in 1870) in 1895.[524]

This pattern of fairly brief management positions ended in 1886, when Sol (Salvator) A. Stephan (1849–1949) was promoted from the post of Head Keeper. He retired in 1937. Born in Ohio to German immigrants, his background was with circus elephants, arriving with Cincinnati's first elephant, an African bull in 1875. In 1902 he became the American agent for Carl Hagenbeck, so that the zoo served as a holding compound for animals imported for eventual sale to zoos and circuses.

His son Joseph Stephan spent several years working for the Hagenbecks in Stellingen, as well as several months working at Theodore Roosevelt's North Dakota Ranch, then becoming Cincinnati Zoo's Assistant General Manager, Assistant General Superinten-dent and finally General Manager at his father's retirement. Joseph's son, Dr. Sol G. Stephan, served as Veterinarian and Curator of Reptiles. Despite his circus background, Sol A. Stephan took his zoo's bird collection seriously, quoted as saying: "You can't have too many birds [...] They're easy to get and easy to keep and breed and there's nothing like 'em to attract the women and children". Documentation of the zoo's work with Carolina Parakeets and Passenger Pigeons, however, left much to be desired.[525]

Although the Carolina Parakeet was exported to Europe at least as early as the 1730s, when a specimen in the Belvedere Menagerie, the Austrian collection of Prince Eugen of Savoy, was figured by Salomon Kleiner, it appears to have had limited popularity during the eighteenth century. In his *Histoire de las Louisiane* in 1758, Antoine-Simon, Le Page du Pratz who spent years with the Natchez, considered this species a poor talker, "similar in this way to the Natives, who speak little. It is doubtless because a silent parrot would not be popular with our women that one does not see them in France."

James Mann, the Boston bird dealer, claimed in 1848 that the "Carolina Parrot" was "much prized in Europe, and in Paris is said to cost more than any other [parrot]", on account of its "beauty and docility". On the other hand, as he himself had "never seen a Carolina Parrot kept in a cage", he found this "a verification of the truth of the old adage: A prophet has no honor in his own country".

This appears to be born out by George Whiting Flagg's "Lady and Parrot", painted around 1835, in the collection of the New York Historical Society. The parrot held by the lady is obviously a Carolina Parakeet, but its beak is dead black, possibly stuffed; a studio prop. By the late Victorian period, attitudes may have changed about the time the species was approaching extinction. Olive Thorne Miller (pen name of Harriet Mann Miller (1831–1918), an amateur ornithologist, pioneering birdwatcher and enormously influential author of bird books, wrote an article on cage birds for *Harper's Bazaar* in 1892. Among her recommen-dations was the Carolina Parakeet, as best choice for "decoration of a room".[526]

Several European zoos kept Carolina Parakeets. The devoted scholar of this species, Daniel McKinley of the State University of New York in Albany, has been able to unearth some long ignored records, partially from the data attached to specimens in museums. The earliest was a bird that died in the Jardin des Plantes, Paris in 1818.

Zoologischer Garten Frankfurt had acquired at least four specimens in the 1860s, of which one was still alive at the end of 1879. Two died at Breslau (now Wrocław, Poland) in 1884. Birds died at Schönbrunn, Vienna in 1877 and 1881. Two died at Sofia, Bulgaria in 1901 (until 1893, when it was opened to the public, this was the private collection of King Ferdinand, who was a well-known aviculturist).

The last bird at Zoologischer Garten Berlin died in 1904, while the last one at London Zoo died in 1902. The last bird in Europe was still alive for the Bronx Zoo's Lee Crandall to see in 1912, at Zoologischer Garten Hamburg, where 39 had been kept the decade following its 1863 opening in the era of Brehm and Dorner. This bird did not die until 1917, the year before Cincinnati's last bird.[527]

Carolina Parakeets at Zoologischer Garten Frankfurt had laid four eggs "as early as 1860". Two eggs were laid at Hannover in 1868 and the Dresden Staatliches Museum held one single egg clutch and two two-egg clutches "from the zoo". Zoologischer Garten Dresden was opened in 1861. No European zoo hatchings are recorded.[528]

On the other hand, breeding occurred in at least five private collections in Europe. Emilius Hopkinson credits the first captive breeding to the German aviculturist Karl Russ, reporting that "a pair in Dr. Russ's bird-room hatched out three young and later on five more. Three to five eggs were laid; both parents sat and fed; nest-down mouse-grey. The young were the same size as the adults, but... a lighter green, ...not till the second years was the yellow or orange of the head fully developed."[529] Daniel McKinley deduces the first of Russ' successes took place just before 1870. In 1877 two French aviculturists, Durand-Gonan in St. Brieue and Verdin in Troyes claimed success, apparently both offering young birds for sale in 1878. A Viennese aviculturist, Dr. Nowotny, kept a pair at semi-liberty in his house. In a detailed account he presented his one successful season in 1879, when ten eggs were laid in three clutches (or perhaps two, as suggested by McKinley).

The first seven were destroyed, either by the parents or in household accidents. The remaining clutch of three, commencing in July 1879, were incubated by the female and all hatched. One chick died at three weeks and the other two at six weeks, a week after Dr. Nowotny was constrained to take them with him on an overnight journey.[530]

The most successful (and ultimately most depressing) results were achieved by Hans Karl Hermann Ludwig, Graf von Berlepsch (1850–1915), a systematic ornithologist and authority on South American birds. At his hereditary estate, where he moved in 1893, he had built a museum, curated by Otto Kleinschmidt. Here 55,000 bird specimens (now the greater proportion of the ornithological collection at the Senckenberg Museum in Frank-furt) were housed. This is probably where he kept two pairs of Carolina parakeets in a "former dove-cote", where around twenty young were reared. This flock was liberated. While frequenting the dove-cote as a feeding station, they took to nesting in lime trees and flocks roamed the countryside. One day all the birds were gone. Years later the Graf was informed they had all been shot thirty miles away.[531]

The earliest clearly documented American captive hatching of a Carolina Parakeet also appears to the first record of any parrot propagated in captivity in the United States. In 1885, eleven years after Philadelphia Zoo opened, one was hatched there on 9 September, from an egg fostered "under a Turtle Dove". The incubation period was reported as fourteen days. Nothing further is recorded of this bird.

McKinley notes that four other eggs, recorded as laid at the zoo in 1885, were received by the Bayard Cutting Arboretum in Oakdale, New York. The zoo held three birds at that time. Seven previously acquired specimens, without histories, were dead "for some years" when a "splendid pair" was obtained in 1884. One of these two, or perhaps a third bird received the same year, came from Smithsonian ornithologist Robert Ridgway and lived until 1898. Apparently no further eggs were laid at Philadelphia, though three birds, collected near Orlando, Florida in 1884 and 1885, were presented to the zoo by their Pennsylvania-based collector in 1897. The last of these three died in 1900.[532]

Robert Ridgway (1850–1929) was the Smithsonian Institution's first full-time Curator of Birds. Three eggs in the Smithsonian collections, respectively laid in 1877, 1878 and 1883, appear to have been produced by a bird in his possession. In 1896 he returned from a Florida expedition with three living Carolina Parakeets, which had survived being shot during the collection of preserved specimens, in the face of the obviously impending extinction of this species. They were maintained in his house in the District of Columbia. An egg laid by one of these birds in 1897 is now in the Harvard Museum of Natural History. Ridgway sold both eggs laid in 1901 to John Lewis Childs (1856–1921), who used the proceeds from starting the first seed catalogue business in the US to build an elaborate private museum on Long Island. Childs was not pleased that one of the eggs was damaged from having been laid off a perch, so Ridgway sent him another laid in 1902.[533]

In November 1902 Ridgway wrote to Childs: "My female Parakeet laid only six eggs the past summer and I shall never get any more, as the bird is now dead. The first she laid is the one I sent you. The remaining five hatched, but I have only two young ones left, a rat having carried off one, another was starved by the parents when half grown, and the third I gave to a friend who had time and disposition to take care of it in order to save it from starving". This appears to be the last recorded captive propagation of the species. Four eggs laid in the collection of a Captain Nicholl in England in 1903 or 1904 were addled. Both the chicks that remained in Ridgway's possession died mysteriously in June 1903, and their remains were consigned to the National Museum of Natural History. They were preceded in death by their male parent in February 1903.[534]

The bird Ridgway gave away in 1902 died in 1914, and for many years was wrongly believed to be the last Carolina Parakeet. This bird was given to Paul Bartsch (1871–1960), who served as the Smithsonian's Curator of Molluscs from 1914 to 1946, starting as aide in the Division of Molluscs in 1896. Born in Silesia (now Poland), Dr. Bartsch also served as Professor of Zoology at George Washington University 1900–1940 and Professor of Histology at the Medical School of Howard University, one of the great African-American Colleges, from 1899 to 1936.

He is considered the pioneering American bird bander, beginning with Black-crowned Night Herons (*Nycticorax nycticorax hoactli)* in Washington DC in 1902. He named the parakeet he hand-raised "Doodles" and wrote an engaging profile of it. Several photographs taken in life are extant.[535] This bird did not spend its entire life as a house pet. For a number of years it lived at the National Zoological Park, in the company of two birds acquired in May 1896. Along with six Clarke's Nutcrackers (*Nucifraga columbiana*), eleven Barn Owls (*Tyto alba pratincola*) and several domestic chickens and turkeys, a single Carolina Parakeet was among the 61 birds transported from the Mall near the Smithsonian museums to the Zoological Park's new site on 15 June 1890. Sadly, both the 1896 birds died in 1913 without reproduction, and presumably after this "Doodles" was returned to Dr. Bartsch, dying in his possession the following year.[536]

The very last documented captive Carolina Parakeet was called Incas. It died at Cincinnati Zoo on 21 February 1918. This bird was one of sixteen collected in Florida and purchased by the zoo in 1886 for $2.50 each. For years it was not clear if these birds ever reproduced at Cincinnati. David Ehrlinger, Cincinnati Zoo's Director of Horticul-ture and resident historian reported: "A number were hatched and raised... Generally, however, the birds were inattentive parents and often tossed their eggs out of their nests".[537] Reproductive attempts appear to have ceased after 1892.[538] Eight remained at Cincinnati in 1911, when two were donated to New York Zoological Park (now Bronx Zoo) on 31 August. These were the only specimens ever held by Bronx Zoo which opened in 1899. The collection had been offered specimens by at least two Florida bird dealers in 1902 and 1903, but its Director, the pioneering conservationist William Temple Hornaday (1854–1937), refused to encourage any further trade in this species.

In 1903, he wrote to one of one of these collectors, Mrs. C.F. Latham: "I would not feel justified in making an offer for Florida parakeets, because the species is so nearly exterminated. We could not hope to induce these birds to breed in captivity, and therefore perpetuate the species, and I think it would not be right to encourage the capture of any of the few living specimens that now remain. The species is so nearly extinct in Florida that unless it is carefully preserved, it will soon totally disappear".[539]

Dr. Hornaday had in mind a disaster perpetrated ten years before: "In 1893, one of the last flocks, of about thirty birds, which nested on the Sebastian River, Florida, was completely destroyed in one night by a local hunter, who captured the entire flock, and sent the birds to a New York dealer, in whose hands all those which reached him alive died in a short time".[540]

Dr. Hornaday's pessimism notwithstanding, it was the intent to attempt breeding the Cincinnati birds at the Bronx Zoo. In 1912, Lee Crandall, acting Curator of Birds, wrote: "If our birds survive until the return of warm weather, they will be placed in an aviary suitable for breeding, and offered every inducement to undertake the task of reproduction, but extreme old age is strongly against the chances for thus increasing the numbers of this vanishing race".[541] As it happened, both died before Paul Bartsch's bird "Doodles", which died in August 1914. By 1916 all but two of the Cincinnati Zoo's Carolina Parakeets had also died. In the summer of 1917, "Lady Jane" died, and the last one, "Incas", died on 21 February 1918. Perhaps because World War I was yet raging, this event did not receive much attention. As already mentioned, Bartsch's "Doodles" was long considered the last bird, or else "Incas's" death year was given as 1914. In 1959, starting his research, Daniel McKinley was informed by the zoo that "no one now employed here has any record of this bird's being on exhibition here".[542]

Similar vagueness attended Martha, the last Passenger Pigeon, which died at Cincinnati Zoo on 1 September 1914. Eleven different years have been presented as this bird's hatching date. A.W. Schorger, devoted scholar of the species, summarized that "the history of the pigeons kept at the Cincinnati Zoological Gardens is hopelessly confused". Though Schorger himself appears to have been unaware of the record, the zoo's first Passenger Pigeons arrived in 1874, a donation of "Frank Louck, Esq., of St. Bernard, Ohio [in the immediate neighborhood of the zoo]". Schorger documented that 22 specimens were present at the zoo's opening in September 1875.

According to various letters, Sol Stephan mentioned 26 birds from Michigan, captured in 1875, seventeen acquired in 1876 and twelve purchased in 1878. Frank Thompson, Superintendent from 1878 to 1885, noted that reproduction commenced in March 1878, and that the birds began breeding in March 1879. There were twenty birds in the collection in 1881.[543]

As previously noted, Passenger Pigeons received in 1830 by London Zoo from John James Audubon and started breeding in 1832. London also received 36 birds bred at the Earl of Derby's Knowsley Menagerie from 1834 through 1850. From 1852 through 1874, six more were purchased or received in exchange. The last Passenger Pigeon acquisition recorded for London Zoo were three "presented by F.J. Thompson, Esq.," on 13 January 1883. Cincinnati Zoo is not mentioned.[544] Of the many contradictory statements emanating from Sol Stephan, one of the better fossilized was published in the *Bulletin of the New York Zoological Society* in 1911. Regarding "Martha", he wrote:

> This bird is now about nineteen years old, and was born [sic] in the Garden in a flock of Pigeons originally received from northern Michigan. The flock was kept in an open cage about twelve feet square, and consisted originally of ten birds. One-half dozen or more birds were hatched from this flock, and it was gradually depleted until in 1910 there were but two birds left. In that year the older of the two birds died, at an age of 26 years, leaving the female which is still alive.
>
> This bird is still active, and for company it had until recently a male Mourning Dove [*Zenaida* macroura]. However, the male mourning dove has been placed in an adjoining cage, because, in spite of the fact that a very good

painting of the bird was placed on the cage, some people had trouble in distinguishing the Passenger Pigeon from the mourning dove. When the flock was originally received the birds were not considered much of a rarity, and no more especial care was taken of the than of other birds. However as the flock decreased in numbers, and the birds became scarcer, greater attention was paid to them, and special attention was paid to their feeding. As a result, we have had good success with them, and I really believe that if we could secure some younger birds our experience would enable us to raise young, and increase the flock from a small beginning.

The last remaining bird has been promised to the Smithsonian Institution: and, while it is hoped that it will be a long time yet before this bird dies, it is hoped that when the end does come it will be in good plumage and condition for mounting. Such was not the case with the old male that died about a year ago. He was moulting at the time and in poor condition, so that it was impossible to secure anything like a good result in the mounting of it.[545]

David Ehrlinger noted that in 1878 (the year 100 million were killed by market hunters at a forty-mile long Michigan nest site), "several pairs" of Passenger Pigeons were purchased for $2.50 a pair. By 1885 fourteen had hatched at Cincinnati, but breeding ceased from that point. From 1899 the zoo promised $1,000 a pair for any further specimens, but to no avail; or so goes this version of the story.[546]

Among the exhibits one may see today at Cincinnati Zoo is a single building remaining from a series of seven, built in 1875, which connected with outdoor flights constituted the zoo's aviaries, and were used as such until 1953, when they became displays for monkeys. In 1977, through the efforts of Cincinnati wildlife artist John Ruthven, journalist George Laycock and zoo director Ed Maruska, this remaining house became the Passenger Pigeon Memorial. Here, along with preserved specimens (none of which lived at the zoo) and photographs, are a collection of newspaper accounts, published when "Martha" died in 1914.

One mentions four pairs obtained in 1877. Another mentions eight, likely the same birds, purchased from a Wisconsin collector for $2.50 each. The first account reports that the largest number ever maintained at the zoo was thirteen, but by 1907 there were only "Martha" and two males, the last of which died in 1910.

What is truly startling among the Memorial's newspaper accounts is that "Martha" actually arrived at Cincinnati in 1902. It is not indicated if this bird was purchased, only that it came from "a private collection in Chicago".[547]

Research aviculture

The Chicago collection belonged to Dr. Charles Otis Whitman (1842–1910) who, among other things, was "Professor and Chairman of Zoology at the Imperial University in Tokyo" from 1879 to 1881, a researcher at Naples Zoological Station (one of the world's oldest aquariums) from 1881 through 1882, the first Director of the Marine Biological Laboratory at Woods Hole from 1888 to 1908, Head of the Department of Zoology at the University of Chicago from 1892 to his death, and "the leader in establishing [1890] the American Morphological Society, which became the American Society of Zoologists".[548]

He is also credited with the first captive breeding in the eastern US of 26 species and subspecies of wild pigeons and doves, including the Passenger Pigeon, as well as such a diversity of birds as the Wonga Wonga Pigeon (*Leucosarcia melanoleuca*), Inca Dove (*Scardafella inca*), Band-tailed Pigeon (*Patagioenas fasciata*), White-crowned Pigeon

(*Patagioenas leucocephala*), Himalayan Snow Pigeon (*Columba leuconota*), European Stock Dove (*Columba oenas*), Wood Pigeon (*Columba palumbus*) and Oriental Turtle Dove (*Streptopelia orientalis*).[549]

These avicultural enterprises of Professor Whitman began in 1895, when he was already 53. After two decades of research on annelid worms, developing embryos and various things involving microscopes, he devoted himself to pigeons and doves, not only wild species, but domestic strains as well. In a way his pigeons killed him. Stephen Jay Gould informs that "on the first chill day of December 1910, Whitman worked all afternoon in his backyard, hastily preparing winter quarters for his pigeons to save them from the cold". As a result, he developed pneumonia and died five days later. F.R. Lillie, once his assistant and then successor at Woods Hole, eulogized: "In his zeal for his pigeons, he forgot himself."[550]

His work was fuelled in part by his determination to validate the controversial theory of orthogenesis, which contends that trends in evolution are largely a product of developmental processes of organisms, and that natural selection acts only as a sort of fine-tuning. Demonstrating the existence of fixed developmental pathways would, he believed, achieve this. He compiled meticulous observations of the feather patterns of juvenile birds and set up intricate experiments in hybridization and inbreeding, producing such mutations as long-tailed Oriental Turtle Doves and African Speckled Pigeons (*Columba guinea*) that looked like Rock Doves (*Columba livia*).

He also bred at least one hybrid between a Passenger Pigeon and a "Ring Dove". At the same time he accumulated and analysed behavioral observations and is considered a pioneering ethologist. His collection of pigeons and doves eventually grew to around 550 birds of about thirty taxa. He maintained the birds with his personal funds at his own residences. Every summer from 1895 through 1903 and in 1907, the entire collection was sent by train from Chicago to Woods Hole, Massachusetts.

He cared for them with the assistance of "one or two maids" and employed two Japanese illustrators, Toda and Hayashi, who produced a stream of minutely detailed illustrations. The work was uncompleted at his untimely death, but three volumes with Toda and Hayashi's beautiful pictures were published posthumously and remain a most interesting document, despite that fact that orthogenesis is generally considered invalid. Eminent evolutionary biologist Steven Jay Gould wrote a useful and sympathetic review of his efforts, pointing out much of value to present-day researchers.[551]

In 1896, a year into his avicultural research, Dr. Whitman obtained Passenger Pigeons, purchasing seven of the fifteen birds owned by Milwaukee aviculturist David Whittaker. Whittaker's flock was descended from a single pair. He obtained four birds, wild-caught in Wisconsin in 1888. One died shortly after its acquisition and another escaped. By 1908, Whittaker had only four males which died between November 1908 and February 1909. Dr. Whitman's birds ceased to reproduce after 1898. He subsequently attributed this to not having fed them earthworms. Two females were left in 1906, of which the last died in 1907. In 1902, the year he transferred "Martha" to Cincinnati Zoo, two other specimens escaped from the Woods Hole aviaries. The last documented shooting of a wild bird had occurred in 1900. While Sol Stephan in 1911 wrote that "Martha" had hatched at Cincinnati Zoo around nineteen years before, in 1907 he had elsewhere explicitly stated that this bird come from Whitman.[552]

Conservation aviculture

In light of the extinction of the Passenger

Pigeon and the Carolina Parakeet, it should not be surprising that William Temple Hornaday, Director of the New York Zoological Park (he despised the name "Bronx Zoo") should have come to believe in the impending extinction of a major portion of the North American avifauna. This appears to have reached obsessive levels; towards the end of his life, he was engaged in "a bitter conservational row" with Dr. Frank M. Chapman, Curator of Birds at the American Museum of Natural History, as to whether "the entire avifauna of the continent was dying out" and writing "sarcastically of Dr. Chapman's beautiful faith in the future of birds".[553]

Opening in 1899, Bronx Zoo never held Passenger Pigeons and only had geriatric Carolina Parakeets. Dr. Hornaday, who served as Director from 1896 to 1926, instead encouraged his staff to focus efforts on other species he thought would likely disappear:The Mourning Dove and Herring Gull (*Larus argentatus*). A few months after "Martha's" death, Lee Crandall, Acting Curator of Birds, wrote: "We are particularly proud of our mourning doves. This is the only large dove now found in the East, and while its non-flocking habits have thus far saved it from the fate of the Passenger Pigeon, its very size makes certain its final reduction. It is therefore a great satisfaction that we have this season reared eight of these birds, two of them being at full liberty with the pair of turtle doves which hatched them. We now have fourteen breeding birds, which should form the nucleus of a valuable flock".[554]

Regarding the 1916 breeding season, Crandall noted: "The mourning doves, that have absorbed a good share of our attention, have reared more than twenty youngsters. Several of these birds killed themselves by dashing about their cages when alarmed by an escaped ring-tailed "cat" [*Bassariscus astutus*], but enough remain to provide a good stock of breeders for next year. A pair of adults is now at liberty in the Park, the male having been free for more than a year".[555]

In the meantime, other activities of Dr. Hornaday were to have a far more profound effect on the Mourning Dove, which is now one of the most common and familiar birds in North America, as well as a well-managed game bird.

Commencing in 1913, he worked tirelessly towards the US/Canadian Migratory Bird Act, which achieved its final form in 1918. His book *Our Vanishing Wildlife*, published in 1913, was tremendously influential, especially after his employer, the New York Zoological Society, distributed 10,000 copies to legislators and other persons of influence across the country. From its incorporation in 1895, the Society was as involved in conservation politics as it was in zoo making. For instance, in 1904 William Beebe, Curator of Birds, and the Society's Secretary Madison Grant spent the night at the base of the torch of the Statue of Liberty, documenting the deaths of songbirds attracted to the light. As a result, the iconic torch was "diminished and rendered indirect". This is richly ironic, as Grant (1865–1937) was a principal force behind the 1924 Immigration Act, which drastically reduced the number and diversity of immigrants to the USA. Grant, a founder of the New York Zoological Society and its President from 1925 until his death, was, alongside his passionate conservation efforts, a notoriously ferocious eugenicist.[556]

Today it seems inconceivable that Herring Gulls could ever have been considered an endangered species, but at the end of the nineteenth century, together with America's other gulls and terns, it was threatened by commercial egging, which entailed not only the collection of eggs from rookeries, but the prior destruction of ones previously laid, to ensure the collected eggs were in no state of development.

In the days before vast commercial poultry

farms, the raiding of sea-bird colonies provided bakeries with oil-rich eggs that were particularly valued for making cakes.[557] In the first decade of the twentieth century, the establishment of a free-flying resident flock of Herring Gulls at the New York Zoological Park was attempted. Dr. Hornaday commented: "In captivity the Gull is badly handicapped. With the primaries of one wing clipped to prevent escape, and without the power of flight, it is not seen at its best; for no Gull is perfect save in flight. Our flock is continually shrieking protests against unlawful detention, and with perfect wings every one would quickly fly away, as did those bred in the park and reared to adolescence with perfect wings. We tried to colonize them, but once away they never came back.[558]

These efforts came to an erupt end, as described by William Beebe and Lee Crandall in 1909: "The [...] Herring Gulls formed a successful breeding colony in the New York Zoological Park several years ago until they were exterminated by wild minks which came down the Bronx River, thus carrying their depredations into the very heart of New York City. The mink danger has now been overcome and a new lot of young gulls has been obtained from Lake Champlain, through the kindness of Mr. Edward Hatch, for the purpose of re-establishing the colony".[559]

Figure 47: Fancy bird cages from George H. Holden's catalogue published in 1883

As with the Mourning Dove, the crusades of Hornaday, Grant and others, resulting in protective legislation passed in the second decade of the century, insured the increase of Herring Gulls, making the establishment of free-flying zoo colonies superfluous.

Around World War I, two other programs to conserve declining American birds through aviculture were initiated, with profoundly contrasting results. Once widespread along the east seaboard, the northeastern subspecies of the Greater Prairie Chicken (*Tympanuchus cupido cupido*), popularly known as the Heath Hen, had been confined the Massachusetts island of Martha's Vineyard since the 1870s.

Here it came close to extinction several times in the 1890s and the first years of the twentieth century, dropping to roughly fifty in 1908. Following the establishment of reserves by the State of Massachusetts, the population exploded dramatically, reaching well over a thousand birds in 1916, but in this year a fire, followed by an unusual abundance of Goshawks (*Accipiter gentilis*) reduced the population by roughly ninety percent. Already before the fire it had been decided to reintroduce Heath Hens to their former range. In the face of the calamitous population crash, it was agreed to attempt the establishment of captive flocks. From the roughly 125 remaining birds, sixteen males and ten females were transferred to facilities in Long Island and Massachusetts. One of these was the aviary of Dr. John C. Phillips, Harvard ornithologist and eventual author of *A Natural History of the Ducks.* It appears no offspring was produced in captivity. The historian of extinct birds Christopher Cokinos finds it curious, in the face of numbers reaching at least 300 by 1921, that no further experiments were attempted. The population crashed again in 1921 due to unseasonal frosts and reproduction did not occur after 1925. By 1928, only three males remained, the last of which was not certainly observed after 6 April 1932.[560]

Of course grouse have never been carefree avicultural subjects, being far more prone to diseases and less amenable to captive diets than other gallinaceous birds. None have been commonplace in collections. Regarding the western subspecies of the Greater Prairie Chicken (*Tympanuchus cupido pinnatus*) William Temple Hornaday noted: "During the first four years of its existence, the N.Y. Zoological Park was able to secure only four living specimens."[561] Of the eleven full species only one, the Ruffed Grouse (*Bonasa umbellus*), appears to have been bred in captivity in America by 1930. Only the development of antibiotics and other pharmaceuticals, as well as an improved understanding of nutritional requirements and other aspects of the species' natural history, have made possible, the captive flock of more than 300 birds which since 1992 constitutes the greater proportion of the Attwater's Prairie Chicken (*Tympanuchus cupido attwateri*) population, the only living subtropical grouse.[562]

As the world emerged from the devastation of World War I, another captive-breeding program was inaugurated. This effort met with success; the species is an excellent and already well-established avicultural subject. The Wood Duck (*Aix sponsa*) has a long captive history. Prince Louis II de Bourbon, whose acquisition of a Cardinal opened this account, received six Wood Ducks for his park in Chantilly in 1663. This species arrived at Versailles around the same time. John James Audubon found it propagated at the Charleston aviaries of his friend Samuel Wilson before 1833. It was being bred in France at least by the 1850s.[563] At the beginning of the twentieth century, this bird was rarely bred in American collections. Beebe and Crandall were not aware of anyone breeding it before the Bronx Zoo. By 1918 the Wood Duck was "virtually extinct" in the wild, with the majority of living birds existing in collections in Britain and the European continent. Four years of war and resulting austerity measures probably threatened the captive population.[564] The import of Wood Ducks from wartorn Belgium was funded by Alain C. White, an internationally known chess master. He achieved notoriety and then celebrity for buying property, often productive farmland, to create nature refuges, eventually donated to the State of Connecticut as the nucleus of its State Park system. The propagation and release of the birds was supervised by Reverend Herbert K. Job, an authority on North American waterfowl. Formerly Connecticut's State Ornithologist, under White's patronage he became Director of the Department of Applied Ornithology for the National Association of Audubon Societies. The term in the title of his book *Propagation of Wild Birds – A Manuel of Applied Ornithology*, "Applied Ornithology", can be translated as aviculture.[565]

Located at White's first preserve in Litchfield, Connecticut, the project initially achieved only limited success until 1922, when at Job's recommendation an English gamekeeper, Henry Bowden, was employed.

Until 1939, when the program was discontinued by the State Department of Fish and Game "after Mr. White's retirement form active direction of his property", 2,579 Wood Ducks were released, as many as 400 in a given year. As these birds were all banded, it could be documented that "the Litchfield birds had reached 15 U.S. states and Ontario in Canada, and had formed the nuclei of breeding colonies form South Carolina up to Massachusetts". At the same time, a total hunting ban of Wood Ducks was imposed nationally from 1928 through 1941. The subsequent recovery of this species has been total and today it is widespread and familiar.[566]

Establishing an aviculture community

In 1909, a decade after the opening of the New York Zoological Park, its Curator of Birds, William Beebe, sought to put in perspective the achievements of his and other American collections. With his recently appointed Curatorial Assistant Lee S. Crandall, he attempted to compile a list of all taxa of birds fully reared in captivity in the eastern United States. Their total came to 105 species and subspecies. They also attempted to document the first breeder of each taxon, not always correctly, leaving out Audubon's list of at least eight species of birds bred by Dr. Wilson in Charleston before 1833, and the propagation of Trumpeter Swans by Adolph Strauch in Cincinnatti before 1880.[567] Still, Beebe and Crandall's list is a very useful record and essentially marks the beginnings of an organized and documented American aviculture. Revisions to this list were published in 1917, 1927 and 1930, all by Crandall, as in 1909 Beebe, whose employment commenced in 1899 when he was 22, ceased to be responsible for the daily operations of the bird collection. This was the result of actions by Anthony Kuser (1862–1929), a New Jersey aviculturist who specialized in game-birds. Kuser endowed the New York Zoological Society with $60,000 to fund the creation and publication of a lavish monograph on the world's pheasants. Beebe embarked on a series of expeditions across Asia, largely to observe the behavior of pheasants in the wild over a course of almost two years. Dr. Hornaday, who himself had a background in fieldwork and collecting museum specimens, wrote to Madison Grant:

> Mr. Kuser is a very wealthy man, amusing himself just now with birds and aviaries. Finding in Mr. Beebe a congenial and admiring spirit, he decides to annex Mr. Beebe; for it is the way of some wealthy men to reach out and take whatever they specially desire. Mr. Beebe is so finely caught that I think he is today ready to fling away a great career as an ornithologist, – if he is denied what he now asks, – and all for the privilege of becoming, for a brief season, only, a rich man's plaything, and making a picture book that he never of his own initiative would have dreamed of making... What, I ask you in all seriousness, will become of the present soldierly discipline of the Zoological Park if Mr. Kuser and Mr. Beebe succeed, *in any measure*, in the effort they are now making?
>
> Shall any man be permitted to inflict serious injury on the Zoological Park by cajoling away from it its best men under the pretext of doing the Zoological Society a 'service'? (If Mr. Kuser gets Mr. Beebe away from me, Mr. Kuser must thenceforward be known as the Evil Genius of the Zoological Park; for no man shall injure this force, and at the same time be credited with having done us a 'favor'.) The value of the Society's imprint on the proposed pheasant book would not counterbalance to the Park the loss of the poorest keeper on the force. Very few persons will see during the next twenty years the costly book that is proposed. The most serious question of the present campaign is the fate of Mr. Beebe. If he goes to the

> orient as proposed, no matter what the *modus vivendi* may be, unquestionably it will end his career as Curator of Birds in the Zoological Park. I never will consent to his going for a longer period than four months; and he never will go for an absence as short as that... The wreck of Mr. Beebe's legitimate career will hurt me personally far more than the mortification of having him triumph over my authority.[568]

The New York Zoological Society, then as now (in its present title as the Wildlife Conservation Society), recognized the values of *in situ* research, overrode the director and enthusiastically supported the project. While Beebe retained the title of Curator of Birds until 1918 and never left the employment of the Society, he devoted his career to exploration and field studies in foreign places, culminating in his historic 3,028 foot plunge into the Atlantic, in the Bathysphere in 1934. He retired as the Society's Director of Tropical Research in 1952.[569]

After Beebe's departure for the Far East, the running of the bird department was effectively taken over by the co-author of the 1909 list. Lee Crandall is best remembered today for the *Management of Wild Mammals in Captivity* (1964), popularly known to at least older zoo professionals, as "The Zoo Man's Bible". It may be a surprise to know that *Wild Mammals* was meant as a practice run.

Until six months before his death in 1969 at the age of 81, Crandall continued working at the zoo five days a week, preparing a never-completed work on the aviculture of all wild birds, though he had retired as General Curator in 1952, 44 years after starting as a student keeper. In his lifetime Crandall "knew birds better than any other zoo man in the world", and managed a collection which was always in excess of 600 species and subspecies, while managing to achieve a respectable breeding record. As an example of the magnitude of the bird collection, on 1 April 1913 the New York Zoological Park held 3,038 birds representing 903 species and subspecies, including 88 galliformes, 72 ducks, geese and swans, 57 pigeons and doves, 86 psittacines, 20 owls, and 378 passerines.[570]

Of the 105 species and subspecies of birds in the 1909 list, 34 are pigeons and doves, 26 credited to C.O. Whitman. Not counting mutations and hybrids, 23 taxa of galliformes are recorded. The one cracid, the Wattled Curassow (*Crax globulosa*), hatched by a private aviculturist, was subsequently "invalidated" by Crandall because the chick did not reach independence. By 1917 full success with the same species had been achieved by another private party, Mrs. F.F. Thompson.

The Ruffed Grouse, which was to be only grouse on any of Crandall's lists, had already been propagated before 1909 by C.F. Hodge of Worcester, Massachusetts, also credited with breeding Bobwhite (*Colinus v. virginianus*).[571] Of the seventeen taxa of pheasants, it is not surprising that Anthony Kuser should be credited with the first eastern breedings of Cabot's Tragopan (*Tragopan caboti*), Black-backed Kalij (*Lophura leucomelanos melanota*), Swinhoe's Pheasant (*Lophura swinhoii*) and Green Pheasant (*Phasianus versicolor).*

The less well-known Dr. George W. Little, of Glens Falls, New York is recognized for breeding the Himalayan Monal (*Lophophorus impejanus*), Lineated Kalij (*Lophura leucomelanos lineata*) and Brown Eared Pheasant (*Crossoptilon mantchuricum*). Along with the Golden Amherst (*Chrysolophus amherstiae*), Silver (*Lophura nycthemera*), Reeve's (*Syrmaticus reevesii*), Ring-necked *(Phasianus cholchicus torquatus)*, and "English" Pheasants *(Phasianus cholchicus cholchicus)*, Beebe and Crandall credited their own zoo with the first eastern US propagation of two further Kalij Pheasants: Nepalese (*Lophura leucomela-nos leucomelanos*) and "Anderson's" (since recognized to be natural hybrid between Kalij and Silver Pheasants). They also credited the zoo with "Red Junglefowl (*Gallus gallus*)", as well as Blue Peafowl (*Pavo cristatus*), though one would suspect someone with an estate or a farmyard had long before achieved the first US breeding of "peacocks".[572]

The 1909 list includes only 22 taxa of ducks, geese, and swans. The zoo is given credit for Canada Goose, Wood Duck (in fact bred by Audubon's (1831–1839) friend Samuel Wilson before 1833), and Mallard (*Anas p. platyrhynchus*). The only other zoo on this list of anatids is the National Zoological Park, credited with Mute Swan. John A. Cox of Fieldstone Farm, East Brewster Massachusetts, is credited with the first eastern US breeding of Mandarin Duck (*Aix galericulata*), Northern Pintail (*Anas a. acuta*), Eurasian Wigeon (*Anas penlope*), Northern Shoveler (*Anas clypeata*), Blue-winged Teal (*Anas discors*) and Green-winged Teal (*Anas crecca carolinensis*). Frederic Gallatin of Noroton, Connecticut is recognized for Bean Goose (*Anser fabalis*), Black Duck (*Anas rubripes*), Indian Spot-bill (*Anas p. poecilorhyncha*), and Redhead (*Aythya americana*).[573]

W.H. Whealton on Virginia's Chincoteague Island (by 1915 also successful with Trumpeter Swans) is listed for Black Swan (*Cygnus atratus*), Whooper Swan and Greater Snow Goose (*Anser caerulescens atlantica*). As earlier noted, the Central Park Menagerie produced a hybrid Whooper X Trumpeter Swan in 1877. Gadwall (*Anas s. strepera*) was attributed to P.K. Hudson of East Norwich Long Island. Although the Canvasback (*Athya valisineria*) was included in the 1909 list, it was subsequently found that W.B. Lawrence of Flushing, Long Island, did not rear the ducklings to independence. Crandall in his 1917 list identified the rearing of three under a foster hen at Rockwood Hall, the estate of William G. Rockefeller (brother of John D. Rockefeller) at Tarrytown, New York, as the first actual success. Typically, Anthony Kuser was credited with two costly imports: The Australian Black Duck (*Anas superciliosa rogersi*) and Chilean Pintail (*Anas georgica spinicauda*). A single Shelduck species, the Ruddy (*Tadorna ferruginea*), appears on the 1909 list, hatched at the Rye, New York, estate of William H. Browning.[574]

The list also credits William Browning with the first eastern breeding of Cockatiels (*Nymphicus hollandicus*). Browning was one the first Americans to maintain an extensive series of psittacines, mentioning thirteen species, along with various softbills and seed-eating birds, in a detailed 1909 description of his aviary. Although he had eggs from Eastern Rosellas (*Platycercus eximius*), the only other complete success he had so far achieved with parrots was Budgerigars (*Melopsittacus undulatus*), of which he bred a "large number". Browning specified: "For breeding places, I first used cocoanut husks, imported from E.W. Harper, Wolverhampton, England, and latterly have been using the No. B logs made by Mr. Herman Scheid, Buren i., Westfalen, Germany, which can be imported for about 50c each". The New York Zoological Park, however, was credited for the first eastern US breeding of Budgerigars.[575]

Only four species of psittacines appear on the 1909 list: The Carolina Parakeet, the afore-mentioned 1885 Philadelphia hatching of a dove-fostered egg being cited, and what was then listed as Masked Lovebird (*Agapornis personatus*), but subsequently determined to be Black-cheeked Lovebird (*Agapornis nigrigenis*), achieved at "Sonneberg", aviaries of Mrs. Frederick Ferris Thompson in Canandaigua, New York. On 21 July 1911, 891 birds of 246 taxa were present at this facility of four buildings occupying around an acre, with a staff of five, open to visitors "every Saturday afternoon from two until five o'clock, and… the second and fourth Fridays of July, August and September". Crandall provided an elaborate 1911 description of this collection's buildings and birds. Passeriformes and psittacines were especially well represented. Of the 600 inhabitants of the "Aviary" and its attached outdoor "flying cage", Crandall considered the rarest specimen to be "undoubtedly the Indian spur-winged plover, (*Hoplopterus spinosus*)... probably unique in America". As Crandall mentions this bird is "common enough throughout the Indian Peninsula", it should be identified as River Lapwing (*Vanellus duvaucelii*).

Other species bred in this "miniature jungle" were Wood Ducks, California Quail (*Callipepla californica*), Bar-shouldered Doves (*Geopelia humeralis*), Diamond Doves (*Geopelia cuneata*) (which Crandall confusingly lists as "Scaly Doves"), Cockatiels, wild-type and yellow Budgerigars, Black-cheeked Lovebirds, Cut-throat Finches (*Amadina fasciata*), Australian Zebra Finches (*Poephila guttata castanotis*), wild-type and white Java Sparrows and Saffron Finches (*Sicalis flaveola*). Along with the Black-cheeked Lovebird, Mrs. Thompson was acknowledged in the 1909 list for the first breeding in the eastern US of the Bar-shouldered and Diamond Dove, the Cut-throat Finch, and the Saffron Finch.[576]

A total of fifteen species of passeriform birds were included in the 1909 list. Somewhat surprisingly, only four were Estrildid Finches; Cut-throat Finch, credited to Mrs. Thompson, Australian Zebra Finch to William H. Browning (who wrote that his birds "bred so abundantly" that they outnumbered the other inhabitants of his finch aviary "three to one"), Chestnut Mannikin or Black-headed Nun (*Lonchura malacca atricapilla*) to the aviaries of C.C. Worthington (whose aviculturist was C.W. Miller) at Shawnee-on-Delaware, Pennsylvania, and Java Sparrow to the ten-year-old Bronx Zoo, whose Bird House was completed in 1905.

The other eleven passerines are a rather odd assortment, reflecting the fact that commercial native bird trapping had ended only three years before. White-breasted Nuthatch (*Sitta carolinensis cookei*) was bred at C.C. Worthington's aviaries. So was the Eastern Song Sparrow (*Melospiza m. melodia*), as well as the European Rock Thrush (*Monticola saxatilis*).

Crandall considered the Worthington and Thompson aviaries the "only two collections of consequence in private hands" of aviary birds – as opposed to game-birds or waterfowl – in the first decade of the twentieth century, and noted: "Both these early collections came to the New York Zoological Park on the deaths of their owners."[577]

Anthony Kuser was credited with the Northern Catbird (*Dumetella carolinensis*). The New York Zoological Park hatched the American Robins (*Turdus m. migratorius*), Eastern Bluebird (*Sialia sialis*), Wood Thrush (*Hylocichla mustelina*), Rose-breasted Grosbeak, the Curacao subspecies of Tropical Mockingbird (*Mimus gilvus rostratus*), and what was then mysteriously referred to as the "Gray Robin (*Planisticus tristis*)", but later identified as *Turdus assimilis*, the White-throated Robin of tropical America.[578] Curacao Mockingbirds reared "one or two broods year after year in a cage only four by five, by eight feet high". Beebe and Crandall noted that "thousands of people pass daily within six feet of the nest without disturbing the sitting bird, or causing the male to interrupt his singing, which he continues long after his mate has begun incubation". The final species was the Saffron Finch, bred in Mrs. Thompson's "miniature jungle" community aviary.

Other orders of birds are represented by a handful of species in the 1909 list. The National Zoological Park is credited for the Florida subspecies of Double-crested Cormorant (*Phalacrocorax auritus floridanus*), as well as the Black-crowned Night Heron. The New York Zoological Park bred

White Ibises (*Eudocimus albus*). The only record for the propagation of any bird of prey was the rearing of two Northern Bald Eagles (*Haliaeetus leucocephalus alascanus*), hatched 18 April 1909 at Buffalo Zoo (started in 1870). The female, caught in Lake Huron's Georgian Bay in 1898, had been paired twice previously, producing two infertile clutches, and one accidentally destroyed by freezing a day short of hatching. Success was achieved with a male, which had arrived from Alaska in 1903. The two chicks were fully parent-reared, leaving the nest at 65 days. The sole gruiform was the Demoiselle Crane (*Anthropoides virgo*), bred by Kuser, and the only charadriiform success was the program for Herring Gulls at the Bronx.[579]

Crandall's three subsequent lists covered records for the entire United States. In 1917 he presented 42 further breeding records. No psittacine taxa were added to the 1909 list. Of the eight new taxa of pheasants, three not surprisingly were the achievement of Kuser: Black-breasted Kalij (*Lophura leucomelanos lathami*) (listed as "*Gennaeus horsefieldi*"), Elliot's Pheasant (*Syrmaticus ellioti*) and Gray Peacock Pheasant (*Polyplectron bicalcaratum*).

A fourth species, Bornean Fireback Pheasant (*Lophura ignita*) was bred by Kuser in 1911, but "through an oversight" it was published only in Crandall's final list. Kuser was also successful with the "Columbia Crested Quail (*Eupsychortyx leucopogon*)" (which, with the Bronx Zoo's "Curacao Crested Quail (*Eupsychortyx cristatus*)" appear to be a subspecies of Crested Bobwhite (*Colinus cristatus*)). The pheasantry of Percy Warner (supervised by Clare Lovett) at Nashville, Tennessee, was obviously of some importance, as the first breeding of Temminck's Tragopan (*Tragopan temmincki*) and Siamese Fireback (*Lophura diardi*) were credited to it. The remaining three pheasants were subspecies of Common Pheasant: C.C. Worthington (who had bred three species of North American native passerines before 1909) is credited with "Mongolian" (Kirghiz) and Prince of Wales Pheasants (*Phasianus colchicus mongolicus* and P. c. *principalis*), while the Bronx Zoo claimed credit for Formosan Ring-neck (*P. c. formosanus*). As far as Crandall could determine in 1917, the hatching of Scaled Quail at Bronx Zoo in 1916 was a US first. The zoo already claimed credit in the 1909 list for propagating California Quail (*Callipepla californica*) and Mountain Quail (*Oreortyx pictus plumiferus*)).[580]

Ratites and tinamous were absent from the 1909 list, but the 1917 list includes "South African Ostrich (*Struthio australis*)" and "North African Ostrich (*Struthio camelus*]", crediting the breeding to "several". (The banning of egret and other native bird feathers through the efforts of Dr. Hornaday and others resulted in great interest in ostrich farms). Bronx Zoo is credited with the Emu, a single chick hatching out of a clutch of six in 1915.[581] William Rockefeller, who was recognized for the first full success with Canvasback, bred Red-winged Tinamou (*Rhynchotus rufescens*) at Rockwood Hall, his Tarrytown estate.

Rockefeller (whose aviculturist Arthur M. Barnes) was also the first US breeder of the Luzon Bleeding-heart Pigeon (*Gallicolumba luzonica*). Four other species of pigeons and doves were added to the 1909 list. A California aviculturist, Harry R. Noack in Oakland, was the first American to breed Cape Dove (*Oena capensis*) and Nicobar Pigeon (*Caloenas nicobarica*). Bronx Zoo hatched a single Picazuro Pigeon (*Columba picazuro*) from a pair received in 1910, which had displayed no nesting behavior until 1916, when one chick was reared.

The Red-billed Pigeon (*Columba flavirostris*), whose range extends into South Texas, was bred by Kenyon V. Painter, whose remarkable Cleveland aviaries will be further discussed.[582]

Only four anatids were added to the 1909 list of 22 ducks, geese and swans. One of them was the success with Canvasbacks at William Rockefeller's estate. With all the records of rare and costly birds achieved by Kuser, it is interesting that he was also acknowledged as the first American to breed Egyptian Geese (*Alepochen aegyptiacus*). A pair of Cape Barren Geese (*Cereopsis novaehollandiae*) began rearing goslings at Bronx Zoo in 1910 and bred every year until 1916, which was exceptionally cold and rainy. Bronx Zoo also achieved the first breeding of the Common White-eye or Ferruginous Duck (*Aythya nyroca*), rearing a single duckling in 1915.[583]

The first North American breedings of three other species of aquatic birds were also achieved by the New York Zoological Park. An African Penguin (*Spheniscus demersus*) hatched in 1915; this was the first penguin reared in the USA. A Brown Pelican (*Pelicanus occidentalis*) reared by its parents in the huge communal "Flying Cage" in 1913 was likewise the first American captive pelican breeding. Laughing Gulls (*Leucophaeus atricilla*) commenced breeding in the same aviary in 1914, but most chicks were consumed by other inhabitants of this aviary, until a system of rock and wire shelters was devised.[584]

The Eastern Snowy Egret (*Egretta t. thula*) was credited to Edward Avery McIlhenny (1872–1949), an ornithologist and pioneering wetlands conservationist, also director of the company which manufactured the famous Tabasco sauce his father invented. Avery Island in Louisiana, where this sauce is made and the same line of peppers are cultivated since the 1860s, is also the site of the earliest truly successful reintroduction of an endangered American bird. In 1892, when the demand for plumes for ladies' hats caused ruthless destruction of nesting colonies, he collected seven (or by some accounts, eight) young birds and placed them "in flight cages built over a lagoon".

Most accounts state that these birds were eventually released, the nucleus of a colony from which a major proportion of the Snowy Egrets now abundant across the country appear to be descended. The inclusion of this program by Crandall implies that some captive breeding was involved, perhaps at semi-liberty. At any rate, "Bird City", the series of bamboo platforms designed by McIlhenny, continue to be maintained and are now a tourist attraction at Avery Island. Jean Delacour informed that in his day, "Monsieur Ned" ran the island as if it were a medieval fiefdom. The people there spoke a form of French Delacour found incomprehensible.[585]

Three species of gruiformes joined the Demoiselle Crane from the 1909 list. The Paradise, Blue or Stanley Crane (*Anthropoides paradisea*) was bred by William J. Mackensen, at Yardley, Pennsylvania some time before 1917. In 1949, Jean Delacour wrote: "For some fifty years Mr. Mackensen, alone in this country, has managed to make a steady living out of his birds, mostly Pheasants, Peafowl, and waterfowl. This is saying much for his skill and also for his honesty. Born in Canada of mixed German and Scottish ancestry, his wife is English, and his son, Richard, who is now finishing a course for a doctor's degree as a Veterinarian, at Philadelphia, married a French girl during the war. He now helps his father whenever he is free to do so, and will eventually take over. He is also an enthusiast of birds, and my usual companion on bird trips in the east. More than twenty years ago I went to Yardley to buy Geese, and when I came to live in New York I got into touch again with Mr. Mackensen. I found him so utterly conscientious and reliable that I asked him to procure or to handle all the birds I acquire for Cleres [...]. I enjoy the proximity of Yardley during the part of the year when most of the birds I see are stuffed specimens, and frequent visits to the Mackensen farm are welcome diversions".[586]

The New York Zoological Society began its long tradition of propagating White-naped Cranes

(*Grus vipio*) in 1916, when a pair imported from Japan in 1910 hatched two chicks and raised one. The first US rail captive breeding occurred at the Bronx, when an "Australian Pectoral Rail (*Hypotaenidia philippinensis australis*)", i.e. Australian Banded Rail (*Gallirallus philippensis mellori*), was parent-reared in an outdoor aviary during the very cold March of 1915.[587]

There are ten species of passeriformes in the 1917 list. One of them, the Blue Jay, had actually been propagated in the 1890s, but it appears Cincinnati Zoo had not communicated this to Crandall until after the 1909 list was published. Cincinnati Zoo was also credited with Strawberry Finch or Red Avadavat. House Finch, which became widespread across the eastern seaboard, was credited to the Bronx Zoo, which also bred two hybrids between two species of Australian Wood Swallows, the Masked (*Artamus personatus*) and White-browed (*Artamus supercilliosus*) in 1915. As remarkable as that achievement was, the White-browed Wood Swallow had already been parent-reared in America in 1912 at the elaborate aviaries of Kenyon V. Painter of Cleveland, Ohio. This collection, rich in softbills, is described in detail by Crandall in an article in the *Bulletin* of the New York Zoological Society.

The first US breeding of the European Blackbird was credited to Kuser.[588] In contrast to people with vast estates or extensive outdoor aviaries, the remaining species of passerines appear to have been bred by aviculturists resident in New York City. C. Campini, at "153 East 33rd St.", is credited with the European Goldfinch. A.G. Ibbekin, at "27 West 124th St." achieved the first US breeding of the Pagoda Mynah (*Sturnus pagodarum*), Shaft-tailed Finch (*Poephila acuticauda*), and Cuban Melodious (*Tiaris canora*) and Yellow-faced Grassquits (*Tiaris o. olivaceus*).

1917 was also when Crandall published also *Pets, Their History and Care*, now an almost forgotten work, but a most interesting document of more than 300 pages, covering birds, mammals, reptiles, amphibians and fishes, as well some aquatic invertebrates and plants. It presents a fascinating profile of the range of birds available to private aviculturists in the decade following the ban on trade with native songbirds.

Despite its title, *Pets* covers a broad spectrum of birds: for instance 28 taxa of pheasants are discussed. Even then, finding pure specimens of Golden Pheasants was "next to impossible", Swinhoe's Pheasants were "not rare", but Elliot's Pheasants were "always obtainable from dealers". Cabot's were the most easily obtained of the three available species of tragopans, though "the price asked is always high". Impeyan Monals, though "generally to be obtained" were "always expensive", costing about fifty dollars each. Six taxa of guineafowl and eight taxa of New World Quails are also presented.[589]

Crandall discusses thirteen species of wild pigeons and doves. He mentions a Band-tailed Pigeon (*Columba fasciata*), which "lived for many years in the New York Zoological Park" and "reared numerous hybrids with various domestic pigeons". Regarding the Palm, Laughing or Senegal Dove (*Streptopelia senegalensis*) he notes: "Curiously enough, while this bird is well known to be perfectly hardy in England, in this country it seems to be unable to endure the slightest cold".[590]

In a book of this kind it is startling to encounter the Whooping Crane, until one finds from a footnote that the seven-page chapter on cranes had originally been an article in the Zoological Society *Bulletin*. The Bronx Zoo's crane paddock, where most of its specimens were kept, is discussed in some detail and ten taxa mentioned. Greater Sandhill Cranes (*Grus canadensis tabida*) produced infertile eggs in 1904 and 1905. Regarding the Whooping Crane, Crandall considered the extinction of "this finest of American birds" inevitable, and thought it probable that "the number of Whoopers in captivity could be counted on the fingers of one hand". He concluded: "It is unfortunate that this splendid crane cannot be induced to follow the example of the wood duck, which is willing to save itself from extermination by breeding freely in captivity." On the other hand, the Siberian Crane (*Grus leucogeranus*) was "one of those species more easily obtained alive, and is brought to this country in some numbers. It is quite hardy and easily tamable".[591] Crandall's book also includes descriptions of 55 taxa of ducks, geese and swans. For some reason the Common Shelduck (*Tadorna tadorna*) was "an exceedingly difficult bird to establish in captivity, and cannot be induced to live for long in this country". The Ruddy Shelduck on the contrary was "a hardy species and not affected by cold. It has been bred frequently in America". The Cape Barren Goose was "so rare, that collectors have to depend upon captivity-bred birds". Strangely, the wild Muscovy Duck (*Cairina moschata*) does not appear in any of his breeding lists, as he states in his book that a "pair of these birds, secured in Colombia, have bred in the New York Zoological Park, the young being colored like their parents". Perhaps these ducklings did not survive to adulthood.[592]

It took years before the protection granted to songbirds and aquatic birds would be extended to most birds of prey, and their possession was in many cases not regulated. Crandall wrote: "It may be noted here that in New York State, at least, while all hawks are without the pale of the law, Horned and Snowy Owls are the only members of their group which legally may be kept in captivity". He discussed only native species. In the days before pesticides, he could endorse Peregrine Falcons (*Falco peregrinus anatum*) and White Gyrfalcons (*Falco rusticolus candicans*).

While he found the Snowy Owl (*Nyctea scandiaca*) "a very satisfactory species in captivity", as long as it was protected from summer heat, he considered the Long-eared Owl (*Asio otus wilsonianus*) and Short-eared Owl (*Asio f. flammeus*) "the most difficult of owls to keep in captivity". He noted: "These owls must have an abundance of natural food if there is to be any hope of keeping them."

He presented 48 psittacines, six lories and lorikeets, six cockatoos (including the cockatiel), five macaws, the African Gray, seven Amazons, eight other New World parrots, five members of the genus *Psittacula*, Australian King Parrot (*Alisterus scapularis*), five lovebirds, three rosellas, and the budgerigar.

From today's perspective some of Crandall's parrot observations are fascinating. Regarding the Galah (*Eolophus roseicapillus*), he noted: "Specimens offered by dealers are invariably very wild and are difficult to tame. Although it is one of the poorest speakers among the cockatoos, it is sometimes sold by the unscrupulous as a 'gray parrot'." Concerning the African Gray Parrot he wrote:

"It appears to be less hardy than Amazons when it comes to enduring dietary abuse, and wrong feeding will quickly upset it. Although it is imported in considerable numbers, the mortality is generally very high among freshly arrived birds, many being infected with psittacosis. Severe losses are beginning to discourage dealers, so that imports of these birds are yearly becoming less". He continued: "Unlike most parrots, the Gray is not difficult to sex. The female is generally smaller than the male, and the bare patch around the eye is rounded posteriorly in the former, but ensd in a point in the sterner sex."[593]

Crandall reports: "The White-fronted or Cuban Amazon (*A. leucocephala*) was at one time the commonest parrot among dealers. Recently, however, its exportation from Cuba has been prohibited, so that it is becoming uncommon." In contrast, he noted:

"The Green-cheeked Amazon (*A. virdigena*), a Mexican species, was once so uncommon here as to be a rarity. Of late, however, it has been imported more freely, at times being the most abundant parrot in the market. Linguistically, it is one of the least gifted, and has not become popular."[594]

The three commonest conures were the Aztec (*Aratinga nana astec*), Cactus Conure (*Aratinga cactorum*) and Brown-throated Conure (*Aratinga pertinax aeruginosa*). Fourth in abundance was the Peach-fronted Conure (*Aratinga aurea*), "occasionally imported in some numbers". The Nanday Conure (*Nandayus nenday*) had until the previous two or three years "always been a rare bird in this country", but was "now more often to be obtained".

The Orange-chinned Parakeet (*Brotogeris jugularis*), then called Tove Parakeet or Beebee, was "imported in greater numbers than… any other American parakeet". While Crandall stated that the Monk Parrot "has been bred in this country by at least one aviculturist", one may surmise, since this species does not show up on any of the first breeding lists, that the chicks did not survive. He also identified the Madagascar Lovebird (*Agapornis cana*) as a "free breeder", though again it appears on none of his lists. It was "imported in considerable numbers" and was "one of the commonest of cage birds". In 1917, it appeared that the blue mutation of budgerigar was in decline and likely to disappear from aviculture.[595]

All of the 85 taxa of "Cage Birds" are passeriformes. Thirty of them are softbills. Asian species form a significant component, with two bulbuls, the Golden-fronted Leafbird (*Chloropsis aurifrons*), three Laughing Thrushes, "Red-billed Hill-Tit" (*Leiothrix lutea*), the White-rumped Shama (*Copsychus malabaricus*), the Common Mynah (*Acridotheres tristis*) and three subspecies of Hill Mynahs, all considered likely to be met with in shops.

Ten European softbills are presented: Skylark, Woodlark (*Lullula arborea*), Blackbird, Song Thrush, Nightingale, Black-cap (*Sylvia atricapilla*), Starling, Jay (*Garrulus glandarius*), Magpie (*Pica pica pica*) and Jackdaw (*Corvus monedula*). Concerning the Black-cap, Crandall noted: "The male is a beautiful songster, held by many to compare favorably with the Nightingale. As a cage bird it certainly is superior to that species, for it is much easier to keep and is more likely to sing. Black-caps thrive on the usual soft food, rich in insects, with a daily allowance of meal worms. Fruit and berries are relished as well Black-caps are inclined to over-eat, and their diet must be restricted."[596]

Crandall was similarly cautious regarding the Slate-colored Solitaire (*Myadestes unicolor*), more familiar to American aviculturists as the Clarino. He reported that it "reaches New York from Mexico in considerable numbers". He considered it "a superb singer, perhaps the finest among birds", noting it "lives well in captivity" but "is of exceedingly greedy habits, and will become monstrously fat if its diet in not carefully regulated. Soft food forms the base, but fruit is the main item, and a liberal allowance must be furnished daily if the bird is to thrive".

Only a handful of other New World softbills were regularly available in American bird shops. The caciques which James Mann found occasionally available in 1848 are not mentioned in 1917. The troupial, on the other hand, was as popular a pet bird in Crandall's time as it was in Mann's. Its close Brazilian relative, the Campo Troupial (*Icterus jamacaii*) appears to have been available as well. Concerning tanagers, "none of them is at all common in the market here, and it is only on very rare occasions that any species is obtainable". The Blue-Gray Tanager (*Tharupis cana*) and Palm Tanager (*Thraupis palmarum*) were "being offered less infrequently than most of the others".

The American Crow (*Corvus brachyrhynchos*),

with some other corvids, was not protected as were other American passerines, so Crandall was able to thoroughly recommend it as a pet for young persons.

Today it is surprising that the Australian Magpie or Piping Crow, as it was called, was commercially available in New York, with both Black-backed (*Gymnorhina t. tibicen*) and White-backed (*Gymnorhina tibicen hypoleuca*) forms to be had. Crandall found them "perfectly hardy", requiring "no artificial heat in winter, so long as they are properly sheltered". He much admired their songs and mimicry.

While no African softbills were common enough to be discussed in detail, of the 55 "Finches" described in his book, 25 are African. Thirteen are estrildines, eight are weavers, three are parasitic whydahs, and two are serins. At this time the Madagascar Weaver or Red Fody (*Foudia madagascariensis*) was also predictably available.[597]

Aside from the Strawberry Finch and Java Sparrow ("no bird could be kept in good condition with less trouble"), all eight of Crandall's Asian "Finches" were members of the genus *Lonchura.* In the spirit of his time, he considered the "Bengalee or 'Japanese Nun' " was "produced by crossing the species of *Urolonchura*, but just which is veiled in mystery". He noted: "The Japanese are given credit for developing the three forms in which they occur: dark brown and white, fawn and white and pure white". It was only by the late 1950s that it became widely accepted that the Society Finch was domesticated entirely from White-rumped Munia (*Lonchura striata*).[598]

Seven Australian estrildines are listed: Diamond Sparrow (*Emblema guttata*), Zebra Finch, Star Finch (*Neochmia ruficauda*), Parson Finch (also called Common Grassfinch) (*Poephila cincta cincta*), Shaft-tailed Finch (of which only the yellow-billed nominate subspecies then appeared to be available), Masked Grassfinch (*Poephila personata*), and both black-headed and red-headed Gouldian Finches (*Chloeba gouldiae*). Crandall stated that Red-headed Gouldians have longer tail feathers than black-headed birds. New World "Finches" are represented by the Saffron Finch, Cuban Melodious Finch, both nominate Cuban subspecies and Mexican Yellow-faced Grassquit (*Tiaris olivaceus pusilla*), and Red-crested and "Red-headed Cardinal (*Paroaria larvata*)" (Pope Cardinal (*Paroaria dominicana*)). The eight European "Finches" discussed in *Pets* are the Goldfinch, Linnet, Bullfinch, Siskin (*Carduelis spinus*), Greenfinch (*Carduelis chloris*), Chaffinch (*Fringilla coelebs*), Brambling (*Fringilla montifringilla*) and Yellowhammer (*Emberizia citronella*).

Crandall went into detail on the trade in trained Bullfinches from Germany, which continued to "reach New York in some number each autumn and are quickly snapped up by dealers" until America's entry into World War I:

> Most of the bird and animal importing trade of this country, exclusive of game birds, is in the hands of two or three New York firms with German connections. It is a much more serious undertaking than might be supposed, a well-organized force of travelers, collectors and caretakers being necessary, and prohibiting the participation of smaller concerns. During good weather and under favorable conditions large shipments are received weekly. These arrivals, chief of which, of course, are canaries, are displayed at the establishments of the importers and then selected by the smaller retail dealers.
>
> The prices at which these birds are sold are fairly uniform and well established, but those asked by the retailers who distribute them to the public are extremely variable. For instance, piping Bullfinches are wholesaled uniformly at $15 each and usually bought by dealers simply on the word of the wholesaler, as these birds

generally will not perform until they feel at home. After each has displayed his own particular degree of proficiency, the price is set at from $25 for the bird that know but a single tune to as much as $200 for one which can deliver three or more. The higher prices, of course, are what might be termed 'fancy' and few care to pay them. The tyro should be on his guard, and should never pay what seems an exorbitant price until he has assured himself that he is not being fleeced.[599]

Pets includes lengthy chapters on Domestic Pigeons, bantams and of course canaries. In a later chapter, Crandall further discussed the bird trade between Germany and America:

> Every one knows that the canary is almost ubiquitous in America, but its derivation is not a matter of common knowledge. Practically all of the birds offered for sale in the bird stores throughout the country are reared by the peasants of Germany. The great majority come from the Hartz Mountains, where canary breeding is a leading industry. The village of St. Andreasburg has become famous as the original home of the sweet-voiced Roller, but now both this specialized bird and the common one are bred throughout this region.
>
> The handling of the birds is a business of considerable magnitude. Most it is controlled by two or three New York firms, the largest of which handles, under normal conditions, about 250,000 canaries yearly. At the close of the breeding season in Germany, expert buyers go about among the peasants and secure the birds, many being contracted for in advance. They are concentrated at suitable points and shipped to New York, in lots of thousands, in charge of experienced men, who spend their entire lives in this branch of the work.
>
> Each bird is placed in a small wooden cage, measuring about five inches wide by six inches long and high. Seven cages are strung on a narrow strip of wood, each unit of seven being known as a 'stick', or 'row'. The cages are placed in large frames with wooden back and bottom, and tight wrapped in canvas during transshipment.
>
> When the canaries reach New York, the cages are placed one above the other in great tiers, and the birds graded according to song. They are then ready for shipment to the thousand to retail dealers who sell them to their final owners.[600]

Regarding the American pet canary, Crandall observed: "The canary thrives in an open brass cage, and is commonly so kept in this country. The box cage is infinitely better, but the brass canary cage is an American institution, which is too firmly established to be uprooted".[601]

Despite the discussion, in *Pets* of 48 psittacines commonly imported to the US in 1917, no new species appeared on Crandall's 1917 list, and the 1927 breeding list adds only one to the four recorded in 1909. The Peach-faced Lovebird (*Agapornis roseicollis*) is credited to the Bronx Zoo. In 1917 Crandall observed that "it has never been a common bird, but quite recently there have been fairly numerous arrivals".[602] In contrast, the period from 1917 through 1926 witnessed the first US breedings of five Australian Finches, all in California. The Diamond Sparrow, Owl Finch (*Poephila bichenovii*), (Black-headed) Gouldian Finch and Masked Grassfinch were propagated by T.F.M. Williamson of Pasadena, while Oakland Judge Mortimer Smith was responsible for Sydney Waxbill (*Neochmia temporalis*). One other passerine record was achieved during this interval: the Black-headed Weaver (*Ploceus melanocephalus*) was bred by C.C. Worthington at his aviaries at Shawnee-on-Delaware, Pennsylvania, which before 1909 was the site of the previously noted first breedings of four other passeriformes.

All ten records for pigeons and doves were achieved in Oakland, California, across the bay from San Francisco. Harry R. Noack, already credited with the first breedings of the Cape Dove and Nicobar Pigeon before 1917, added the Scaly Pigeon (*Patagioenas squamosa*), the Indian subspecies of Spotted Dove (*Streptopelia chinensis suratensis*), Ruddy Ground Dove (*Columbina talpacoti rufipennis*) (listed as "Red-winged Gound Dove (*Chaemepelia rufipennis*)"), Brush Bronze-wing (*Phaps elegans*) and Plumed Dove (*Geophaps plumifera plumifera*). The other five columbids were all island endemics and the achievement of Edward Winslow Gifford (1887–1959), one of the most remarkable figures in research aviculture. Though renowned as an anthropologist, he began his career in 1904 at the California Academy of Sciences, working with molluscs and birds. During the Academy expedition in 1906 to the Galapagos, he was the first person to observe tool-use by Woodpecker Finches (*Camarhynchus pallidus*). In 1912 he became Assistant Curator of the Museum of Anthropology at the University of California, Berkeley. There he rose in the ranks as Associate Curator in 1915, Curator in 1925 and Director in 1947. Despite having no formal education after the age of seventeen, he also taught at Berkeley as an Associate in 1938, then Full Professor from 1945. In 1942, Jean Delacour lamented that "my old friend, Mr. E.W. Gifford, has given up his doves, a decision that we all regret". Gifford did not end his avicultural work in order to relax, but he plunged himself into anthropological research, embarking on expeditions to Fiji in 1947, New Caledonia in 1952 and Yap in 1956.[603]

Four of Gifford's breeding records were for the genus *Gallicolumba*, whose members are scattered across the Western and Central Pacific: the Shy Ground Dove (*G. stairi*) of Fiji, Tonga, Samoa and other islands, Jobi Ground Dove (formerly called Amethyst Pigeon) (*G. jobiensis*) of New Guinea and many smaller islands, Buff-hooded Ground Dove (*G. xanthonura*) of the Marianas and Marquesan Ground Dove (*G. rubescens*). Although Gifford published a 1934 account of his experiences with the Buff-hooded Ground Dove, this data was overlooked. Thus, when in the course of establishing a captive population of this threatened bird, hatchings commenced in several US zoos in the first decade of the twenty-first century, it was widely believed they were the first in captivity.[604]

On the other hand, Gifford's work with the Marquesan Ground Dove remains well known, given his extensive notes remain the standard reference for it. Until 1922, the bird was known only from a plate in an 1814 edition of Adam Johann Von Krusenstern's account of the Russian Imperial circumnavigation 1803–1806. It was rediscovered by the American Museum of Natural History's Whitney South Seas Expedition, led by Roland Beck, who had led the 1906 Galapagos expedition, in which a teen-aged Gifford had participated.

Beck had 31 living specimens delivered to Gifford in Oakland in January 1923. Gifford recorded and published extensive behavioral notes. Reproduction began two months after the birds arrived. By 1925, Beck had distributed specimens to a number of other collections.[605] Jean Delacour observed: "Some were sent to England and to France, where Mme. Lecallier raised some during the following years. These curious little doves are extremely quarrelsome and vicious, fighting constantly, even mates of the same pair, the male bullying the hen, sometimes even killing her. This bad temper has caused this once well-established species gradually to disappear from aviaries, the last specimens in California having died about 1945".[606] This species, now listed as vulnerable by the IUCN, appears to never again have been brought into captivity.

Different results were achieved with the Socorro Dove (*Zenaida graysoni*), declared extinct on its island in the 1970s, subsequently found to persist in private collections. It has been suggested that there is some genetic contribution from more recent imports, but it is generally agreed that the birds now living are at least partially descended from the seventeen birds Gifford received from a California Academy of Sciences expedition in 1925. Again he published extensive observations. Breeding commenced at his aviaries in 1926 and soon after at other collections, to which he sent wild-caught birds.

At present there is a well-coordinated effort not only to maintain a self-sustaining, genetically pure population, but to re-introduce it to the Mexican island of Socorro. More than 25 institutions and organizations are involved.[607]

Of the three pheasant species in the 1927 list, two were the achievement of Anthony Kuser: White-crested Kalij (*Lophura leucomelanos hamiltoni*) and Cheer Pheasant (*Catreus wallichi*). He was also credited with the first US breeding of the Painted Quail (*Coturnix chinensis*), long before that species became domesticated. The third species of pheasant was the Satyr Tragopan (*Tragopan satyra*), bred by J.V. de Laveaga near San Francisco, who was later to achieve North American first breedings of fourteen species of anatids before 1931, none however mentioned by Crandall.

The Bare-faced Curassow (*Crax fasciolata*) (listed as "Banded Curassow (*Crax sclateri*)") was bred at the Bronx Zoo.[608] TheBronx is also credited with the Greater Magellan Goose (*Chloephaga picta leucoptera*) (listed as "Magellan Upland Goose (*Choephaga magellanica*)"). Four other anatids US first breedings during 1917–1926 are listed. The Lesser Snow Goose (*Anser coerulescens coerulescens*) was credited to H.J. Jager of Owatonna, Minnesota; the Barnacle Goose (*Branta leucopsis*) was bred at William Rockefeller's estate. Crandall also credits him with the first breeding of the Rosy-billed Pochard (*Netta peposaca*), but the International Wild Waterfowl Association credits that species to Henry Cook of Woodbury, Long Island, by 1915. The New Zealand or Paradise Shelduck (*Tadorna variegata*) was bred by T.A. Havemayer of Mahwah, New Jersey.[609]

The remaining species on the 1927 list were zoo breedings. The North American Black Vulture (*Coragyps atratus atratus*) and Silver Gull (*Chroicocephalus novaehollandiae*) were bred at the Bronx, while the American Coot (*Fulica americana*) was bred at the National Zoological Park.

In 1930, Lee Crandall published his final list of US first breedings, presenting "the rather astonishing total of 170" taxa hatched until 1929 (as compared to 105 in 1909).[610] No novel passeriformes were hatched in the period 1926–1929, but of the 23 taxa that were first propagated in this interval, eleven were psittacines.

Considering that the total number of parrot species prior to 1926 was five, of which four were bred by 1909, it is obvious that a new era in American aviculture was at hand. The first taxon of lory or lorikeet hatched in North America was Swainson's or Australian Rainbow Lorikeet (*Trichoglossus haematodus moluccanus*), a record credited to the San Diego Zoo, a relative upstart established in 1916; by 1930 fourteen taxa of psittacines had hatched there.

One of the two cockatoo records was the achievement of William Browning in Rye, New York, who had achieved US first breedings of Zebra Finches and Cockatiels before 1909. His success was with Leadbeater's Cockatoos (*Cacatua leadbeateri*); a breeding pair had been obtained in

1905. In 1940, "after a lapse of nine years, [they] nested and reared a splendid youngster".

The other cockatoo was the Galah, an achievement of British-born Francis H. Rudkin (1861–1956) and his son Harold F. Rudkin in California's San Fernando Valley. Founded in 1912, the Rudkin Aviaries flourish today, still under family management, specializing in Australian Cockatoos (including Galahs).[611] Although William Browning had eggs from Eastern Rosellas before 1909, Crandall did not credit him with a US first breeding until he published his 1930 list, in which he also recognized Browning for rearing Red-rumped Parakeets (*Psephotus haematonotus*).[612]

While Rudkin eventually established the blue phase of the Masked Lovebird, Kenyon V. Painter , earlier noted for his remarkable propagation of the White-browed Wood Swallow, was responsible for the first US breeding of the Masked (*Agapornis personata*), as well as Fischer's Lovebirds (*Agapornis fischeri*). In fact, it was Painter who introduced both species to aviculture, as related by David Seth-Smith in 1925, Curator of Birds at London Zoo:

> Living examples of two very rare species of Lovebirds, never before imported alive to Europe, have recently passed though London *en route* for the U.S.A., and it is as well that a record of this event should be kept. Mr. Painter of Cleveland, Ohio, is a very enthusiastic aviculturist, as well as a great traveler and big game hunter. He has recently made a trip through East Africa, and on to Cairo via the Nile. At a place called Nagara Nagara in Tanganyika Colony, at an altitude of some 6,000 feet, he obtained several birds, amongst which were two species of Lovebirds, of which a number were secured. A journey of one thousand miles by motor-car and another two thousand miles by river steamer while confined in traveling-boxes greatly reduced their numbers, and four only of the Lovebirds reached London. I am indebted to Mr. Painter for the privilege of seeing and identifying these. There were three examples of the Masked Lovebird... Of Fischer's Lovebird (*Agapornis fischeri*) there was but one example... Both of these species were described and figured by Dr. Reichenow in 1887 in the *Journal für Ornithologie*, but very few skins exist in any museums.[613]

Intrigued by this 1925 incident, I have discovered that from 1907 until his brief incarceration in 1935 for "abstracting and misapplying funds" from the Union Trust Bank, of which he was both Director and the largest stockholder, Painter (who died in 1940) conducted 31 expeditions, collecting specimens for the Cleveland Museum of Natural History, as well as his enormous trophy room. Having first visited Tanganyika in 1907, he invested millions of dollars, purchasing 11,000 acres of land outside Arusha to grow coffee and constructing the first post office, church, hospital, a coffee research center and the famed New Arusha Hotel in 1927. He was in the habit of returning from a safari with living specimens as well. In 1921 Lee Crandall noted: "On September 10, Mr. Kenyon V. Painter presented the [New York Zoological] Society with an unusually fine South African ground hornbill (*Bucorvus leadbeateri*)."[614]

According to avicultural historian George Smith, in the spring of 1926 Painter was able to achieve the first global captive breeding from the Masked Lovebirds he imported in 1925. As he had brought only one Fischer's Lovebird to Cleveland, his success with that species was contingent upon importation of further specimens in commercial quantities by the famed British animal dealer George Chapman. Within an eighteen-month period, Chapman was also the first commercial importer of Masked Lovebird and Nyassa Lovebird (*Agapornis lilianae*). The latter species was first bred in the US before 1929 by Karl Plath (1886–1970) of Chicago, whom Crandall also credited with the Green-rumped Parrotlet (*Forpus passerinus*).

After establishing an international reputation both as aviculturist and artist, Plath became the first Curator of Birds at the Chicago Zoological Park (more popularly known as Brookfield Zoo), a position he held from 1935 to 1961, when among many other things the first US breeding of the Princess Parakeet (*Polytelis alexandrae*) and the first captive breeding of Goldie's Lorikeet (*Trichoglossus goldiei*) were achieved.[615]

The Scarlet Macaw (*Ara macao*) is credited to Chris Holmes of Santa Barbara. An eccentric millionaire, he gave Alex Isenberg his first Greater Bird of Paradise (*Paradisaea apoda*) and a pair of White-bellied Go-away Birds (*Corythaixoides leucogaster*) that bred for the first time in captivity in 1935. Isenberg achieved an amazing series of softbill breedings from the 1930s to the 1960s. Holmes donated to San Diego Zoo its one and only Sumatran elephant in 1926, and left at his death in the 1940s a collection of animals in Hawaii, which became the nucleus of the mammal collection of the resurrected Honolulu Zoo.

The final parrot in Crandall's list from 1926 to 1929 was Petz's Conure (*Aratinga canicularis*), a species that was to become immensely popular in America in the 1950s and 1960s as the "Halfmoon Conure". Its US first breeding is credited to a certain Palmer of whom nothing more is known.[616]

In contrast to the parrots, little had occurred with game-birds or waterfowl in the 1926–1929 interval. As noted, Kuser's propagation of the Borneo Fireback, appearing for the first time in the 1930 list, had actually occurred in 1911. While New York Zoological Park was credited in the 1909 list with breeding the "Red Junglefowl (*Gallus gallus*)", in the 1930 list the National Zoological Park was recognized for the "Red Junglefowl (*Gallus bankiva*)", which today would be recognized as the Javan Red Junglefowl (*Gallus gallus bankiva*). The 1930 list credits the Bronx Zoo with the Gray Junglefowl (*Gallus sonneratii*). The only anatid mentioned is the "Blue Goose (*Chen caerulsecens*)", credited to the National Zoo, now recognized as a color phase of the Lesser Snow Goose, then recognized as *Chen h. hyperborea,* credited to H.J. Jager of Owatonna, Minnesota, by 1926.

Five new species of pigeon and dove breedings occurred. It appears E.W. Gifford was interested in comparing *Gallicolumba* species, with which he was previously successful, with ecologically similar New World species, propagating the White-tipped Dove (*Leptotila verreauxi*), Key West Quail Dove (*Geotrygon chrysia*) and Blue-headed Quail Dove (*Starnoenas cyanocephala*). The Himalayan Bar-tailed Cuckoo-dove (*Macropygia unchall tusalia*) was bred by someone called Greenshields. Mrs. A.R. Black, who maintained Green-naped Pheasant Pigeons (*Otidiphaps nobilis nobilis*) in her Gardena aviaries in California three years before breeding any, not only wrote an account of her success, but was much admired for propagating this specie for at least seven years.[617]

The forty-year-old National Zoological Park was responsible for the other four species in 1930. All but one had taken place much earlier and somehow been left out of previous lists. American Egret (*Ardea alba egretta*) had been hatched in 1906, as had Sacred Ibis (*Threskiornis aethiopicus aethio-picus*), while American White Pelican (*Pelicanus erythrorhynchos*) first hatched there in 1914. There is no mention in Sally Tongren's centennial compilation of any national zoo hatching before

Figure 48: Exotic birds depicted in the American bird-dealer George H. Holden's book on Canaries and other cage birds. From the left a male Pin-tailed Whydah, a Society Finch (or Bengalese Finch), Diamond Finch and a Spice Finch

1974 of Sandhill Crane, but "Sand-hill Crane (*Grus mexicanus*)" is mentioned in a communication "supplied by Dr. William Mann, Director of the Park", which Crandall included in 1930. Four, without date, are included in a list of birds bred at the National Zoo by 1928.[618]

The age of documentation

Beebe and Crandall's original list of records was published in the *Bulletin* of New York Zoological Society, as were Crandall's lists of additions in 1917 and 1927. Crandall's final revised list, however, appeared in *Aviculture*, the journal of the Avicultural Society of America, organized in 1927, with William Browning as its first president. For two years the Society published articles in the now forgotten *Pet Dealer*, but in 1929 it started its own publication.

The first number was issued as *The Avicultural Magazine*, but hurried correspondence from across the Atlantic revealed that the Avicultural Society had been publishing a journal of that name since 1894, and this prompted a change to *Aviculture*. The advent of the Great Depression almost terminated the Society at its start, resulting in a transfer of operations from New York City to southern California; the confusing number of "series" in which *Aviculture* was published testify to various financial troubles. Yet the resulting body of avicultural literature is magnificent. American aviculture reached the point of self-documentation with the publication of the journal.

Exploring back issues of this magazine is thoroughly recommended, as is reading *A.F.A. Watchbird*'s republication of Jean Delacour's notes from his 1936 visit to America, first published in *The Avicultural Magazine* in 1937, Captain Delacour's *Avicultural Magazine* notes on 1949 visits, and Rae Anderson's personal retrospective in *A.F.A. Watchbird*, a wonderful parade of personages of the 1930s until the 1950s, which, if presented here, would double the size of this book and its bibliography.[620]

Aviculture ended publication in 1951, when Jean Delacour, President of the Avicultural Society of America, just starting his duties as Director of Arts, History and Sciences for Los Angeles County, merged it with *The Avicultural Magazine*, resulting in an enhanced international focus for the latter journal, which has been maintained ever since. The Avicultural Society of America started its own publication again when the *Avicultural Bulletin* appeared, edited by the erudite

Eliott Morgan, Head of Research at MGM Studios, in 1962.[621] The inauguration in 1974 of *AFA Watchbird*, journal of the American Federation of Aviculture, has resulted in another magnificent repository of avicultural documentation.

One might expect American aviculture to languish during the difficult years of the 1930s and 1940s, but the evidence provided by *Aviculture* and Delacour's notes shows the contrary. The keeping of cage birds, tropical fishes and cultivation of such house plants as African violets greatly expanded among American households during the Depression.

As later seen in the Soviet Union and countries of the former Warsaw Pact, general conditions of austerity seem to encourage the cultivation of small living things in homes, perhaps in compensation for the inability to practice more conspicuous consumption. There was also a certain degree of "cottage industry" propagation as a source of supplementary income in a sparse economy. At the same time, the prosperity of the 1920s had put in place establishments for mass production of fishes, plants and birds. In 1928 e.g. J.C. Edwards, a charter member of the Board of Directors of the Avicultural Society of America, "raised and disposed of some 15,000 young [budgerigars], including many Blues, Blue-breds, Cobalts, Mauves, and other color varieties" at his Los Angeles establishment. The reader is reminded that in 1917 Crandall considered blue budgerigars a dying breed and otherwise mentioned only the "yellow" mutation.[622] At any rate, Jean Delacour found "a tremendous change in the state of things" between his 1927 and 1936 visits to America. In 1927 he found "very good public collections of live birds in different zoos, but only a very few private ones and some commercial breeding establishments".[623]

In Southern California in 1936 private aviculture flourished. Delacour observed that "the prices of imported foreign birds are generally higher than with us: but Japanese, Filipino, and Australian birds are about the same or cheaper. Mexican, Central American, Colombian, and Venezuelan birds, rather cheap. Indian and, still more, African birds are five or six times dearer than they are in Europe. Pheasants and waterfowl have more or less the same value.

Among the smaller birds bred in California, the rarer Parakeets, Parrots, and Finches fetch good prices: but Lovebirds, Zebra, and other Common Finches are very cheap, fetching about one-half of European prices only." Delacour noted birds "one sees almost everywhere" in the Southern Californian collections: "With the exception of the Firetails and Painted Finches, all Australian [finch] species are numerous and bred in large numbers.

Gouldians do extremely well, and also Diamond Sparrows. There are only a few Parrot Finches, and all African Waxbills are rare, although they do well. Zebra Finches are common and very cheap: a silver and a white strain has been evolved. White and grey Java Sparrows are widespread. Other seed-eaters are scarcer than in Europe. Doves of all sorts are more numerous than they are with us, and do exceedingly well. Quite a few Amazon and Grey Parrots, Ringnecks, and Conures are bred each season, and also *Eclectus* and *Tanygnathus*, and cockatoos. Grass Parakeets, Lorikeets, and Broadtails in general are reared in numbers... Thousands of Masked, Fischer's, Nyassa, Peach-faced, and Black-cheeked Lovebirds are bred, as well as Budgerigars, and all have a low value."

At the end of 1940, Delacour found himself living in America, appointed Technical Advisor to New York Zoological Society. Until 1947, he served in equal capacity with Lee Crandall, by then General Curator of Mammals and Birds, John Tee-Van the ichthyologist, and Fairfield Osborn, President of the Society (and himself an aviculturist), as a "planning committee" of an otherwise director-lacking Bronx Zoo.

The zoo flourished during those years and American aviculture in general was not severely disturbed by World War II. While rationing of gasoline and other petroleum-derived products made things difficult, America's vast agricultural resources meant that the rationing of foodstuffs was not nearly as rigorous as in Europe, and ended as soon as the war was over. By 1949, Delacour found private and zoo aviculture prosperous and expanding.

The profound alteration of lifestyles in the post-war years encouraged bird breeding by individuals.[624] Delacour, in his capacity as Director of the Department of History, Sciences and Art of the County of Los Angeles, maintained aviaries there from 1952 until 1960, and stayed there at least part of each year for the rest of his 95 years. He wrote: "It is idle to compare life in Los Angeles with that in other large cities in the United States or even in the rest of the world. There is no real center of activities. Distances being always tremendous, walking in the streets is rare, and a sort of suburban life prevails everywhere, with its advantages and its drawbacks. One stays at home a great deal, and, to be happy, a hobby such as gardening or bird-keeping is essential. I indulged in both extensively."[625]

Such suburbanization would follow elsewhere in the US, with attending opportunities for backyard aviculture. In the midst of increased activity in the post-war years, there was much that did not happen. While Honolulu Zoo bred Southern Pied Hornbills in 1953, no hornbills were bred anywhere else in the USA until the 1970s. Neither were any toucans, the breeding of other softbills was infrequent and until the 1960s did not result in any kind of program. In retrospect, a basic problem was that birds were cheap.

The advent of air cargo shipments made birds available to zoos and collectors in a way they had never been before, but at prices that were not conducive to the creation of captive self-sustaining populations. As an example, Jungle World, Inc., operating out of Miami, made the following offers in its 1956 price list: Caribbean Flamingoes (*Phoenicopterus ruber*) were $50 each. So were Humboldt's Penguins (*Spheniscus humboldti*). $30 was the price for a Sulphur-breasted (*Ramphastos sulfuratus*) or Cuvier's Toucan (*Ramphastos tucanus cuvieri*), while for $35 one could purchase a juvenile King Vulture (adults cost $85).

Screamers, Trumpeters (*Psophia* sp.), Hill Mynahs and Sun Bitterns (*Eurypyga helias*) could be had for $25. Jabiru Storks (*Jabiru mycteria*) cost $40, while American Wood Storks (*Mycteria americana*) were $20, and Roseate Spoonbills (*Platalea ajaja*), presumably captured in South America, were $15. A Motmot, most likely Blue-crowned (*Momotus momota*), cost $10. "Blue", "Green" and "Scarlet" Tanagers were $3 each and "Foreign Hummingbirds" $10. Because of the psittacosis scare parrots cost more. Adult, wild-caught Scarlet and Blue-and-gold Macaws (*Ara ararauna*) ranged in price from $50 to $100. The most expensive birds on the list were Andean Condors (*Vultur gryphus*), offered for $250.

Such prices definitely had an influence on the breeding programs of zoos. While the occasional propagation of softbills was certainly interesting, going to any great trouble or expense to create self-sustaining populations of birds that cost less than the wages of their keepers was difficult to justify before a municipal park commission or a zoological society's board of directors.

The first volume of the *International Zoo Yearbook* documents 113 taxa of birds hatched in 1959 among the ten zoos that participated in the survey. Anseriformes with 31 taxa, galliformes with 25 taxa and psittaciformes with 20 taxa dominate this list. All of the thirteen species of passeriformes are seed-eating birds, and all but one, the

Yellow-faced Grassquit, are estrildines.[626] The only softbill in this list is the Red-billed Streamer-tail (*Trochilus polytmus*), hatched at Cleveland Zoo.

As this was the first fully successful captive breeding of a hummingbird outside South America, it was well documented, but remained the only zoo hatching until 1970, when San Diego Zoo and Tierpark Berlin hatched Sparkling Violet-ears (*Colibri coruscans*). The Red-billed Streamer-tail was not hatched in any zoo again until 1977, when Zoo Wuppertal in Germany was successful, and not hatched again in the USA until 2000, when it was bred at Florida's Butterfly World in Cocoanut Grove.[627]

In 1970 San Diego Zoo recorded 1,126 species and subspecies of birds, the largest number ever kept in a single American collection. At the end of the year, 1,097 taxa were present. The total number of bird taxa fully reared at San Diego in 1970 was 34. That same year, 290 taxa were hatched among 98 US collections: 78 were anseriformes. Next in representation were psittaciformes with 55 taxa; 25 of the 34 taxa reared at San Diego were parrots, including the first captive-breeding of Western Iris Lorikeet (*Trichoglossus i. iris*), as well as such rarities as Swift Parakeet (*Lathamus discolor*), Black-cheeked Loverbirds, and Malabar Parakeet (*Psittacula columboides*)). The third most well-represented order was galliformes with 42 taxa. There were only sixteen taxa of pigeons.[628]

In contrast to 1959, when all but one of the passeriformes were estrildines, and all were seed-eaters, of the 38 passerines hatched in 1970, fifteen were estrildines, ten more were other seed-eating birds, but the remaining thirteen were softbills. While only a single non-passeriform softbill hatched in 1959, thirteen species were bred in US zoos in 1970. None were hornbills and these achievements represented six orders and nine families, including the first toucans.[629]

The final compilation of breeding records in the *International Zoo Yearbook* is for 1996: 520 taxa were bred among 97 participating US collections. The number of anseriform taxa hatched that year was 71, seven less than the figure for 1970.

Galliforms also underwent a decline, with 39 taxa hatched in 1996 as opposed to 42 in 1970. On the other hand, the number of parrot taxa grew from 55 in 1970 to 72. The increase in pigeon breeding was remarkable: 47 in 1996 as compared to sixteen in 1970. No members of Fruit Dove genus *Ptilinopus* or Imperial Pigeon genus *Ducula* were hatched in 1970. In 1996, eleven species of *Ptilinopus* were hatched among eighteen US collections and four species of *Ducula* among nine zoos.[630]

Aviculturally, Fruit Doves and Imperial Pigeons are considered softbills, and the above statistics for these genera are but part of the profound advancement in softbill husbandry that occurred in American aviculture in the last four decades. In contrast to 1970, when no hornbills were hatched in the US, fifteen taxa were bred in 1996. In 1970, four turacos of three taxa were hatched among three collections. For 1996, the figure was 81 birds (of which 59 were reared) of ten taxa among nineteen collections. The first breeding of toucans in the US occurred in 1970, when two Ecuadorian Crimson–rumped Toucanets (*Aulacorhynchus haematopoygus sexnotatus*), out of three hatched, emerged from the face of a large and ugly wooden "Tiki" at the Los Angeles Zoo. In 1996, 21 ramphastids of four taxa were hatched among eight collections. While only thirteen of the 38 passeriform taxa hatched in zoos in 1970 could be classified as softbills, 80 of the 114 taxa hatched in 1996 were softbills.[631]

This dramatic shift is due to several factors. The first was the sudden imposition in August 1972 of a government ban on bird imports, in reaction to a 1971 outbreak of Exotic Newcastle's Disease. [623]

When imports resumed, it was only under the condition of a one month quarantine in a handful of approved facilities; the situation stands to this day.

The result was a dramatic explosion of prices. In 1972, with much effort I cajoled my parents into purchasing Cut-throat Finches, Strawberry Finches and Red-eared Waxbills (*Estrilda troglodytes*) from Woolworth's and White Front stores, at $3.95 a pair. Two years later, the same birds were entirely beyond my means, retailing at $40 a pair.

The previously mentioned prices of softbills correspondingly changed. Compounding this reduction in availability, countries such as Colombia, Costa Rica, Angola, Thailand and India, which had been major sources of commercial bird trade during the 1960s, ceased to export and others drastically reduced their quotas. The passage of the US Endangered Species act in 1973 and implementation of the appendices created by the Convention on International Trade in Endangered Species (CITES) in 1975 further restricted the variety of imported birds. While the effects of CITES on US bird import initially concerned the species listed in Appendix I, implementation of the Wild Bird Conservation Act in 1993 prohibited commercial import of birds in all three appendices, essentially ending the mass import of parrots to the US, as well as such species as Peking Robin (*Leiothrix lutea*) and many of the commonest African finches. The establishment of self-sustaining populations, previously considered only for endangered species, increasingly became a priority in zoo management.

When comparing the number of taxa fully reared and the number exhibited between 1970 and 1996 for eighteen American collections, a profound decrease in overall species holdings immediately becomes apparent. In 1970 six of the zoos had collections in excess of 400 avian taxa. At the end of 1996 only one did, San Diego, with considerably less than half the bird species and subspecies it held at the end of 1970. Of these eighteen collections, in 1970 fourteen held more than 200 taxa of birds, while at the end of 1996 only six did.

In some cases, this reduction led to dramatic increases in the number of taxa successfully bred. This can be clearly seen for San Diego Zoo, augmented by the achievements of San Diego Wild Animal Park, still under construc-tion in 1970. Similar increases in propagation occurred at San Antonio, the National Aviary at Pittsburgh and Busch Gardens. Yet some institutions show a marked decline in the number of taxa fully reared. In the case of Bronx Zoo, National Zoo and St. Louis, this is due to the general decline of American zoo pheasant and waterfowl collections, demonstrating a decrease in the number of anseriform and galliform taxa propagated between 1970 and 1996. The programs for these birds were to some extent victims of their own success. The startlingly smaller number of species produced by Los Angeles Zoo can in part be attributed to the remarkable success of a single project.

Defeating extinction

Although only five taxa of birds were fully reared at Los Angeles Zoo in 1996, one of them was the California Condor (*Gymnogyps californianus*), of which seventeen hatched and fourteen survived that year. Considering that in 1987, the last wild specimen was taken into captivity, the total population of this species stood at 27, it is entirely understandable that efforts towards increasing its numbers would take precedence over any other activity. Not all of these birds came from eggs hatched in Los Angeles. Eggs laid at San Diego Wild Animal Park were transferred for incubation and hand-rearing of the resulting chicks with puppets, to avoid imprinting.

The *International Zoo Yearbook* records that on 1 January 1996, there were 43 birds at Los Angeles and 24 at the Wild Animal Park. In 1996 breeding began at the Peregrine Fund's World Center for Birds of Prey in Boise, Idaho. The population had reached an all-time low of 22 in 1982, when only three males were in captivity. In 1998 the population exceeded 150. Each of the 27 birds existing in 1987 had produced offspring by 1999. In 2009, there were more than 300 living California Condors with slightly more specimens in the wild than in captivity.[633]

While the first captive breeding of California Condor did not occur until 1988 (following the hatching of wild-collected eggs starting in 1983), serious efforts to establish an avicultural program for this species were initiated in the 1940s. Encouraged by sustained success with Andean Condors, San Diego Zoo (which hatched nine from 1942 through 1952) applied in 1949 for permission to trap a single pair of young California Condors. The permit was granted in 1952, but rescinded in 1954 before any birds were captured, due to the opposition of several prominent zoologists and the Audubon Society to this project. The last captive bird had lived at San Diego from 1929 to 1940, and not until 1967 another bird was brought into captivity, the male Topatopa, who lived by itself at Los Angeles Zoo for years, finally siring offspring in 1993.[634]

Popular imagination would instead be captured by attempts to propagate the Whooping Crane in the 1950s, though only limited success was achieved and it would ultimately prove irrelevant to the present status of the species. The grotesque circumstances surrounding the first full captive rearings of this species have been documented in painful detail. Suffice it to say that the last representative of the Louisiana non-migratory population, brought into captivity in 1941, came to be considered the property of the Park and Recreation Department of the City of New Orleans. Although this bird, Josephine, had produced a subsequently predated chick in 1950 at a government-administered facility in Texas, following its sudden forced return to New Orleans with its mate in 1951, no further eggs were laid until 1955. From 1949 to 1965, Josephine produced 47 eggs, of which fifteen hatched, but only four raised to adulthood. The last of these birds died in 1980 without any contribution to the current gene pool. From a low of 23 specimens in 1941, the total population of this species did not reach fifty until 1964.[635]

That fact that any Whooping Cranes were reared at all in New Orleans is due to the vigorous intervention of a number of persons. A highly diplomatic Jean Delacour and an increasingly antagonistic Dillon Ripley visited and proffered avicultural advice. William G. Conway, then at the beginning of his illustrious and highly influential career with the New York Zoological Society, was on hand for the first complete success in 1957, accompanying George Scott, the Bronx Zoo's Head Keeper of Birds, who was "sent on loan" for two breeding seasons at the Audubon Park Zoo.[636] Consistently successful propagation of Whooping Cranes eventually became a reality in the 1980s at the US governmental Patuxent Wildlife Research Center and the International Crane Foundation. There are presently roughly 500 birds, with about 150 in captivity, in several breeding facilities and an increasing number of US public exhibits.[637] A result of the publicity around the hatching of Whooping Cranes in New Orleans in the 1950s was a renewed focus on saving endangered animals through captive breeding. Such programs became an increasing priority for American public zoos, and by the 1970s had become a primary mission for many collections.

Managing Bali Mynahs

One program almost became a victim of its own

Figure 49: African Gray Parrot (*Psittacus erithacus*)

success. The Bali Mynah (*Leucopsar rothschildi*), described to science only in 1912, first reached aviculture in 1928 when Alfred Ezra received five birds from Walter Goodfellow. Ezra achieved the first captive breeding at Foxwarren Park, Surrey in 1931.

Although the Bronx and San Diego Zoos received specimens in the 1940s, the first US hatching did not occur until 1962 at San Diego, which was the only American collection to breed this species until 1965, when the Pittsburgh Conservatory Aviary was successful. Through the 1960s four US zoos produced at least sixteen birds. From 1970 to 1979, at least 785 hatched in US zoos, of which 438 were reared.

The total for the 1980s was 891 hatched and 531 reared. From the records of the *International Zoo Yearbook*, the peak year for zoo propagation was 1981, when 152 were hatched of which 66 survived among 23 collections.[638] In 1981, the Bali Mynah Species Survival Plan (SSP) was established under the aegis of the Association of Zoos and Aquariums, one of the inaugural SSPs and the first for any bird. Its goals were twofold; to maintain a self-sustaining captive population and supply birds for re-introduction in Bali. The explosion of US zoo propagation coincided with the calamitous collapse of the wild population. This was not entirely coincidental.

Until the abrupt end of Sukarno's 21-year presidency in 1967, the export of Bali Mynahs from the Republic of Indonesia was under strict control. For instance, K.C. Lint, San Diego's longtime Curator of Birds, described his efforts to obtain this species in 1961: "This bird is loved and widely protected by the Balinese. A personal trip to the government headquarters located at the famous Bogor Botanical Gardens and Institute for the Conservation of Nature in Indonesia was necessary before special permits could be issued by Director T. Made Taman of Kepala Lembaga Pengawetan Alam (name of the Gardens). The permits obtained allowed the exportation of four of these birds from the Surabaja [sic] Zoo in Java".[639]

The first years of the 1970s saw a startling increase in the number of Bali Mynahs hatched each year, and the number of collections hatching them. In 1970, three years after the (US supported) fall of Sukarno's Government, more than 100 birds were commercially imported to the US. Comparable numbers were exported elsewhere, and trade within Indonesia became widespread.[640]

While the arrival of large numbers to the USA was a brief phenomenon, curtailed by the imposition of the 1972 Newcastle import ban and further restricted by the US Endangered Species Act in 1973, exploitation of wild birds continued with disastrous effect.

It has been estimated that when this species was described in 1912, the wild population did not exceed 1,000. Although 1976 estimates, based on the observation of 127 wild birds, put the wild population at between 500 and 1,000 birds, by 1984 the estimate was fewer than 200 and less than 100 by 1988. Continued poaching reduced the population to around fifteen in 1990. This number was increased by a series of releases of captive-bred birds (including ones from the American SSP), but stood at twelve in 1999, the same year 39 freshly imported captive-bred birds were stolen at gun-point from the holding facility at Bali Barat National Park.[641]

Thus the mission of the Bali Mynah SSP became increasingly crucial to the survival of the species. In its first decade an obvious problem, in common with many zoo breeding programs, was lack of documentation of the ancestry of much of the population. In some cases, determining origins was impossible. Following the establishment of a regional studbook, it was determined that a significant proportion of the US zoo population was descended from a handful of especially prolific breeding pairs, while many wild-caught birds had few, if any, documented descendants.

In order to insure maximum genetic diversity, the officers of the SSP designated only the propagation of pairs selected for optimal genetic value, and deliberately controlled the number of offspring produced to reduce the effects of over-representation of some bloodlines, and prevent the further production of birds of uncertain ancestry. It quickly became apparent, as someone soon observed "that there were reasons unrepresented founders were unrepresented". It was indeed frustrating, when excellent results had been achieved by "tossing some birds in a cage and watching them breed", that the results of setting up pairs chosen for maximum genetic value were "all too often, nothing at all".[642]

The reduction in annual numbers of birds hatched was eventually reflected in a declining US zoo population, documented by Robert Webster, Curator of Birds at Toledo Zoo, who spent years monitoring the status of passeriform birds in US collections. From a 1985 population of 395 specimens, the zoo population fell to 274 in 1990. Although it rose to 313 in 1991, it fell to 253 in 1992 and has since never exceeded 300; the population declined to 217 in 2001. By the end of 2009, the International Species Information System (ISIS) documented 180 birds distributed among 59 American zoos. The separately managed European and Japanese populations respectively numbered 273 and 140 birds. Naturally, this decline has resulted in concern and criticism, especially among American private aviculturists, among whose collections this species is held in small numbers. It is estimated that without the addition of any further birds to the US zoo population, an at least 90 per cent genetic diversity will be present fifty years from now. The Bali Mynah SSP considers the current situation to be a satisfactory result of several decades of strict management; it is intended to increase the US zoo numbers to 250 and maintain them at that level.[643]

Rescuing birds from Guam

While the Bali Mynah had an avicultural history starting decades before the inauguration of its SSP, the Species Survival Program for the Guam Kingfisher (*Todiramphus c. cinnaminus*) was established only four years after the first specimens were brought into captivity in 1984. Although the Kookaburra (*Dacelo novaeguineae*) was by then well established in American zoos, and some success had been achieved in propagating two species of African Kingfishers, the idea of creating a self-sustaining population of a Pacific island kingfisher in America's zoos would have seemed to verge on science fiction, had not the only alternative been its complete extinction. The precipitous destruction of the avifauna on Guam, and the only partially successful attempts to prevent its extinction, is well documented in Mark Jaffe's *And No Birds Sing* (1994).

In 1981, the population of Guam Kingfishers was estimated at 3,000. In 1984, when 21 (the first in captivity) were collected, it had declined to less than fifty. In 1988, the year the SSP was established, the last wild specimens were observed. This collapse was due to the accidental introduction of the Brown Tree Snake (*Boiga irregularis*) sometime before 1952. This reptile's population exploded on Guam in the 1970s, resulting in the complete extinction of three taxa of birds and the reduction to captive populations of Guam Kingfisher and Guam Rail (*Gallirallus owstoni*).

In addition to the 21 Guam Kingfishers distributed among a consortium of American zoos in 1984, a further eight were collected in 1986. From these 29 birds, sixteen became founders of the current population. Breeding began in 1985, when birds were hatched in Bronx, National and Philadelphia Zoos. Results were initially encouraging, with 65 specimens distributed among 21 collections by 1990.

However, the population declined to less than 60 by 1992 and remained so through 1998. By 1993, only three of the wild-collected birds remained alive, and the situation was further jeopardized by the discovery of tuberculosis in seven specimens among four zoos by 1992. Beginning in 1999 this situation grew less precarious, with consistent increase in numbers, accelerating after 2003. In October 2005, the population stood at eighty birds distributed among eleven collections on the US mainland and a facility on Guam. At the end of 2009, ISIS indicated that 109 were present among eighteen US collections, and 24 had been hatched over the previous twelve months.[644]

Guam became a US territory in 1898 following the Spanish-American War and the neighboring Northern Marianas a US Commonwealth after World War II. As the Brown Tree Snake appears to be present on the island of Saipan and, despite vigilant precautions, its invasion of other Marianas islands remains a continuous threat, US zoos have instituted programs for other taxa endemic to these islands.[645] In 1968, when the population of Guam Rails was estimated at 80,000, ten were sent to San Diego Zoo. From 1968 to 1975, this species bred there repeatedly and offspring were sent to other collections. Yet when this species was discovered to be nearing extinction in the 1980s, the captive bloodline had died out. All specimens now living are descended from fifteen wild-caught birds. From 1983 to 1985, when the species went extinct in the wild, a total of 21 specimens were collected as adults, chicks, or eggs.

Despite difficulties with forming compatible pairs, 150 birds were in captivity by 1989. Breeding in US zoos was curtailed by the SSP in 1994 and did not resume for several years, until it was determined that programs reintroducing the species to Guam and introducing it to the nearby island of Rota were not successful. In 2009, the ISIS-listed US zoo population numbered 52 among sixteen collections, with no breeding indicated for the preceding twelve months. Roughly 120 captive birds are maintained on Guam.[646]

Expanding Marianas program

Following widespread zoo propagation of

Ptilinopus doves in the 1990s, it was decided to establish a US captive population of the Marianas endemic species *Ptilinopus roseicapilla*, the Mariana Fruit Dove.

Propagation began in 1995 at Philadelphia Zoo. In 2009, ISIS listed 62 birds in fifteen collections, with six bred in the last twelve months. This species is currently classified by the IUCN as endangered. Several years ago, zoo-sponsored expeditions to Saipan collected Buff-hooded Ground Doves (*Gallicolumba xanthonura*) and Golden White-eyes (*Cleptornis marchei*).

As noted, E.W. Gifford began hatching *G. xanthonura* in the 1920s and San Diego Zoo had bred from his stock in the 1940s. This captive bloodline was gone and largely forgotten by the time the species was re-introduced to captivity. The first offspring produced by the new birds hatched at Louisville Zoo in 2006. In November, 2009, ISIS listed 56 specimens among nine collections, with nine hatched in the last twelve months. This species is classified as near-threatened by the IUCN.

The Golden White-eye is the sole member of its genus, found only on the islands of Saipan and Aguijan. An enigmatic bird, it was classified as a honey-eater (*Meliphagidae*) until the 1980s, when research suggested it to be a White-eye (*Zosteropidae*). In the 1980s it was "common to abundant" on Saipan. In 2004 it was categorized by the IUCN as critically endangered, entirely due to the presence of Brown Tree Snakes. The present population still counts tens of thousands, but the object-lessons of Guam's extinctions are all too apparent.[647]

A pair of Golden White-eyes were collected by Marquis Yamashina in 1930 and kept in Tokyo, in a traditional Japanese cage "less than a cubic foot square in size, with only two perches, less than five inches apart" until 1939, when they were liberated in Marquis Hachisuka's thermally-heated conservatory in Atami. This pair nested and on 19 March 1940 laid eggs which did not hatch. This remained the only record of the species in captivity until the first decade of the twentieth century, when an American zoo consortium collected a number. In 2009, 38 were distributed among six US collections, and ISIS recorded one hatched in the previous year at North Carolina Zoological Park.[648]

The efforts which culminated in the current populations of Guam Rails and Guam Kingfishers were actually initially targeted towards three now extinct taxa of small passerines: Guam Broadbill Flycatcher (*Myiagra freycineti*), Guam Rufous Fantail (*Rhipidura rufifrons uraniae*) and Guam Bridled White-eye (*Zosterops conspicillatus conspicillatus*). Although avicultural experiments were initiated, all were gone before any serious work could begin. In the face of the potential spread of Brown Tree Snakes across the Marianas, zoo programs have been established for related taxa.[649]

While there has been little to show in US zoo work with the Rota Bridled White-eye (*Zosterops rotensis*), better results have been obtained with the Saipan and Tinian subspeces of Bridled White-eye (*Z. conspicillatus saypani*), which was collected at the same time as Buff-hooded Ground Doves and Golden White-eyes. In 2009, nearly thirty birds were held between Louisville Zoo in Kentucky and Sedgwick County Zoo in Wichita, Kansas, with hatchings at the latter collection that year. In the same year, Bridled White-eyes were transferred from Tinian to the tiny snake-free island of Sarigan.

This was a project of the Mariana Avifauna Conservation Program, administered by the Division of Wildlife of the Commonwealth of the Northern Marianas, the United Fish and Wildlife Service and the Association of Zoos and Aquariums. In conjunction with these efforts, two species of Marianas birds were transferred to American zoos. While the 38 species of the Fantail genus

Rhipidura have an enormous range from Pakistan and Southern China into the Pacific, to New Zealand and Samoa, this genus has until now scarcely factored in American aviculture. Four specimens of the Tinian Fantail (*Rhipidura rufifrons saipanensis*) were sent to Honolulu Zoo in 2009 and two were alive in early 2010. It is planned to involve more US zoos in their care.[650]

There are 31 full species in the genus *Monarcha*, distributed from Australia, New Guinea and Wallacea, east to the Solomons and the New Hebrides. Again their representation in American aviculture has been minimal. In 2009, 24 Tinian Monarchs (*Monarcha takatsukasae*) were divided among Honolulu, Memphis and Riverbanks Zoos. Although "all of those birds were moved from the mist nest to each zoo with no losses" there were subsequent mortalities due to aspergillosis and head trauma "from males trying to escape their cage to attack the other males". By the spring of 2010, several were doing "extremely well on exhibit" and had constructed a nest in Memphis.[651]

Diverging paths

This sort of aviculture is very far from most of the attempts at breeding endangered birds in US zoos in the 1960s and 1970s. Previously, the emphasis was on pheasants, waterfowl and psittacines. The propagation of pheasants by US public zoos was practiced with far more enthusiasm by public zoos than it is today. The larger number of bird taxa reared at Bronx, St. Louis and National Zoos in 1970, as opposed to the number bred by the same collections in 1996 reflects this; the statistics were earlier dominated by pheasants and waterfowl.

This emphasis on "game birds" represents the background of many staff members. The maintenance of such collections as a hobby often begun in childhood. The fact that this could be done economically and the production of offspring could produce revenue, rather than drain it, made this kind of aviculture attractive to zoo administrations. In 1970, 78 taxa of anseriformes were hatched in US Zoos, compared to 71 in 1996. The 1971 numbers for galliformes was 42, while 39 hatched in 1996.

Their overall numbers continue to decline; large collections of waterfowl and comprehensive series of pheasants that were widespread zoo features across the country are far less common today.[652] Many staff members are more likely to have an ornithological focus from their college classes rather than a background in aviculture from their youth. Keepers (and curators) in today's zoos are inclined to find raking pheasant cages less rewarding than dealing with the challenges of hand-feeding the chicks of softbills out of the egg or meticulously collecting and cataloguing behavioral observations.

This trend is well illustrated by comparing and contrasting the breeding results of San Diego Zoo for 1959 and 1996, the first and last years that the *International Zoo Yearbook* published records of animals bred in captivity. In 1959, a significant portion of the 74 taxa hatched were species also well represented in private aviculture; 22 were galliformes. Four of the six pigeons and doves were species established in private collections and the Galapagos Dove (*Zenaida galapagoensis*) would soon join them for some years.

The nineteen psittacines include several popular Australian species. All but one of the eleven passerines were estrildines. There were no non-passeriform softbills at all. If one does not include the nine *Ptilinopus* fruit doves and the one *Ducula* nutmeg pigeon or the charadriiform waders, the 1996 San Diego breeding list still includes eleven non-passerine softbills including four hornbills. Of the fourteen taxa of psttacines, none were widely-kept Australian endemics. Only two of the 24 passerines were estrildines.

At the end of the first decade of the twenty-first

century, the propagation of softbills remains a significant component of US zoo aviculture. For example, five species of Birds of Paradise were maintained as the Association of Zoos and Aquariums sanctioned Population Management Plan (PMP) taxa. Recent achievements have far exceeded what any zoo aviculturist would have imagined thirty years ago. Although at least sixteen Andean Cocks of the Rock (*Rupicola peruviana*) were hatched among five zoos and one private collection in the USA, from an undocumented date in the 1950s to 1989, only five lived to independence. Fifteen of the 23 specimens hatched at The Dallas World Aquarium since 2007 were alive in 2010.[653]

More than 150 members of the toucan family, representing thirteen taxa, were fully reared (either by hand-raising or parent-rearing) at that same collection in 2009 alone. There is a vigorous community of American private aviculturists breeding softbills, some active for well over twenty years, and various species of captive-bred aracaris, hornbills and touracos, among others, are usually available from these individuals.

As an example of an "endangered bird" once widespread in US zoo aviculture which has fallen from favor, the Swinhoe's Pheasant serves as an excellent illustration. In the 1960s, the ornithologist and conservationist James Fisher proclaimed this species to have "the largest aviculture bank, in Europe and North America, of any endangered pheasant, with an estimated captive population of nearly 600, of which about half are in at least 65 public zoos".[654]

The *International Zoo Yearbook* notes that Swinhoe's Pheasants hatched among a total of 36 zoos in the USA every year from 1960 to 1986, as well as in five other collections. From 1972 to 1980, no fewer than six US zoos bred it each year, with nine collections doing so in 1976 and 1977, but in 1981 the number fell to two collections, never again rose above four, and eventually petered out entirely.

By the end of 1999, the International Species Inventory System listed only four US public exhibits of Swinhoe's Pheasants clustered in Utah, North Dakota and Kansas. By the mid-1990s, some zoo aviculturists could not recognize this species. At of the beginning of 2010, six US zoos exhibit twelve specimens, with no recent propagation indicated by ISIS. In contrast, ISIS lists 33 zoos in the United Kingdom and continental Europe collectively holding over a hundred specimens.

On a 1967 visit to San Diego Zoo, New Yorker Magazine's reporter Emily Hahn was told by K.C. Lint that the Swinhoe's Pheasant was extinct in "Formosa" (Taiwan), but that there were "enough in the hands of breeders to repopulate Formosa if the people there want them". Around the same time James Fisher only considered them nearly extinct.From this perspective, it was pleasantly surpising to find this bird considered neither vulnerable nor endangered under the criteria of the Mace-Lande Threat Categories, but instead classified as safe, as long as then current levels of habitat protection were maintained in Taiwan, and thus in no need of inclusion in 1995–1999 Conservation Action Plan for pheasants, drafted by the International Union for the Conservation of Nature. It appears no real field surveys were conducted in Taiwan until the 1970s, when it was estimated to number close to 10,000. In 2010 the IUCN listed Swinhoe's Pheasant as near-threatened, as it appears populations outside of protected areas may be declining, though within protected areas numbers appear stable.[655]

At the same time American zoos were producing large numbers of Swinhoe's Pheasants, other "endangered pheasants" were also being propagated in numbers and have since undergone a similar decline in American zoos. The Western Hume's Pheasant (*Syrmaticus h. humiae*) was one of the last pheasant taxa to be exported from Asia in 1961, and bred for the first time in captivity in Britain during 1962. Propagation in American public collections began at Tracy Aviary in Salt Lake City in 1968, and until 1980, 158 birds of which 78 were reared, hatched among eight US zoos. None have been hatched in US zoos after 1980 and the species has been non-existent in these collections for more than a decade. In 2010, eleven UK and continental European zoos collectively held 31 birds. This situation was not effected by the American Zoo Association's Galliformes Taxon Advisory Group's 1999 recommendation that ten pairs of Hume's Pheasants be maintained among participating institutions.[656]

The Galliform TAG's 1999 recommendations also included the maintenance of ten pairs of Edward's Pheasants (*Lophura edwardsi*) in American zoos. In 1999, ISIS listed only two US collections, both in Utah, each holding a single male. In 2010 ISIS catalogued six specimens among four US zoos; nearly 150 were held among 38 zoos in the UK and continental Europe. This bird was introduced to aviculture by Pierre Jaboulle and Jean Delacour in 1923 and began breeding at Delacour's park in Cleres in 1925. All specimens in Europe and North America are descended from Delacour's exports before 1939. A number of US aviculturists propagated it by Delacour's 1936 trip to America, and he found a healthy American stock after World War II. In the 1980s, shipments of American birds across the Atlantic were effective in quelling such effects of inbreeding as the reduction of the white crest. [657] Delacour had presented Bronx Zoo with an Edward's Pheasant Cock before the war, and this species was present at the inauguration of St. Louis Zoo's redesigned pheasantry in 1954. However, the bird was not hatched in a US public zoo until 1970, when St. Louis was successful. From 1970 to 1993, the *International Zoo Yearbook* recorded 131 hatched of which 91 were reared, among six US zoos, with the above-noted result of an American zoo population of two males at the end of the twentieth century.[658]

In the 1990s, when it appeared that Edward's Pheasant was extinct in the wild, one might have thought vigorous efforts to conserve the US zoo population might have been pursued. Such efforts were, however, partially discouraged by the fact that this species was available in American private aviculture for prices that verged on the trivial. For instance in advertisements in a December 1999 number of *The Gamebird Gazette*, nine individuals offered Edward's Pheasants with prices ranging from $100 to $175 a pair, mostly tending toward the lower sum. Prices for a pair of Hume's Pheasants varied from $100 to $120. A private zoo director has informed me that Hume's may still be purchased in Texas for these prices. Other traditional "endangered pheasants" were similarly valued. Elliott's Pheasants ranged only from $75 to $85 the pair. (Five specimens were collectively held by three US zoos at the beginning of 2010, while in continental Europe and the UK 62 were distributed among 21 institutions).

The two individuals offering Mikado Pheasants (*Syrmaticus mikado*) in *The Gazette* priced them at $140 and $150 a pair. The only US specimens listed by ISIS at the beginning of 2010 were a pair at the Bronx Zoo, while seventeen were held among nine European zoos. Returning to Swinhoe's Pheasant; though eight persons offered this species in *The Gazette*, only two prices were quoted, either $50 or $65 per pair.

The perception that cheap prices have some bearing on the choice of species in US zoo programs is only supported by considering Palawan Peacock Pheasant (*Polyplectron napoleonis*). This bird has a similar history to Swinhoe's Pheasant in that over the period covered by the *International Zoo Yearbook*, at least 644 were hatched in US zoos, with 463 surviving.

Hatchings occurred every year from 1972 to

1996. From 1979 to 1994, no fewer than four US zoos hatched it each year. Twelve places hatched it in 1983, and in three other years nine collections did so. At the end of 1999, 67 specimens were distributed among 28 US zoos, with one hatching indicated by ISIS for the previous year. Twenty years later there were 39 specimens among 22 US zoos, with three hatched among three zoos the previous year. While not quite following the 1999 recommendation of the Galliform TAG that the population continue to be maintained at that year's level, two decades of management resulted in the Palawan Peacock Pheasant being the fifth most widely kept zoo pheasant (if one goes by number of institutions) at the beginning of 2010. ISIS also inventoried 39 "Ring-necked Pheasants", distributed among only fourteen zoos.

The Palawan Peacock Pheasant is designated as vulnerable to extinction by the IUCN, and certainly much effort has been put into its propagation, especially by San Diego Zoo, the only IZY-listed US collection to hatch before 1972; it has produced at least 165 chicks, with 106 hatched there from 1980 to 1988 alone. This reversal of fortune for a species which was not bred at all in US zoos between 1961 and 1972 was effected by San Diego's import of birds, unrelated to any other US bloodlines, from Hong Kong Botanical and Zoological Gardens.[659]

Such efforts would not necessarily guarantee a constant place in US zoo collections for a pheasant. The Zoological Society of San Diego imported eight species of pheasants from the People's Republic of China from 1980 to 1986 and produced hundreds of chicks from this stock. While this addition of new bloodlines of Elliott's, Reeve's, Blue-eared and Brown-eared pheasants was certainly appreciated by private aviculturists, among whom specimens were thoroughly distributed, it did not prevent their decline as American zoo animals. In 2010, ISIS inventoried five Elliot's Pheasants among three collections. Only ten US zoos exhibited Reeve's Pheasants, holding a total of 23 birds, with one hatched the previous year. One of the more famous of the traditional "endangered pheasants", the Brown-eared Pheasant, was represented by seven birds at New York's Central Park Wildlife Center and one at Bramble Park Zoo in Waterton, South Dakota. In the 1999 *Gazette* advertisements, five persons [660]offering Palawan Peacock Pheasants priced their birds at either $400 or $450 a pair. In comparison three purveyors of Brown-eared Pheasants offered them at $175, $220 and $250 a pair.

For the thirteen listings of Blue-eared Pheasants, prices began at $70 and did not exceed $100. There were also thirteen persons advertising Reeve's Pheasants; prices ranged from $25 to $50. The number of Silver Pheasants listed by ISIS for US public zoos in 2010 was almost identical to the figure for Reeve's Pheasants, 24 distributed among ten collections. Thirteen breeders offered the birds in *Gazette* with prices per pair ranging from $25 to $45.

This is not to say that great availability among the private sector automatically sentences a given species to scarcity in American zoos. The public expects to see peafowl and for most visitors, they continue to rank among the most recognized and admired birds. It is not at all surprising that more than 900 specimens, mostly at semi-liberty, should be present among roughly 100 US zoos in 2010. Likewise, the Golden Pheasant, though identical in price to the Silver Pheasant, is met with such enthusiasm by visitors that 37 US zoos collectively exhibited 113 (with six hatched in 2009). Fewer zoos exhibited more "Red Junglefowl" (most of doubtful purity) and almost all at semi-liberty; 173 listed among nineteen collections.

In the USA, 25 zoos held 45 Amherst Pheasants, with no recorded breeding. If the three most well-represented US zoo pheasants (by number of institutions) at the beginning of 2010 were species also common in private aviculture, fifth place was occupied by one of the most coveted. No Malay Argus Pheasants (*Argusianus argus argus*) were advertised in the 1999 *Gazette*, and this species remains one of the most expensive for private aviculturists, in part because of the high demand for unique feathers from trout fly tiers. Over 32 of the 38 years of breeding records catalogued by the *International Zoo Yearbook*, fifteen US zoos collectively produced at least 213 chicks, of which around 150 reached independence. In 2010, 67 specimens were distributed among 23 US collections, with five indicated to have hatched the previous year.

As magnificent as its courtship displays are, this is a species often overlooked by the public, so its continued presence in zoos is due to the enthusiasm of zoo staff. The fact that a normal clutch consists of only two eggs partially explains both its persistence in zoos and its scarcity in private aviculture. All Peacock Pheasants also produce two egg clutches. In contrast, Swinhoe's Pheasants routinely produce six to a dozen eggs, Hume's around ten and one Elliott's clutch may consist of up to twenty eggs.[661]

In today's American public zoos, exponential population growth can be a liability. Zoo aviculturists of a certain age remember species that once bred abundantly, were acquired as captive-bred specimens by many collections, subsequently considered "over-produced", had their propagation curtailed, then fell into precipitous population declines and eventual extinction as zoo animals. From 1982 to 1986, the *International Zoo Yearbook* documented 251 Red-and-White (or White-bellied) Crakes (*Laterallus leucopyrrhus*) hatched among fourteen US zoos, of which 157 were reared to independence. 68 were hatched in 1983 alone over this five-year period; Chicago Zoological Park (Brookfied Zoo) hatched 49 and reared forty. ISIS listed 43 specimens present among seventeen US collections on 30 June 1988, with at least 67 per cent indicated captive-bred, and seven hatched in the previous twelve months. For 30 June 1997, ISIS listed five specimens distributed among four collections. Three years later there were three specimens among two collections and several years later it was gone from US zoos.

From 1980 to 1989 the *International Zoo Yearbook* recorded 268 Crested Barbets (*Trachyphonus vaillantii*) hatched among fourteen US zoos, and 200 of these were reared to independence. From 1981 to 1988, The Pittsburgh Conservatory Aviary (since then re-named National Aviary at Pittsburgh) hatched 32 and reared 31. North Carolina Zoological Park at Asheboro hatched 42 of which 40 survived from 1983 through 1988. 1988 was the most successful year for this species, when seven US zoos hatched a total of 52 and reared 41. In 1992 and 1993 only one hatched each year. The three hatched at Kansas City in 1994 were the last recorded for a US zoo.[662]

Improvement of management of individual zoo animals contributed to curtailing the propagation of "overly prolific" species. In the 1990s it became routine for zoo veterinarians to insist upon comprehensive pre-shipment examinations and tests for every animal leaving the collection, even when not requested by the recipient. Such expenses were added to the price of outgoing birds, so that both the production and acquisition of "$50 pheasants" and waterfowl became unattractive to the administrations of many zoos, especially as the higher levels of zoo management became increasing staffed by persons with public relations, marketing or business administration backgrounds, rather than purely an interest in captive animals.

At the same time, the sale of birds to private individuals and the purchase of birds from the same sources became an increasingly sensitive issue in the age of animal rights activists, so that many public zoos work with only a limited number of "approved vendors", who undergo intense examination and checking of references before being accepted.

Along with the establishment and perpetuation of zoo populations of conservation-significant animals, the prevention of further instances of "over-production" and subsequent population collapse of other species has been the aim of Taxon Advisory Groups established under the supervision of the Association of Zoos and Aquariums (formerly the American Association of Zoological Parks and Aquariums).

The TAG for passeriform birds was established in 1992, and subsequently expanded to include caprimulgids, hummingbirds, trogons and mousebirds. By the end of the 1990s, TAGs had been established to cover all birds. These groups meet at several year intervals to determine which species should be approved for zoo programs, ranging from display, without propagation (especially of native species) through "display, education and research populations" and "population management plans" to Species Survival Programs (SSP). Existing programs are re-evaluated at these meetings, also target populations and annual recommended numbers of offspring produced each year are adjusted. This has by necessity limited the number of taxa maintained and displayed by participating institutions, especially where administrators do not themselves have an avicultural background and are less likely to question TAG recommendations.

Thus one sees a divergence between the sorts of birds maintained by American zoos and private aviculturists. American private aviculture documented in its infancy by Crandall in the first decades of the twentieth century, and approvingly observed by Delacour in the 1930s and 1940s, underwent profound changes in the recent decades. The mandatory quarantine period that has existed since 1972, in response to the threat of Exotic Newcastle's Disease, resulted in a profound increase in the prices of all birds.

An immediate reaction was an initial genuine concern that private aviculture could be legislated out of existence to appease agricultural interests. The American Federation of Aviculture was born out of this atmosphere in 1973 in the living room of California aviculturist Jerry Jennings. AFA rapidly evolved from a political action committee into a leading avicultural society. Its journal *A.F.A Watchbird,* created in 1974, has since served as a major document of the development of American aviculture both in zoos and the private sector.

One can follow through its volumes how species are pulled back from the brink of extinction and popular mutations proliferate through the captive populations of others; how the captive breeding of Touracos and African Hornbills becomes a routine procedure and species such as Sun Conures (*Aratinga solstitialis*) and Goldie's Lorikeets (*Trichoglossus goldiei*), unobtainable at the magazine's founding, become established as staple species among private breeders. Surgical sexing, Feather and Beak and Pacheco's Disease, hybridi-zation controversy, pelleted foods and the Wild Bird Conversation Act are vigorously addressed on its pages, which have recently been digitally archived.

The sudden increase in the cost of imported birds resulting from the mandated quarantine procedures was actually a stimulus to breeding cage, aviary and companion birds in USA. Until then not only did breeders find it difficult to interest buyers in species available as wild imports, but they also faced stiff competition from overseas breeders. For instance, though notoriously fragile, Japanese-bred, Society Finch-fostered Gouldian Finches flooded the market well into the 1960s. The robust strains of Gouldians, bred in US outdoor aviaries only became widely available after the Japanese birds stopped coming.[663]

At the same time, as it became profitable for Americans to propagate psittacines and softbills, innovations in sexing procedures made the establishment of true pairs a far simpler procedure than it had ever been before. Surgical sexing via a fiber-optic laparascope, an innovation of the 1980s, was followed by non-invasive DNA sexing in the 1990s. When imported birds again became regularly available, many persons purchased multiple pairs of the species they were interested in, being assured that they indeed had true pairs.[664]

Beginning in the late 1970s, there was a proliferation of facilities designed to produce a maximum number of psittacines as hand-reared companion birds. While some of these were very large commercial enterprises, many more were "cottage industries", often literally established in a garage, basement, attic, barn or shed, where suspended metal cages were maintained in rows. Similar installations appeared out-of-doors in places with suitable climate. British aviculturist Rosemary Low became a vigorous critic of these techniques, earning a degree of animosity from sectors of American private aviculture for her efforts. In particular her 1996 paper entitled "Aviculture at the Crossroads", expressing her concern that parent-rearing of macaws and cockatoos appeared to be neglected in favor of hand-rearing, which would result in a future collapse of captive populations, drew criticism of a similar nature to the protests later directed against the Obama administration, resulting in vigorous discussion among aviculturists.[665]

Another contentious issue has been the pervasion of mutations among the stock of an increasing number of species, to the extent that wild-type specimens became scarce or even regionally not obtainable. In the first decade of the twentieth century there had been an explosion of interest in establishing mutations of new world parrots and in a short time it was thoroughly achieved with Pacific Parrrotlets (*Forpus coelestis*), Lineolated Parrots (*Bolborhynchus lineola*), Monk Parrots and Green-cheeked Conures (*Pyrrhura molinae*). By the 1990s, the black-headed morph of Gouldian Finch, as well as the Turquoisine Grass Parakeet (*Neophema pulchella*) and Bourke's Parrrot (*Neopsephotus bourkii*) had become scarce.

From the mid-1970s until 1993, when the Wild Bird Conservation Act put an end to the commercial import of wild-caught psittacines, a startling variety of South American parrots arrived in the USA, many of them previously rare or unknown in aviculture. Many American private aviculturists and zoo professionals still remember the 1980s as the time they first saw such birds as Red-bellied Macaws (*Orthopsittaca manilata*), Blue-headed Conures (*Aratinga acuticauda*), Austral Parrots (*Enicognathus ferrugineus*), Slender-billed Parrots (*E. leptorhynchus*), Bronze-winged Parrots (*Pionus chalcopterus*), Dusky Parrots (*P. fuscus*) and Alder (Tucuman) Amazons (*Amazona tucumana*).

At the same time, a bewildering assortment of Indonesian parrots arrived. Several species, previously known only as treasured zoo specimens, became available to private aviculturists. Among them were the Goldie's Lorikeet, Central Stella's Lorikeet (*Charmosyna papou goliathina*) and Pesquet's Parrot (*Psittrichas fulgidus*). Others, such as the Red-flanked Lorikeet (*Charmosyna placentis*) and Desmerest's Fig Parrot (*Psittaculirostris desmarestii*), were complete novelties in America. In retrospect, the sheer volume of psittacine imports to the US during the 1980s can only be considered appalling. Rosemary Low noted that 77,908 macaws arrived in the USA between 1980 and 1984, and in 1985 alone 29,849 Indonesian parrots were imported.[666] For several years more than 6,000 Salmon-crested or Moluccan Cockatoos (*Cacatua moluccensis*) were collected and exported annually from Seram.

The Convention on International Trade in Endangered Species (CITES) documented 66,654 arriving at its signatory countries between 1983 and 1990. In 1988, the Fauna Preservation Society included this species on its annual list of the ten animal and plant taxa most likely to go extinct and it is currently classified as vulnerable by the IUCN. In January 1990, legal international trade ended and an immediate result was a last flurry of imports by American animal dealers in the months before the law went into effect. Multiple pairs were purchased by aviculturalists across the country.[667]

The majority of offspring were destined to be hand-reared as companion animals. Male cockatoos thus raised have often been found to be useless as breeders, attacking females, destroying chicks or otherwise exhibiting pathological behavior. The Salmon-crested Cockatoo has traditionally been a standard exhibit in American zoos. The world first breeding, a hand-rearing, was achieved at San Diego Zoo in 1951.

Over the years documented by the *International Zoo Yearbook*, 85 (with 59 surviving) were hatched among fourteen institutions from 1969 to 1996. Although a PMP was designated for this species and a studbook maintained, there has been little recent effort among zoos to establish a stable population. At the end of 1997, ISIS listed 119 distributed among 52 zoos. In 2010, seventy were present among 42 collections, and no breedings were indicated for the previous year. Five zoos collectively hatched fifteen and reared twelve in 1994. A number of American cockatoo breeders are currently producing parent-raised birds of some species.[668]

Among other parrots whose conservation status deteriorated as a direct result of commercial trade, were two species of macaws, both endemic to Bolivia, and both unknown in aviculture before the 1970s. The Red-fronted Macaw (*Ara rubrogenys*) was not exported alive until 1973. It appears the first shipments to the US were sixteen birds in 1977; 82 were imported in 1978 and 125 in 1979. Figures are incomplete for 1980, but in 1981, 210 arrived in the USA.[669] Its legal trade ended with a CITES I classification in 1983.

Classified by the IUCN as endangered, the wild population is estimated at less than 5,000, but in 2010 it was not difficult to find hand-reared birds sold as pets. ISIS listed 46 distributed among nineteen zoos, but with no propagation indicated for the previous year. The *International Zoo Yearbook* documents US zoo breedings each year from 1983 through 1996. Seven places collectively hatched 171 and reared 150. Eighty-nine, of which 81 were reared, hatched at the New York Zoological Society's St. Catherines Wildlife Survival Center, off the coast of Georgia, where breeding occurred each year except for 1995. As with pheasants, the propagation of parrots that are conservation-significant, yet widespread in private aviculture, is no longer a priority in US zoos.

No specimens of the Blue-throated Macaw (*Ara glaucogularis*) appears to have been kept in captivity before 1977, when one became part of the encyclopedic parrot collection at Walsrode Bird Park in Germany. Only a few years before, its existence as a taxon was in question, regarded by many as an ornithological Loch Ness Monster or Sasquatch. This macaw's range was not scientifically established until 1992, long after the major portion of its population had been removed from the wild. Until it was placed on CITES Appendix I in 1983, it appears over 1,200 were captured for export. In 2010, the wild population was thought to number no more than 350, and it is classified as critically endangered by the IUCN.[670]

Rosemary Low was aware of four Blue-throated Macaws that had arrived in the USA by 1980 "via a large quarantine station" in California. One of these was on loan to Los Angeles Zoo in 1980. In 1981, two years before the CITES listing, fifty were exported to the USA and ten to Europe. Population estimates at the time ranged from 500 to 1,000, but in either case, such a reduction in a single year could only have negative effects.[671]

Hand-raised birds, offered as pets by private breeders, are today easily located. While prices are generally between $2,000 and $2,500, many birds can be found for considerably less. Of the 52 specimens ISIS catalogued in 2010, 23 were distributed among seven public zoos, while 29 were held at Natural Encounters, the off-display breeding center maintained by Steve Martin, the pioneering creator of conservation-oriented birds shows for zoos and related facilities. During the period covered by the *International Zoo Yearbook*, St. Catherines Island was the only US zoo-related collection to breed this species (then listed as "Caninde Macaws (*Ara caninde*)").

From 1989 to 1996, fifteen hatched there and fourteen were reared. The St. Catherine Island parrot collection was dispersed in the first decade of the twenty-first century. In 2005, Natural Encounters received fifteen Blue-throated Macaws, the nucleus of its current breeding group. Steve Martin informed me that parent-raised chicks have proven as amenable to hand-reared birds for training to appear in his remarkable free-flight demonstrations.

One of the first parent-reared St. Catherine Island-hatched Blue-throated Macaws, a female, was sent to Houston Zoo, where it was paired with a captive-bred male, also parent-reared. In 2000 this pair produced its first chick, which remained with its parents for almost a year. Ten years later this bird and one of two hatched in Houston in 2001 resided at The Blues Conservatory, a Texas facility operated by non-profit Bird Endowment, incorporated in 1998, to insure a parent-reared US population of Blue-throated Macaws.[672]

The Endowment's founder, Laney Rickman, had not kept birds prior to 1990, when she was given a "Halfmoon Conure". Her fascination with this bird led her to volunteer at Houston Zoo in 1992 and 1993, where she became acquainted with Blue-throated Macaws and grew deeply concerned for their avicultural future. She acquired her first breeding pair in 1992.

At the end of 2009, The Blues Conservatory held nine bonded pairs, six unpaired juveniles maintained in a flock and three young birds still with their parents. As the various founder pairs had come from a variety of facilities, where in some cases all the eggs they had previously produced were pulled for hand-rearing, some presented frustrating challenges in establishing them to consistently raise their own chicks, and hand-rearing has been a last resort at various times. However, successes have increased and from 2006 to 2009, fifteen chicks were parent-raised to adulthood.[673] The Bird Endowment has also marshalled the efforts of many sponsors in engaging *in situ* programs in Bolivia, in conjunction with Asociacion Armonia and various international organizations. Land has been purchased and an on-going program for erecting nesting sites is being carefully monitored. Eco-tourism is promoted and education programs have instilled pride and concern among local people. While population estimates vary, there is optimism that the wild population may begin to increase.[674]

As with public zoo efforts to conserve Guam Kingfishers, the idea of private aviculturists establishing foundations to preserve a macaw, almost unknown to science until the seventh decade of the 20th century, might seem a fantasy, except that such things are, in fact, being done.

Thus, American aviculture is making a vigorous entrance into the twenty-first century, continuing the work of many remarkable people. Viewing what is yet to be achieved in the light of what has so far been accomplished can only benefit it.

Chapter 12. Hummingbirds in European Aviculture

Christoph Hinkelmann

"The family of hummingbirds attracts ornithologists more than any other birds. Only a few other birds resemble them in beauty, intelligence, individuality, and elegance, which is why so many nature researchers observed and studied them, and for a long time bird-keepers dream of keeping them alive in captivity", wrote the famous French-American ornithologist Jean Delacour in 1924.[675] From here it was only a small step to risk attempting to try keeping hummingbirds in captivity. The idea was not developed in the twentieth century, but much earlier, but there are only very few reports on early attempts of hummingbird care in captivity. They all originate in the New World; all of the around 330 recent species are limited to the Americas, occurring between Alaska, Tierra del Fuego and the Caribbean.

First attempts of hummingbird care

The earliest known records of hummingbirds kept in captivity tell us about the Aztecs in Central America. Later reports were written in the eighteenth and nineteenth centuries and originate from the southern parts of present-day United States of America, Jamaica as well as other unspecified places in the Americas. From these accounts it is mostly impossible to determine the species of hummingbird. All agree that the birds stayed alive for only a short time.[676] Hummingbirds are food specialists and feed on nectar of blossoming plants, a liquid containing quickly re-absorbed natural sugars, amino acids and lipids. The only other nutrition components are small insects and spiders. Early hummingbird keepers knew of this diet basis and offered various mixtures of sugar and water, and occasionally insects. Today we know that this artificial food could not cover the requirement of nutrients and resulted in the premature death of the hummingbirds. In 1754, Paul H. Tesdorpf reported that an Englishman living in southern North America was feeding his hummingbirds with honey spread on small twigs. No bird survived this food conversion longer than one month. Other authors mention using, in addition to sugar and water compositions, also diluted honey and biscuit, but none of these resulted in the birds living longer. Artificial food was the most important requirement for keeping hummingbirds in captivity, but finding the right combination was not always easy.

Nineteenth-century reports from the USA mention hummingbirds in cages with small bowls filled with a sugar-and-water composition in a syrup-like concentration, with flowers with sepals of conspicuous colors. The birds survived this care for a period of a few months to a half year. However, when kept in a larger cage with gauge allowing small insects to get inside, and fed by diluted honey or dissolved sugar, they lived for

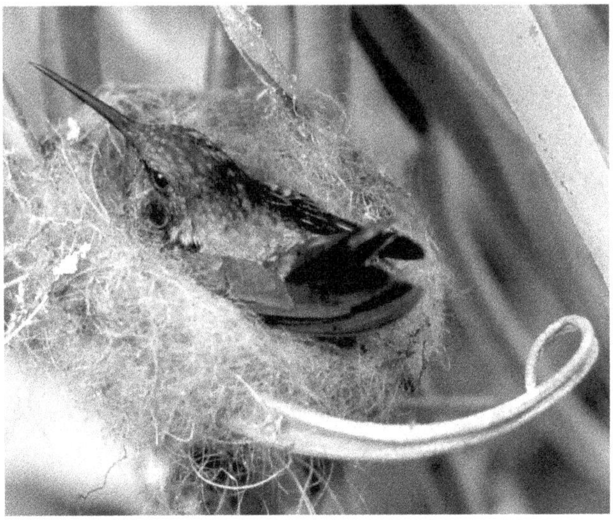

Figure 50: A nesting Sparkling Violetear (*Colibri coruscans*)

twelve months. In Peru, the Swiss explorer Johan Jakob von Tschudi observed a native girl feeding a Violet-fronted Brilliant (*Heliodoxa leadbeateri*) with finely crushed pieces of sugar offered between her lips. Probably the bird would survive this nutrient-poor food, only because it could use the natural supply of flowers.[677]

The history of hummingbird breeding and care was mainly written in Europe. The most important challenge was the transportation across the Atlantic. In the periods preceding intercontinental flights, the journey took weeks, depending on the power of sailing ships or steam vessels. The first reports of hummingbird transports are probably the result of fantasy, but the first documented imports seem to be almost unbelievable as well. In 1837, an Englishman found a breeding hummingbird just a short time before returning to Europe and spontaneously decided to take it with him. It was reported that the female continued to sit on her two eggs, apparently accepting the kidnapping. She was fed on honey and water. The mother died soon, but both young hatched and arrived in England, where they were kept for a short period.[678]

Reports of the last decades of the nineteenth century are more trustworthy. Dozens of hummingbirds were brought to England and France; some of them survived for several months. They were fed on syrup and animal protein, but there is no indication on the origin of the latter. Harro Strehlow reported that the first humming-bird imported alive to Germany arrived in Hamburg in 1879 and could be admired in the famous Carl Hagenbeck menagerie. Unfortunately, notes on the species of hummingbirds as well as the time these birds stayed alive are lacking.[679] The better documented reports of hummingbird transports to Europe are known from the first decade of the twentieth century. A Sparkling Violetear *(Colibri coruscans)* presented in London in 1905 was the first hummingbird in a European zoological garden. A total of thirteen different species lived there until 1927. In 1913, five of ten Cuban Emeralds *(Chlorostilbon ricordii)* delivered from Cuba, arrived alive in Hamburg, probably the first hummingbird import for sale in Germany. They were fed on sugar and water compositions, honey and small insects. Attempts were also made to keep hummingbirds alive with mixtures of honey, evaporated milk without sugar and *Mellin's Food*, an infant food from England dissolved in boiling water. Delacour noted: "Anyway, under natural conditions hummingbirds are probably feeding on insects and they suck the nectar from the flowers only for digestion and in addition to their food. This is known by experience, and confirmed by observations of these birds in liberty as well as in captivity."[680]

Heyday and a new start

There were no regularly imported hummingbirds before World War I. Jean Delacour reported 1924 on three birds, Cuban Emeralds *(Chlorostilbon ricordii)* and Bee Hummingbirds *(Mellisuga helenae)*, brought alive to Hamburg in 1923.

Unfortunately very little was documented of these arrivals. In 1932 many hummingbirds came to Germany. They were sold to the zoological gardens of Berlin, Cologne, Leipzig, Halle/Saale, Hamburg and Frankfurt am Main. No fewer than seventy birds arrived in Berlin Zoo and were offered for sale: "The board of directors is willing to supply a few hummingbirds on moderate terms."[681] The hummingbirds were nourished following the instructions of the importer, Heinrich Danisch, for a successful transport:

"One third of the food is composed of *Liebig's Fleischextrakt* [a meat extract from Germany] dissolved in hot water [...] In addition, four teaspoons of evaporated milk stirred cold, and five to six spoonfuls of honey from tree flowers, four heaped spoonfuls of *Mellin's Food*, eight teaspoonfuls of cane-sugar.

Every seven to eight days a point of a knife of medical carbon will be added to the mixture. Mr. Danisch considered it as appropriate to offer the acclimatized hummingbirds only honey and water composition from 5:00 pm to 5:00 am. The feeding of this liquid takes place in tubular glass bottles of about 10 cm length, closed above by a cork stopper. Just above their bottom, these bottles have a spout of about 15 cm length, slightly tilted upwards, where the hummingbirds suck their food."[682]

Publications of 1932 indicate as well the first sales to private hummingbird keepers in Germany, but without species determination. Reports on their care were provided by the Berlin and Cologne Zoos, mentioning details of keeping and nouris-ment as well as many observations, just in the same manner as we are used to present day publications in bird-keeping magazines, presenting experiences to ordinary people. Although records are lacking, we can be sure that hummingbirds were regularly imported to Germany in the fourth decade of the twentieth century, and many of them were bought by private individuals.[683]

Many interesting experiences were made in Breslau Zoo (today Wrocław in Poland) with five species of hummingbirds, among them Jamaican Mangos *(Anthracothorax mango)*, Black Jacobins *(Florisuga fusca)* and Swallow-tailed Humming-birds *(Campylopterus macrourus)*. Some of them lived for as long as two years. They were fed by a special nutriment solution created by keeper R. Wiesner (honey, Indian cane-sugar, *Nestlé's Infant Meal* from Switzerland, unsalted meat extract, warm water and living *Drosophila* flies). The keeper discovered that hummingbirds in the European winter conditions require longer days, i.e. elongated light periods, to have sufficient time to feed.[684]

World War II interrupted the encouraging improvements in European and in particular German hummingbird care. The zoological garden in Wuppertal received hummingbirds from a German living in Recife, Brazil, and by 1952 Frankfurt/Main Zoo could exhibit hummingbirds again. The interruption of experience in hummingbird care prevented for a long time reaching former successes, in particular extending the period of keeping birds alive in captivity. Once again most imported hummingbirds survived no more than a few weeks or months.

Longer periods occurred, but remained rare exceptions; the director of Copenhagen Zoo, Axel Reventlow, reported 74 birds of 14 species having stayed alive for many months, some of them even years.[685] In the 1950s, hummingbird enthusiasts could only visit zoos to see their favorite birds in Germany, or read reports from their American homes. In Vienna, however, the team of the Biological Station Wilhelminenberg caught up much more quickly with the pre-war situation. Otto Koenig presented in 1959 a large overview on their experiences in the collective care of hummingbirds and sunbirds "since many years". He reported on behavior, feeding, size and technical requirements of cages and aviaries, temperature, procurement, and many other topics. For the first time it was strongly recommended to use evaporated milk without any sugar, but never fresh or condensed milk with sugar in the food composition.[686]

The successful care of hummingbirds cannot be determined merely by the period an individual bird or a certain species survives in captivity. The challenge was to successfully breed them under controlled conditions. Given the problems of reliable artificial food and all the questions, the breeding of hummingbirds in captivity appeared utopian. Still, in 1938 Georg Steinbacher reported that a female Swallow-tailed Hummingbird

(Campylopterus macrourus) laid a single, naturally not fertilized egg in the Berlin Zoo. The first successful breeding of a hummingbird in captivity probably happened in 1959 in Cleveland Zoo, Ohio, where a young Red-billed Streamertail *(Trochilus polytmus)* fledged.[687]

No earlier than in the 1960s, did hummingbirds regularly come into the hands of private keepers. These persons (some of them became successful scientists) took over the leadership in hummingbird breeding and care. After only a couple of years, zoos, institutes, and bird parks profited from the amateur experiences.

The enthusiasts dedicated their spare time to the development of optimal keeping conditions for hummingbirds.[689] The publications of Walter Scheithauer in 1966 signified a milestone in this process. He reported on all facts hitherto known, and presented a well-tried food composition extrac-ted from a great variety of recipes from literature, zoos and private hummingbird keepers. He also provided practical experience and recommendations, as well as superb color photographs of many species in captivity, and concluded: "I consider hummingbirds to be perfect cage-birds."

As a consequence, it came as no surprise that many bird-keepers in the late 1960s took on hummingbird care. Bird dealers and importers offered them regularly. Due to export controls in many American countries, the hummingbirds mainly came from the Andean states of northwestern South America. Although there were still many losses, in particular within the first months of keeping, the regular supply provided enough birds. As an example for the experiences of many hummingbird keepers of those times, Manfred Wittmann wrote: "I kept most [...] hummingbird species no longer than a few months, some of them even only for a few days. My personal record was an *Eriocnemis luciani* [Sapphire-vented Puffleg] that stayed alive for one year and seven months."[690]

Regular keeping conditions

The breakthrough to a "normal", regular and successful keeping of hummingbirds came from one man, Karl-Ludwig Schuchmann. A student in Frankfurt/Main, he started hummingbird care in the late 1960s and suffered from the same failures as probably all other keepers. He would not accept this situation and collected experience reports from all over the world, dissected dead hummingbirds to find out the causes of death, and tested new food compositions.

In 1972 he presented the first results on hermit hummingbirds, a group of hummingbirds extremely difficult to keep alive in captivity. In the same year, Schuchmann published a study including scientific data of life-spans of hummingbirds

Figure 51: Ruby topaz (*Chrysolampis mosquitus*)

in captivity, resulting in the statement that mistakes in diet caused the preliminary death of many birds. This data collection was a second milestone, calling for improvements.[691]

The study, however, was not complete. The author focused primarily on the experiences of private keepers and zoological gardens in Germany, Great Britain, the Netherlands, Denmark, France and the USA. A very important contributor to this data collection was Anthony Mobbs from England, who also published his experiences in the *Avicultural Magazine* in the 1970s. He later summarized his results in a book.[692]

Schuchmann's study lacks data of many private hummingbird keepers (presumably they did not provide precise information), and in particular experiences published in the pre-war literature. Many hummingbird friends misunderstood and wrongly criticized Schuchmann's study, for instance Dieter Poley, who yet rightly pointed to the neglected experiences of many years in Copenhagen Zoo.[693]

As a result of a reader's question, Schuchmann published again in 1972 the most important facts of his findings on hummingbird diet. These were avoidance of any ingredients containing milk, the suitable relationship and concentration of sugar (disaccharides), vitamins and minerals, as well as the use of tinted tubular feeding bottles made of glass to reduce the decomposition of B-vitamins caused by light.[694]

Based on these results, Schuchmann developed an artificial food which could be dry-stored and easily prepared by stirring. It was produced and sold commercially, a large breakthrough in the care of hummingbirds. Now it was possible to feed the birds adequately and in agreement with their natural needs, and thus hummingbirds survived for ten and more years in captivity. Besides, breeding success was achieved under man-made conditions. Scheithauer reported in 1975 on a successfully reared hybrid of a female Fork-tailed Woodnymph *(Thalurania furcata)* and a male Amazilia Hummingbird (*Amazilia amazilia dumerilii*).

It was probably the first breeding success by a private hummingbird keeper in captivity, at least in Germany. The first successfully reared hummingbird in a Continental European zoo was a Red-billed Streamertail *(Trochilus polytmus)* in Wuppertal 1977.[695] In the following years, private keepers as well as zoological gardens reported on breeding of several species, a remarkable, almost incredible success with regard to the history of hummingbird care. Criticizing the new development, Koenig commented that hummingbirds "appear to become the new fashion trend of record effort in animal care in captivity".[696]

Scheithauer preferred keeping hummingbirds in a glass case or large conservatory or greenhouse with many tropical plants, and many hummingbird keepers followed his recommendations.[697] It took some time before breeders realized that several or many hummingbirds living together in a single room, independently of the cage or aviary size, generated a lot of stress between the birds, and often resulted in an untimely death of individuals.

Keeping hummingbirds in a community required daily observations to remove single, subordinate birds. To avoid the abandonment of keeping many hummingbirds in large aviaries, the owners increased plant density, hoping to reduce the inter-individual aggressiveness, by improving opportunities for the birds to hide. Letting hummingbirds live alone in large cases or aviaries proved to be the best solution to keep them alive for years.

Scheithauer, Poley, Schuchmann and Folger provided information and practical advice for the optimal care of hummingbirds in such captivity. As an essential pre-supposition, in addition to diet and size of their homes, "clean accommodation plus fresh air" was considered important.[698]

If a keeper was lucky to possess male and female of a certain species, he had to look for opportunities to get them into direct contact during the mating season. Due to the limited number of countries permitting exports, there were only a limited number of species available, and it was easier to find both sexes of a given species.

Experienced hummingbird keepers knew the exact time to remove one of the partners, if there was no harmony between them. Even some hermit hummingbirds, difficult to keep, bred and raised young in captivity.[699] The late 1970s and 1980s can be regarded as the second heyday of hummingbird care in captivity in Germany, and probably in other European countries as well. New birds arrived in small numbers and regularly, the major problems of keeping were solved and many hummingbird friends reported publicly on their experiences.[700]

The general situation changed when the European countries enacted new legal regulations as a consequence of the CITES agreement of 1976. For instance, in West Germany, the federal species protection decree of 1986 covered all hummingbird species in general and put them on the list of completely protected species of birds. Private hummingbird keepers were obliged to declare to the responsible nature conservation authority that their birds were bred in captivity. For imported birds, the majority in private collections, this demand could not be fulfilled. Hummingbirds in a personal estate had to be registered and were permitted, but further imports were almost impossible.

The only exceptions to legally imported hummingbirds were those permitted for scientific investigations. Regular supply was stopped. If there were still hummingbirds in private care, their owners in many cases could not fully prove their descent, and the keepers avoided reporting on their care experiences. Scientific ornithology thus lost a major source of information on hummingbird biology and behavior provided by amateur researchers. Due to recent tightening of legal regulations, e.g. the generally prohibited import of birds not bred in captivity to the European Union in 2007, there is little chance that there will be another heyday in keeping and breeding hummingbirds. Keeping hummingbirds in European private collections is an almost closed chapter of aviculture.[701]

Chapter 13. Gaolbirds

The Interconnected Lives of Prisoners and Birds

Benedict Taylor

> In that garden of despair there was a little aviary of singing birds – 'gaol birds.' One of the prisoners looked up as he passed that cage and spoke to his fellow-prisoners, also caged.
> *The Sun,* 1926

> One feels like some rare exotic bird, trapped in an intricate gilded cage; a metaphor not inappropriate as the hammer beamed roofs frequently resound to the flapping of real birds, curious and unfortunate enough to have found their way into these vast basilicas of human discontent.
> *Peter Wayne, 1994*

Londoners fluent in rhyming slang have long referred to time in prison as *bird*. The term is an abbreviation of *birdlime*, a sticky substance used to capture birds. 'Birdlime' rhymes with 'time' and thus, convicts 'did their bird.'[702] We use expressions such as this – *gaolbird* especially – so often that it is easy to forget that we are speaking metaphorically. It is not just a metaphor that goes unnoticed, though; few realize that prisons have provided habitats for a great many real birds, both kept and free.

Given how much food and refuse pass through them and the shelter they offer, prisons have presumably provided habitat for birds for as long as they have for humans. Prisoners kept birds of their own initiative at first, but also as part of official rehabilitation programs in recent decades. Many of the stories of prisoners and their birds are poignant and even touching.

They show how care and intimacy can flourish in even the most hostile environments, and do much to dispel monolithic stereotypes of hardened, heartless prisoners, or "gaolbirds". Images of desperate villains are difficult to reconcile with descriptions of men who tenderly care for small feathered creatures.

The presence in prison of animals of any kind has gone almost unnoticed by historians. Joseph Berkovits, a rare exception, wrote that when Canadian prisoners "were at their loneliest" they tamed rats, birds, and insects.[703] Phillip Priestly noted the solace that prisoners in Victorian England found in "the indigenous fauna of the prison cell"[704], but the matter has not been treated in its own right.[705]

Environmental historians have also neglected the subject, although some have begun to explore what Nigel Rothfels calls the "unnatural histories" of animals. By this Rothfels means animals' "lives and afterlives not in their 'native haunts' but in such human environments as museums, books, circuses and zoos."[706] Intriguing lines of research flow from thinking about the history of animals in this way, including the history of the prison as a habitat for birds. Prisons can, after all, be understood as an unnatural haunts or unusual ecosystems, comprised of unforgiving niches.

In this light, the bonds formed between prisoners and birds appear as sensible adaptations to difficult environments. But birds have not only been prisoners' consoling pets, they have also helped inmates understand, and come to terms with, the experience of imprisonment and the nature of being free.[707]

The Big House

Inside ordinary houses, birds are regarded as omens of bad fortune or even heralds of death. In the 'Big House', however, where all the inhabitants already consider themselves unlucky, and who have often been regarded as suffering a civil "death", birds inside have been small blessings and eased the trauma of incarceration. Even when not kept as pets or companions, they have been a comfort to prisoners. The famous nineteenth-century London journalists Henry Mayhew and John Binney wrote of prisoners at Pentonville Prison listening eagerly to "the voices of the birds telling their love of liberty in a very spasm, as it were, of melody".[708] Over a century later, on his release from Long Bay Prison in Sydney in 1964, a man said that prison "at night isn't as you might think, filled with the clanking of locks and keys or the yells of prisoners…D Hall is silent except for the birds on the adjoining roofs. Men stand on their beds to look through the 3ft x 1ft window with the horizontal bars to look for hours on end at the birds".[709]

'Angelo', a man who spent many years in the prisons of California in the late twentieth century, thought there was not a prison in the world without a flock of visiting birds. "I and many others have come to enjoy these visitations", Angelo wrote, "but I also find them puzzling: why with the option to go virtually any place they want do the birds spend their time hanging out with us? But whatever, the entertainment value of the birds is sometimes priceless as they cavort like comedians, competing for whatever edibles they can snatch up or steal from one another. So each yard has groups of bird lovers who accumulate bags of bread and lunch meats to dispense to what quickly becomes an armada of feathers and noise".[710] Frank Maritt, an English prisoner, was one such bird lover. In 1986 he wrote to his friend Jim Little: "After twenty two years in jail I'm a little sick of male company. There's a little garden outside my cell and I fed the birds during the cold spell. I found out that blackbirds, wagtails will eat bread. I thought they only ate live worms etc. There's also two robins I see regularly. They are the first I've ever seen in my life."[711]

The history of prisoners keeping birds themselves dates back at least as far as the 1840s, when a prisoner at the Norfolk Island Penal Station stole a much-loved parrot with its cage from one of the island's official residents. Interestingly, although the thief was known, he was not punished and the bird was left in his possession.[712] Keeping birds was a common practice in England by the 1880s, when Michael Davitt, a prisoner at Portland Prison, kept a bird in his cell. Davitt, an Irish nationalist, dedicated his *Leaves from a Prison Diary* to Joe, his Blackbird (*Turdus merula*)::

"To the memory of the little confiding friend whose playful moods and loving familiarity helped to cheer the solitude of a convict cell."[713] Davitt recalled how Joe would "stand upon my breast as I lay in bed in the morning and awaken me from sleep. He would perch upon the edge of my plate and share my porridge […] He would 'fetch and carry' as well as any well-trained dog. Towards evening he would resort to… the post of the iron bedstead, and there remain, silent and still, till the dawning of another day, when his chirrup would again be heard, like the voice of Nature, before the herald of civilization, the clang of the prison bell at five o'clock".[714]

When observers ventured inside the prison, they were often struck by the care prisoners bestowed on their birds and by how well-trained birds like Joe were. "It is curious how the convict loves a pet", wrote a reporter for the Melbourne *Argus* in August 1920: "Several canaries chirped happily in the captivity they shared with their masters, and a blackbird fluttered on to one prisoner's wrist." [715]

But while outsiders were impressed by the ways in which prisoners trained their pets, prisoners themselves often noted the way in which a bird could change a fellow-inmate's behavior. From 1896, in New South Wales (NSW), Australia, prisoners were permitted to care for birds sent by friends or relatives.[716] Grant Hervey, a journalist with a colorful history, was a prisoner at Goulburn Prison during World War I. There he met a man known as 'Twenty' after his prison number. Twenty was "a down-to-date Sydney larrikin", possessed of a foul mouth. The two men worked together in the boot shop at Goulburn and after a short time Hervey began to tire of Twenty's language. Another prisoner, known as Sparrow Jack, also worked in the shop and had a number of tame Canaries (*Serinus canarius*). Hervey persuaded Jack to give to him a Canary, which Hervey then gave to Twenty on the condition that he "cut out the language".[717] The plan worked like a charm: "It was wonderful to see that little bird living its merry life in gaol…

Figure 53: Canary with broken bones. Drawing by Robert Stroud

Whistle! How that bird whistled! I have repeatedly seen the whole shopful of human nondescripts momentarily give over working, just to look at Twenty and his bird. And when the little fluffy, yellow-coated creature would hop down off his perch […] with his head on one side and looking so cunning, whenever the canary did that Twenty's happiness was complete […] Deep down beneath the ununderstandable [sic] of Twenty's nature there was something that was good. It required the canary to bring it out."[718]

Hervey was a man with grand ambitions and he felt the stigma of imprisonment sharper than most. He had deeply personal reasons to seek to humanize prisoners through his writing, but his story of Twenty and the canary is only one of a number of such accounts. Vance Marshall was one of Hervey's fellow-prisoners at Goulburn, and also a writer. He too was struck by effect of birds on prison life. They lent "a gleam of sweetness to a life of monotony, gruelling and grovelling", he wrote. In the workshops their "trill rang the day long above the racking clang of the anvil, the sullen drone of the grindstone, and the dreary rip of the saw". Prisoners are accustomed to regarding timepieces as oppressive devices, yet at Albury Prison (NSW), where Marshall was confined for a time, a pair of Welcome Swallows (*Hirundo neoxena*) would ride the hands of the main clock, "with laughable unsteadiness", freeing the watching prisoners briefly from the mirthless tyranny of the clock. And just as the prisoners felt an affinity with their fellow gaolbirds, so too did the birds seem to prefer the prisoners over others. An Australian Magpie (*Gymnorhina tibicen*) perched near the main gates at Albury and regularly swooped down on those prisoners, newly arrived, who had not yet donned their prison uniforms.[719]

Marshall marveled, as Hervey had, at the effect the little feathered creatures could have on hardened characters. When a Willie Wagtail (*Rhipidura leucophrys*) was seen "skipping on the sill of the workshop window" then burglars and forgers and "even the old 'lifer', who long ago killed his mate" would lay down their tools and say "Hush!... Let him be, and he might come back again."[720]

Marshall was touched by the care prisoners bestowed on their birds; he took it as a kind of badge of honor: "When I have watched them take the fluffy young in the palms of their tainted hands, and have seen the tiny things croon responsively content [...] I am proud to think that MY name, MY finger-prints, and MY record stand inscribed in the annals of crime with theirs. It warms my heart to still recall the innate goodness which they were not, and still are not, ashamed to express." Jack Garvey was another who knew the inside of NSW's gaols in the early twentieth century. Unlike Hervey and Marshall though, he spent time at the dreaded Berrima Prison, infamous for holding the worst of the state's prisoners and for driving them mad. But birds softened the brutality of life at Berrima, too. "I have known a sparrow perched on the window-sill to help stay the craving for love in a convict's heart", Garvey wrote.[723]

Individual prisoners came and went, but the presence of birds in and around prisons remained a constant throughout the years, smoothing – for some at least – the rough edges of the aesthetics and experience of imprisonment. In 1960, a journalist for the *Sunday Telegraph* inspected a number of cells at Goulburn Prison and noted how the prisoners were "encouraged to make them as homely as possible". One of these cells, occupied by a man serving a life sentence, was decorated with numerous paintings and photographs, and "an even more homely touch – he kept a budgerigar!"[724] The McLeod Prison Farm in Victoria was located on French Island in Westernport, an island famous for its rich and varied bird life. In 1965, an inmate at McLeod attended so anxiously to three young hatchlings that another inmate was prompted to write that "the new arrivals are assured of at least one doting grandparent".[725]

Robert Stroud, Birdman

The best-known prison bird-keeper was an American, Robert Stroud (1890–1963). A murderer known to many as 'the Birdman of Alcatraz', Stroud actually never kept birds at Alcatraz. It was at Leavenworth Penitentiary in Kansas, where he served most of his sentence that he became enchanted with the creatures. In solitary confinement, he bred and raised hundreds of Canaries and sparrows and eventually became a self-taught expert on bird pathology. From the moment he rescued a pair of injured baby sparrows from the exercise yard in 1913, Stroud's life revolved around birds. According to his biographer, Stroud "crowded his solitude with the hottest and most amazing vitality on earth – that of living birds". He "tended them like a father", watched "his little prison family" thrive, and felt "some part of him that had been starving was richly fed". Birds filled his isolation: "Since he was excluded from people, his people were his birds. Birds were his family, wife, children and afternoon matinee."[726]

Stroud became obsessed with finding cures for the ailments of his birds. He read hungrily on the subject and conducted numerous experiments. He contributed to bird enthusiasts' magazines and kept up a lively correspondence with other breeders. He even published two books on bird pathology; the second, *Stroud's Digest on the Diseases of Birds*, became a standard work on the subject.

Under the heading 'Post-Mortem Examina-tions', in an otherwise dry and practical manual, he shed the language of science and plainly said what his birds meant to him: "Years of work, of study, of careful observation; the lives of literally thousands of birds, the disappointment and heartbreaks of hundreds of blasted hopes have gone into these pages [...] have had birds die in my hand when their death brought me greater sadness than that I have ever felt over the passing of a member of my own species. And I have dedicated all this to the proposition that fewer birds shall suffer and die because their diseases are not understood."[727]

Stroud's biography was widely read in Australian prisons. Reviewing the book in *Neptune News*, the magazine produced by inmates at the McLeod Prison Farm, one prisoner thought it "little wonder" that Stroud had tried to commit suicide after his birds were taken from him.[728]

Birds continue to console those in the unnatural environment of the prison to this day, and not only prisoners. Tim Duggan worked as an interrogator at Abu Ghraib prison in Iraq in 2003 and 2004 and was featured in the documentary made by Eroll Morris about the abuses that took place in the prison after the Americans assumed control. Dugan was an experienced interrogator, yet he found the practices at Abu Ghraib deeply disturbing. Struggling to find something of a natural pattern in his environment, he took to watching the birds roost in the date palms outside the prison walls.[729]

Birds and rehabilitation

Images of prisoners raising birds or teaching them tricks were good for prisons departments' public relations. Even better were those occasions where prisoners seemed to have genuinely changed for the better through their contact with birds. From the early twentieth century Australian penal systems, like others around the world, established prison farms where inmates worked on their honor, without surrounding walls, often raising poultry and other birds.[730] Later, bird-rearing programs were introduced into conventional, walled prisons, and aviaries became increasingly common prison features.

In 1949, a prisoner who had been sentenced to ten years in prison for a violent crime was released from Emu Plains Prison Farm near Sydney after serving only two and half years. He declared his intent to settle on the land and establish a chicken farm. "I got the inspiration for chicken farming at Emu Plains", he said. "Most of my work was taken up with feeding and looking after the farm's fowls". His time at the prison farm was, he said "the best part of my imprisonment [...] I spent most of my spare time with the chickens – cross breeding and studying them. It was fascinating work".[731]

In 1964, at Palen Creek Prison Farm in southern Queensland, Superintendent L.C. Robson, offered a favorable assessment one of his long-term prisoners to the parole board, making special reference to the inmate's interest in the farm's poultry. Robson informed the parole board the prisoner displayed "a keen interest in Veterinary matters [...] I do not know any reason that would suggest that he would not honour any obligations imposed should parole be granted".[732] A short film produced by the NSW Department of Prisons not long after the poultry-farming prisoner was released from Emu Plains depicts inmates milking cows and handling newly-hatched chicks at Emu Plains. The prisoners are, we are told, "learning the love of animals for the first time".[733] The image of the young bird enclosed in the hand of the prisoner was clearly a powerful one. For the prisons department it was evidence for the positive affects of a benevolent penal system, while for Vance Marshall it had spoken, equally powerfully, to the gentleness in prisoners that remained unbroken despite their imprisonment.

Not all observers saw the relationships between birds and prisoners as entirely healthy. Some perceived an ironically cruel twist in prisoners keeping and caging small animals. In 1968, Eric Price, the secretary of the South Australian Offenders Aid and Rehabilitation Services, described a number of the innovations at Cadell Prison Farm, near the Murray River. Prisoners were allowed to "tend native birds, which with the consent of authorities they have robbed of their freedom and keep in cages which permit of no flight".[734]

Despite the rhetoric, "learning the love of animals", was not always sanctioned, and could occasionally actually lead prisoners into further trouble. In 1931, inmates from the Bombala Prisoners' Afforestation Camp in southern NSW took with them native animals, including parrots, that they had raised whilst confined at the camp. Others, upon being transferred to other prisons, left their pets in the care of their fellow-prisoners, hoping to reclaim them later. These practices, the Comptroller-General insisted, "must cease". The area around the camp was a proclaimed sanctuary under the terms of the Protection of Fauna and Flora Act, and the prisoners' actions had come to the attention of the local police. Those caught "had been arrested, fined, and had their licence revoked". Their fondness for their pets had lead them back to prison.[735]

Beginning in the United States, animals gradually became incorporated into official prison rehabilitation programs from the 1970s, as authorities sought greater control over prisoner-animal interactions. Most of these programs involved birds. Interestingly, this development began through one prisoner's initiative. In 1975, an inmate at the Lima high-security hospital for the criminally insane in Ohio, found, as Robert Stroud had, an injured sparrow in the yards. Animals were not permitted in the facility, but the prisoner hid the bird in a broom closet and together with the others in his block – the most troubled individuals in the institution – fed and nursed the bird back to health. Dramatic improvements in the inmates' behavior and happiness were observed, which prompted a year-long study on the benefits of animal therapy in the facility. Two wards of the same size run along the same lines, and the same types of inmates were compared. In one, the inmates were given pets, while in the other they were not. Those with pets needed only around half the level of medication required by those without pets. The ward with pets also saw less violence and there were no suicide attempts. Eight suicide attempts took place in the ward without pets during the trial period.[736]

Australian prisons were slow to initiate their own animal therapy programs. The partnership beginning in 1994 between the Won Wron Prison Farm in Victoria and a local animal sanctuary was the first. At Won Wron, prisoners in the 'Raptor Release' program cared for injured birds of prey and prepared them for their return to the wild. As Rod Wise, the Director of Victoria's Department of Prisons noted, "the parallels between the situations of the birds and the prisoners who looked after them, rehabilitating them and then setting them free, were lost on no-one".[737] For the first four years of the program, not one of the prisoners involved returned to prison. Deputy-Governor of Won Wron, Steve Bergervoet wryly noted: that "if we had that kind of success rate across the prison system, officers like us would be out of a job".[738]

A particularly successful prison bird-rearing project was developed at Pollsmoor Prison in South Africa in 1998 and is still in operation. Prisoners hand-raise the birds, mostly Peach-faced Lovebirds (*Agapornis roseicollis*) and Cockatiels (*Nymphicus hollandicus*), which are then sold to the public or given to children in hospital. The architect of the scheme, Wikus Greese, was the deputy director of prisons, and a founding member of the Western Cape Parrot Breeders Association. Greese reportedly formed the idea after reflecting on the role of birds in his own life (not after watching *The Birdman of Alcatraz*, the film based on the life of Robert Stroud, as some newspapers reported).[739] "I have never met a person who doesn't need love or respond to nature", said Greese.[740] Warders reported that violence amongst inmates in the so-called 'bird block' dropped steeply, while according to inmate Johannes Adams, the experience of being part of the project was a transformational one. He no longer swore, whereas he had, he said, been "full of swear words" before. After 33 years in prison, he had found, as Stroud, a surrogate family in the birds. "More and more the love of birds took a place in my life", he said.[741]

Even where aviculture was not a formal part of a prison's rehabilitation program, aviaries became increasingly incorporated into prison design during the late twentieth century. The results, however, have not always been happy. Plans for a new Remand Centre in Melbourne drawn up in the mid-1980s featured caged white birds in the visitors' contact yard as "a metaphor for the function of the establishment".[742] An aviary of native Australian birds was installed at the Woomera Detention Centre, too. Situated in the middle of the South Australian desert, Woomera's purpose is to hold asylum seekers while their claims are processed. Nevertheless, it is a prison by any other name, with individuals locked against their will behind walls, bars and fences, often for up to several years. Najaf Mazari, a refugee from Afghanistan, was held there from 2001–2002. He was puzzled by the aviary: "Perhaps keeping birds is a hobby of one of the security guards, or perhaps the people in charge of Woomera think it is pleasant for the refugees to study these birds. But when we study these birds, we have only the one thought, and the thought is this: How wrong to keep these birds from the life they were made by God to live! They fly from one end of the cage to the other, or hang by claws of their feet to the wire and cry out."[743]

It was made clear to the detainees at Woomera that any attempt to interfere with the aviary would be regarded as a serious matter. Nevertheless, a number of them smashed the lock from the cage and opened the door. Later, Mazari joked that he and his fellow detainees had granted "visas" to the birds.[744] Deprived of any real relationship with the birds, the inmates found the aesthetics and the metaphors of aviculture confronting. After his claim to be a refugee was recognized by the Australian authorities, Mazari was released from Woomera and settled in Melbourne. Each Saturday he visited a local market, where a vendor sold caged Canaries. On each visit, Mazari bought one of the birds and took it back to his shop, where he re-enacted the release from Woomera: "First I watch the bird jumping about in the cage, and I whistle too amuse him, or maybe it is a her. Then I go to the back of my shop and wish the bird good fortune. I open the cage and let the bird fly away. It is so good for my heart! I see the bird fly swiftly to a tree or to the top of the fence, then he looks around as if he can hardly believe that he is free. Off he flies, I hope to a good life."[745] Mazari acknowledged the risk that the Canary would face that "another bird will have him for dinner", but felt sure "he would choose freedom, even if he knew the danger".[746]

Birds and the nature of imprisonment

When prisoners felt that they were given the opportunity for meaningful interaction with birds, the consequences were often profound and went well beyond companionship, therapy or training. In the outside world, birds have often been "instruments of epiphany" and it is an observation that rings true in prison, too.[747] Birds feature frequently in prison literature. Kept and free, they provided a spectrum of comforting tropes. James Montgomery, imprisoned in 1790 in York castle for defamation, ruminated on his circumstances in a series of poems inspired by birds, including *The Captive Nightingale* and

Soliloquy of a Water-Wagtail.[748] Many prisoners felt imprisonment to be such a de-humanising experience, that they were forced to imagine themselves as birds in order to understand it.[749] The years spent observing the effects of captivity on his birds eventually prompted Robert Stroud to turn his attention to penology. He completed a monumental history of the United States penal system, only to see it suppressed by the Bureau of Prisons: "Writing about jailbirds in steel cages was less welcome in a prisoner than writing about pet birds in wire ones."[750]

A handful of prisoners saw birds as harbingers of a world without crime and punishment, a place where they could be free not only of the prison walls, but of the stigma of being a 'gaolbird'. For Vance Marshall, writing in Goulburn prison, the prisoners' love for their birds indicated to him their "unashamed nature love", a sign of an unbroken spirit, one that remembered the good in the world despite its own sins and misfortune: "It is not confined to birds, this yearning tenderness, but extends to the hills, the rivers, the trees – all that is no longer part of our lives." In nature prisoners found "the only friend in whom our faith remains unshaken beyond these walls which crush our manhood from us", a friend oblivious to the past.[751]

Quoting the words Oscar Wilde wrote in a cell in Reading Gaol, Marshall concluded: "Society, as we have constituted it, will have no place for me, has none to offer; but Nature, whose sweet rains fall on unjust and just alike, will have clefts in the rocks where I may hide, and secret valleys in whose silence I may weep undisturbed. She will hang the night with stars so that I may walk abroad in the darkness without stumbling, and send the wind over my footprints so that none may track me to my hurt: she will cleanse me in great waters, and with bitter herbs make me whole."[752]

At the end of the twentieth century, birds could still represent transcendent hope. Former prisoner Leonard Peltier wrote of a pigeon, "unfettered as the wind", that visited his cell window: "To me this jailbird, this pigeon is as holy a messenger as the eagle. He tells me of the Sky World beyond the steel and cement and razor wire."[753] If birds represented an escape from society and its judgments, they also helped prisoners hold on to a sense of a normal, social world. Indeed, birds have long stood as emblems of human qualities. Anthropologist Claude Lévi-Strauss noted that despite obvious differences between human and bird communities, birds are often given common human names while animals such as cats and dogs are not. Bird wings, feathers, eggs and ability to fly set them apart from humans, but as Lévi-Strauss suggested, this very independence means that the world of birds appears to us like another society, homologous to that in which we live. Birds love freedom; they build themselves homes in which they live a family life and nurture their young; they often engage in social relations with other members of their species and they communicate with them by acoustic means recalling articulated language.[754] If birds stand in the human imagination for freedom, family, home and speech, it is not difficult to understand why they have become so important in the lives of prisoners, from whom all these things have been taken. "I have always loved birds", Najaf Mazari wrote. "They are intelligent, they work hard to make a living, and they find time to sing."[755]

Conclusion

Prisoners are often casually condemned as "animals" but behind the walls, real animals, especially birds, have for centuries been a rare locus of care and selflessness in an otherwise bleak environment. Birds enabled prisoners to imaginatively escape the prison and made the raw facts of captivity a little less grueling. Caged, birds were captives, as prisoners wished to see themselves

– blameless and born for freedom. They were a way for some prisoners of taking responsibility for something, when all other responsibilities had been taken from them.

For others, birds took the place of friends or family. Free birds were a tangible link with the outside world, a world beyond the prison that still held a place for them. Prisoners forged their connections with birds long before official animal rehabilitation programs were introduced; this was, in a sense, a "bottom-up" movement in penal reform. For a few prisoners, keeping birds and writing about them, offered more than just solace or companionship, these became ways to grapple with the experience of imprisonment. These stories of gaolbirds caring for animals of other species add to our understandings of the "unnatural histories" of birds and humanize our understandings of imprisonment.

References

Chapter 1

Adams, H. G. (1854): *Cage and Singing Birds: How to Catch, Keep, Breed and Rear Them.* London.

Albin, E. (1738): *Natural History of Birds* vol. 1. London.

Albin, E. (1740): *Supplement to the Natural History of Birds.* London

Alves, R. R. d. N. & E. E. G Nogueira, (2010): Bird-keeping in the Caatinga, NE Brazil. *Human Ecology* 38: 147–156.

Aristotle, (1991): *History of Animals. Book* VII–X. Cambridge, MA.

Barton, F. T. (1912): *Pheasants in Covert and Aviary.* London.

Bechstein, J. M. (1795): *Naturgeschichte der Stubenvögel: oder Anleitung zur Kenntniß und Wartung derjenigen Vögel, welche man in der Stube halten kann.* Gotha.

Belda, A., J. E. Martínez-Perez, C. Martín, V. Peiró & E. Seva, (2010): Plants used to capture and sustain wild **finches** (Fringillidae) in Southeast Spain. *Economic Botany* 64: 367–373.

Belon du Mans, P., (1555): *L'histoire de la nature des oyseaux.* Paris.

Bender, M., D. P. Sponenberg & D. Bixby, (2000): *Taking Stock of Waterfowl: The results of the American Livestock Breeds Conservancy's Domestic Duck and Goose Census.* Pittsboro, N.C.

Berg, G., (1937): Lejon, kameler, apor och andra exotiska djur i det gamla Stockholm. In *Svenska kulturbilder.* Ny följd, vol. 5, eds. S. Erixon & S. Wallin. Stockholm. Pp. 313–336.

Berg, G., (1980): The collection of wild birds and nestlings in Sweden. In *Heiðursrit til Sverra Dahl 70 ár 31. Mars 1980,* eds. H. D. Joensen, J. P. Joensen & J. H. W. Poulsen. Tórshavn. Pp. 132–157.

Bergman, S., (1960): The smallest parrots of the world. *Avicultural Magazine* 66: 209–215.

Bergström, E., (1933): *Fågelliv i städernas parkdammar och vattendrag.* Hudiksvall.

Bezzel, E. & R. Prinzinger, (1990). *Ornithologie.* Stuttgart.

Bielfeldt, H., (2008): *Kanarien: Gesangkanarien, Farbenkanarien, Positurkanarien, Mischlinge.* Stuttgart.

Bigwood, J. M., (1993): Ctesias' Parrot. *The Classical Quarterly* N.S. 43: 321–327.

Birkhead, T., (2004): *The Red Canary: the Story of the First Genetically Engineered Animal.* London.

Birkhead, T. R., K. Schulze-Hagen & R. Kinzelbach, (2004): Domestication of the canary, *Serinus canaria* – the change from green to yellow. *Archives of natural history* 31: 50–56.

Blake, W., (1989): *The Complete Poems.* Edited by W.H. Stevenson (= Longman Annotated English Poets 9), London.

Blechman, A. D., (2007): *Pigeons: the fascinating saga of the world's most revered and reviled bird.* St. Lucia.

Boehrer, T. B., (2004): *Parrot Culture: Our 25000-Year Long Fascination with the World's Most Talkative Bird.* Philadelphia.

Boettger, C. R., (1958): *Die Haustiere Afrikas.* Jena.

Borowski, O., (1998): *Every Living Thing: Daily Use of Animals in Ancient Israel.* Walnut Creek, CA.

Boyd, B. W., (1987): The death of Corinna's parrot reconcidered: poetry and Ovid's *Amores. The Classic Journal* 82: 199–207.

Brehm, A. E., (1872) : *Gefangene Vögel. Ein Hand- und Lehrbuch für Liebhaber und Pfleger einheimischer und fremdländischer Käfigvögel* vol. 1.Leipzig.

Brooks, D., (1995): *Love Birds: Unlocking the Secrets of The Emperor's Bird Scroll.* New York.

Brusewitz, G., (2001). De hjälpsamma djuren. In *Människan i naturen: etnobiologi i Sverige* vol. 1. Stockholm. Pp. 119–126.

Chaiklin, M., (2005): Exotic-bird collecting in early-modern Japan. In *JAPAnmials: History and Culture in Japan's Animal Life,* eds G. M. Pflugfelder & B L. Walter. Ann Arbor MI. Pp. 125–160.

Chekhov, A. (1922). *The Tales of Chechov,* vol. 12. London.

Darwin, C., (1861): *On the Origin of Species by Means of Natural Selection.* New York.

Davis, K. (2001): *More Than a Meal: The Turkey in History, Myth, Ritual and Reality.* New York.

Donkin, R. A. (1989): *The Muscovy Duck* Carirna moschata domestica: *Origins, dispersal and associated aspects of the geography of domestication.* Rotterdam.

Donkin, R. A. (1991): *Meleagrides: an historical and ethnogeographical study of the guinea fowl.* London.

Drews, C., (2001): Wild animals and other pets kept in Costa Rican households: incidence, species and numbers. *Society and Animals* 9: 107–126.

Dundes, A., (1982): Releasing caged b**irds** in **Iran.** *Midwestern Journal of Language and Folklore,* 8:116–119.

Dundes, A. ed., (1994): *The Cockfight A Casebook.* Madison.

Dyson, C. E. (1878): *Bird Keeping: A Practical Guide for the Management of Singing and Cage Birds.* London.

Edwards, G. (1747–51): *A Natural History of Birds* vol. 1–4. London.

Erkes, E., (1942): Vogelzucht im alten China. *T'oung Pao* 37:15–34.

Farrar, C. D., (1910): *Birdroom and Aviary: Trials and Triumphs of a Yorkshire Parson.* London.

Festing, S. (1988): Menageries and the landscape garden. *Journal of Garden History* 8: 104–107.

Finn, F., (1915): *Garden and Aviary Birds of India.* Calcutta.

Finsch, O. (1867): *Die Papegeinen* vol. 1. Leiden.

Fuller, R. A., J. P. Carroll & P. J. K. McGowan, (2000): *Partridges, Quails, Francolins, Snowcocks and Guineafowl.* Cambridge.

Gillmore, R. M., (1950): Fauna and ethnozoology of South America. In *Handbook of South American Indians* 6. Washington D.C. Pp. 345–463.

Gould, J., (1865): *Handbook to the Birds of Australia* vol. 2. London.

Green-Armytage, S., (2003): *Extraordinary Pigeons.* New York.

Grindol, D. & T. Roudybush, (2004): *Teaching Your Bird to Talk.* Hoboken, N.J.

Grundel, W., (1993): *Brieftauben.* Stuttgart.

Gunda, B. (1969): Die Jagd und Domestikation des Kranichs bei den Ungarn. *Anthropos* 63–64:473–496

Hams, F. (2008): *Old Poultry Breeds* (=Shire Library). Princes Risborough.

Harper, J., (1972): The tardy domestication of the duck. *Agricultural History* 4: 385–389.

Hesse, B., (1980): Archaeological evidence for Muscovy Ducks in Equador. *Current Anthropology* 21: 189–190.

Holden, G. H., (1888): *Canaries and Cage-Birds: The Food, Breeding, Diseases and Treatment.* New York.

Holland, G., (2007): *Encyclopedia of Aviculture.* Surrey, B.C.

Hornell, J., (1947): Egyptian and Medieval pigeon houses. *Antiquity* 21: 182–185.

Humboldt, A. v., (1821): *Reise in die Aequinoctial-Gegenden des neuen Continents during the years 1799–1804,* vol. 5. London.

Ives, P., (1951): *Domestic Geese and Ducks.* New York.

Jackson, C. E., (1997): Fishing with cormorants. *Archives of Natural History* 24: 189–211.

Jenyns, S., (1930): The Magpie Robin as a cage bird. *The Hong Kong Naturalist* 1: 51–52.

Jepson, P., (2010): Towards an Indonesian bird conservation ethos: reflections from a study of bird-keeping in the cities of Java and Bali. In *Ethnoornithology: Birds, Indigenous Peoples, Culture and Society,* eds. S. Tidemann & A. Gosler. London. Pp. 313–330.

Jepson, P. & R. J. Ladle, (2005): Bird-keeping in Indonesia: conservation impacts and the potential for substitution-based conservation responses. *Oryx* 39: 442–448.

Jerolmack, C., (2009): Primary groups and cosmopolitan ties: rooftop pigeon flyers of New York. *Ethnography* 10: 435–457.

Kawata, K., (2012): Passerine birds in Japanese zoos 2009. *International Zoo News* 59: 181–197.

Keller, O., (1913): *Die Antike Tierwelt* vol. 2. *Vögel, Reptilien, Fische, Insekten.* Leipzig.

Killermann, S., (1921): Zur Geschichte der Einführung der Papageien. *Naturwissenschaftliche Wochenschrift* N.F. 20: 545–550

Kinzelbach, R. & J. Hölzinger, (2000): *Marcus zum Lamm (1544 –1606). Die Vogelbücher aus dem Thesaurus Picturarum.* Stuttgart.

Kinzelbach, R., (2004): The distribution of the Serin (*Serinus serinus* L., 1766) in the 16th century. *Journal für Ornithologie* 145: 177–187.

Kisling, V. N. (1998). Colonial menageries and the exchange of exotic faunas. *Archives of Natural History* 25: 349–368.

Koppenol, J., (2007): Noah's Ark disembarked in Holland: animals in Dutch poetry 1550–1700. In *Early Modern Zoology: the Construction of Animals in Science, Literature and the Visual Arts,* edited by Karl A.E. Enenkel and Paul J. Smith. Leiden, Pp. 451–528.

Laidlay, C., (1914): Some notes on ornamental water-fowl, *The Amateur Menagerie Club. Year Book,* 1914: 197–205.

Laufer, O., (1939): *Singvögel als Hausgenossen im deutschen Glauben und Brauch.* Berlin.

Lazenby, F. D., (1949): Greek and Roman household pets. *The Classical Journal* 44: 299–307.

Layton, L., (1991): *Songbirds in Singapore: the Growth of a Pastime.* Singapore.

Lindemann, K., (1994): *Der Papagei: Seine Geschichte in der Deutschen Literatur.* Bonn.

Low, R., (1972): *The Parrots of South America.* London.

Low, R., (1984): *Endangered Parrots.* London.

Low, R., (2006): *A Century of Parrots.* Mansfield.

Lutwack, L., (1994): *Birds in Literature.* Gainesville, FL.

Malkin, B. (1962): *Seri Ethnozoology.* Pocatello, ID.

Möller, D. (2015): *Dumb beasts in hallowed tombs: Swedish funerary poetry for animals 1670-1760.* Frankfurt am Main.

Morris, S., (2001): Bird trade in Hanoi. *Bulletin of the Oriental Bird Club* 33:34–35.

Muse, C. & S. Muse, (1982): *The birds and birdlore of Samoa/O manu ma tala'aga o manu o Samoa.* Washington.

Naefe, B., (2001): Die Kormoranfischer vom Erhai-See: eine südwest-chinesche Wirtschaftsweise im Wandel. Köln (= Kölner Ethnologische Beiträge 1). Cologne.

Parker, A. J., (1988): The birds of Roman Britain. *Oxford Journal of Archaeology* 7.

Parson, J.J., (1987): The origin and dispersal of the domesticated canary. *Journal of Cultural Geography* 7: 19–34.

Peleman, B., (1922): *De Belgische Waterslager.* Antwerpen.

Pliny the Elderly, (1983): *Natural History in Ten Volumes. Libri* VIII–XI. Cambridge, MA.

Pluis, J. & E. Stupperich, (1986): *Die Lachtaube. Eine historische und volkskundliche Untersuchung.* Vreden.

Ptak, R. (2006): *Exotische Vögel: Chinesische Beschreibungen und Importe.* Wiesbaden.

Raethel, H-S. & J. R. Mancini, (2005): *The Duck Handbook.* New York.

Rahmani, A.R., (1999): The growth of ornithology in India and the contribution of indigenous knowledge. In *Proceedings of the 22 International Ornithological Congress Durban,* ed. N. J. Adams & R. H. Slotow. Johannesburg. Pp. 1339–1343.

Rea, A. M. (2007): *Wings in the Desert: A Folk Ornithology of the Northern Pimans.* Tuscon.

Robbins, L. E., (2002): *Elephant Slaves and Pampered Parrots: Exotic Animals in Eighteenth-Century Paris.* Baltimore.

Roberts, S., (1972): *Bird-Keeping and Birdcages: A History.* London.

Schlenker, R., (2004): *Bibliographie der deutschen vogelkundlichen Literatur von 1480 bis 1850.* Stuttgart.

Schulze-Hagen, K., F. Steinheimer, R. Kinzelbach, C. Gasser, (2003): Avian taxidermy in Europe from the Middle Ages to the Renaissance. *Journal für Ornithologie* 144: 459–478.

Serpell, J., (1986): *In the Company of Animals: A Study of Human-Animal Relationships.* Cambridge.

Shurtleff,, L. L. & C. Savage, (1996): *The Wood Duck and the Mandarin: The Northern Wood Ducks.* Berkeley, CA.

Shiu, H. & L. Stokes, (2008): Buddhist animal release practices: historic, environmental, public health and economic concerns. *Contemporary Buddhism: An Interdisciplinary Journal* 9: 181–186.

Sick, H., (1981): Zur frühen Darstellungen neotropischer Papageien. *Journal für Ornithologie* 122: 73–78.

Simakov, G. N., (1989): Hunting with raptors in Central Asia and Kazakhstan. In *Ecology and Empire: Nomads in the Cultural Evolution of the Old World,* ed. G. Seaman. Los Angeles, pp. 129–133.

Sossinka, R., (1982): Domestication in birds. *Avian Biology* 6: 373–403.

Sparrman, A., (1786): *A Voyage to the Cape of Good Hope: towards the Antarctic Polar Circle, and around the World, but chiefly into the country of the Hottentots and Caffres, from the Years 1772–1776,* Vol 2. London.

Ståhlberg, S. & I. Svanberg, (2012): Wild animals in Russian, Siberian and Central Asian households according to eighteenth-century travel reports. *Journal de la Société Finno-Ougrienne* 93: 353–383.

Stoner, C. R. (1948): Fishing with the Indian Darter

(*Anhinga melanogaster*) in Assam. *Journal of the Bombay Natural History Society* 47: 746–747.

Stresemann, E., (1914): Die Vögel von Seran. *Novitates Zoologicae* 21: 25–153.

Strunden, H., (1984): *Papageien einst und jetzt. Geschichtliche und kulturgeschichtliche Hintergrunde der Papageienkunde.* Bomlitz.

Svanberg, I., (2007a): »*Deras mistande rör mig så hierteligen«: Linné och hans sällskapsdjur.* Uppsala.

Svanberg, I., (2007b): Golden Pheasant (*Chrysolophus pictus*) in Sweden in the 1740s. *Der Zoologische Garten* N.F. 77: 24–28.

Svanberg, I., (2007c): »Kunna lära att hwisla och tala«: Linnés och andras papegojor i 1700-talets Sverige. *Uppland: årsbok* 2007: 115–150.

Svanberg, I., (2008): Towards a cultural history of the Bengalese Finch (*Lonchura domestica*). *Zoologische Garten* N.S. 77: 334–344.

Svanberg, I., (2009a): Färöische Riesenalken *Pinguinus impennis* (L.) in Gefangenschaft. *Ornithologische Mitteilungen* 61 (1):13–20.

Svanberg, I., (2009b): Undulatens historia. *Fågelhobby* 39: 214–215.

Svanberg, I.,(2011): *Kanariefågelns svenska historia.* Uppsala.

Svanberg, I., (2012): Exotic birds at Skansen zoo, Sweden, before 1920. *The Bartlett Society Journal* 22: 5–14.

Svanberg, I. & S. Ægisson, (2005): Great Northern Diver (*Gavia immer*) in circumpolar folk ornithology. *Fróðskaparrit: Annales Societatis Scientiarum Faeroensis* 53: 51–56.

Svanberg, I. & S. Ægisson, (2006): The Black Guillemot (*Cepphus grylle*) in Northern European folk ornithology. *Scripta Islandica* 56:104–110.

Svanberg, I. & J. Strand, (2012): Giovanni Pietro Olinas beskrivning av kanariefågeln av år 1622. *Svenska Linnésällskapets Årsskrift* 2012: 21–29.

Ticehurst, N. F., (1957): *The Mute Swan in England.* London.

Toynbee, J. M. C., (1973): *Animals in Roman Life and Art.* London.

Tyrwhitt-Drake, G., (1912): Herr Carl Hagenbeck's Ostrich Farm at Stellingen, *The Amateur Menagierie Club. Year Book* 1912: 74–82.

Tschirren, B. & E Postma, (2010): Quantitative genetic research in Zebra Finches: where we are and where to go. *Emu* 110: 268–278.

Verrill, A. H., (1915): *Pets for Pleasure and Profit.* New York.

Vriends, M. M. & T. M. Heming-Vriends, (2002): *Hancock House Encyclopedia of Estrildid Finches.* Surrey, B.C.

Wai, C. S., (2006): *Religious release of birds in Hong Kong.* The University of Hong Kong. (MA thesis).

White, C. M. N., (1975): The problems of Cassowary in Ceram. *Bulletin of British Ornithologist's Club* 95: 165–170.

Wille, L. & D. Spormann, (2008): *Buchfink und Mensch: Geschichte der Finkenliebhaberei im Harz.* Benneckenstein.

Wiese, V., (1896): Tropefuglenes liv i fangenskap. In *Haandbog i fuglens røgt, pleje og opdræt* vol. 2. Aarhus.

Yamashina, Y., (1961): Quail-breeding in Japan. *Journal of the Bombay Natural History Society* 58: 216–222.

Yapp, W. B., (1982): Birds in captivity in the Middle Ages. *Archives of Natural History* 10: 479–500.

Zaunschirm, T., (1984–85): Affe und Papagei: Mimesis und Sprache der Kunst, *Kunsthistoriker: Mitteilungen des Österreichischen Kunsthistorikerverbandes* 1–2: 14–17.

CHAPTER 2

Allchin, B. & R. Allchin (1982): *The Rise of Civilization in India and Pakistan.* Cambridge.

Allin, M., (1998). *Zarafa: A Giraffe's True Story, from Deep in Africa to the Heart of Paris.* New York.

Auboyer, J., (1965): *Daily Life in Ancient India.* New York.

Breasted, J. H., (1906–07): *Ancient Records of Egypt.* Chicago.

Clagett, M., (1955): *Greek Science in Antiquity.* New York.

Collins, B. J., (1989): *The Representation of Wild Animals in Hittite Texts.* New Haven: Yale University. [Ph.D. Dissertation]

Contenau, G., (1964): *Everyday life in Babylon and Assyria.* London [Reprint of 1954 edition, translated by K. R. and A. R. Maxwell-Hyslop]

Dalley, S., (1993): Ancient Mesopotamian Gardens and the Identification of the Hanging Gardens of Babylon Resolved. *Garden History* 21 (1): 1–13.

Erman, A., (1971): *Life in Ancient Egypt.* New York. [Reprint of 1894 edition translated by H. M. Tirard]

Fairservis, W.A., Jr., (1971): *The Roots of Ancient India*. New York.
Finkel, I.L., (1988): The Hanging Gardens of Babylon. In *The Seven Wonders of the Ancient World*, eds. P. A. Clayton & M. J. Price. London. Pp. 38–58.
Frederick II., (1943): *The Art of Falconry* (Casey A. Wood and F. Marjorie Fyfe, Editors and Translators). Stanford.
George, W., (1969): *Animals and Maps*. Berkeley.
Glacken, C. J. (1967): *Traces on the Rhodian Shore: Nature and Culture in Western Thought From Ancient Times to the End of the Eighteenth Century*. Berkeley.
Gowers, W., (1947): The African Elephant in Warfare. *African Affairs* 46 (181): 42–49.
Graham, D., (1938): *Chinese Gardens*. New York.
Hawkes, J., (1973): *The First Great Civilizations: Life in Mesopotamia, the Indus Valley and Egypt*. New York.
Huff, T. E., (1993): *The Rise of Early Modern Science: Islam, China and the West*. Cambridge.
Hughes, J. D., (1975): *Ecology in Ancient Civilizations*. Albuquerque.
Jennison, G., (1937): *Animals for Show and Pleasure in Ancient Rome*. Manchester.
Kees, H., (1961): *Ancient Egypt: A Cultural Topography*. Chicago.
Kisling, Jr., V. N., (2001): *Zoo and Aquarium History: Ancient Animal Collections to Zoological Gardens*. Boca Raton.
Kosambi, D. D., (1969): *Ancient India: A History of Its Culture and Civilization*. New York.
Laufer, B., (1928): *The Giraffe in History and Art*. (=FMNH Anthropology Leaflet 27.) Chicago.
Lloyd, J. B., (1971): *African Animals in Renaissance Literature and Art*. Oxford.
Loisel, G., (1912): *Histoire des Ménageries de l'antiquité à nos jours* 1–3. Paris.
Luckenbill, D. D., (1926–27): *Ancient Records of Assyria and Babylonia*. Chicago.
Marsh, G. P., (1965): *Man and Nature: Or, Physical Geography as Modified by Human Action*. Cambridge.
Menzies, N. K., *(1994): Forest and Land Management in Imperial China*. New York.
Montet, P., (1962): *Everyday Life in Egypt*. London.
Moynihan, E. B., (1979): *Paradise as a Garden: In Persia and Mughal India*. New York.
Naville, E H., (1894–1908): *The Temple of Deir el-Bahari*. London.
Needham, J., (1954): *Science and Civilisation in China*. Cambridge.
Oppenheim, A. L., (1964). *Ancient Mesopotamia: Portrait of a Dead Civilization*. Chicago.
Polo, M., 1934: *The Travels of Marco Polo* (Manuel Komroff, Editor and Translator). New York.
Possehl, G. L.,(1982): *Harappan Civilization: A Contemporary Perspective*. New Delhi.
Rao, S. R., (1973): *Lothal and the Indus Civilization*. New York.
Schafer, E. H., (1968): Hunting Parks and Animal Enclosures in Ancient China. *Journal of the Economic and Social History of the Orient* 11: 318–343.
Scullard, H. H., (1974): *The Elephant in the Greek and Roman World*. Ithaca, N.Y.
Snow, P. (1988): *The Star Raft: China's Encounter with Africa*. New York.
Toynbee, J., (1973): *Animals in Roman Life and Art*. Ithaca, N.Y.
Tyldesley, J., (1996): *Hatchepsut: The Female Pharaoh*. New York.
Wheeler, R. E. M., (1968): *The Indus Civilization: Supplementary Volume to the Cambridge History of India*. 3rd Edition, Cambridge.

Chapter 3

Armstrong, E.A. (1975): *The Life and Lore of the Bird: in Nature, Art, Myth, and Literature*. New York.
Ægisson, S. (1996): Ísfygla/Aves Islandicae. Grenjaðarstaður.
Ægisson, S. (2014): *Icelandic Trade with Gyrfalcons: From Medieval Times to the Modern Era*. Siglufjörður.
Beebe, F. L. (1976): *Hawks, Falcons and Falconry*. Saanichton. Hancock House Publishers
Bernström, J. (1959): Falkar. In *Kulturhistoriskt lexikon för nordisk meddeltid* vol. 4. Malmö. Pp. 142–150.
Berridge, W. S. (1934): *All about Birds*. London.
Brown, L. & Amadon, D. (1968): *Eagles, Hawks and Falcons of the World*. Vol. 2. Feltham.

Brown, L.H. & Newton, I. (1985): Falcon. In Campbell, B. & Lack, E.: *A Dictionary of Birds*. Calton. Pp. 202–203.

Cade, T. J. (1982): *The Falcons of the World.* Ithaca.

Cade, T. J. (1999): The Gyrfalcon's Adaptations for Survival in Arctic Regions. In *Gyrfalcon,* ed. E. Ford. London. Pp. 55–68.

Christensen, N. H. (1995): *De store falke fra de danske konger*. Niels Hesselbjerg Christensen, Aalborg.

Ciesielski, L. C. (2007): *Der Gerfalke, Falco rusticolus*. Hohenwarsleben.

Clum, N. J. & Cade, T. J. (1994): *Gyrfalcon (Falco rusticolus).* Ithaca.

Cramp, S., & Simmons, K. E. L. (eds.) (1980): *Handbook of the birds of Europe and the Middle East and North Africa.* Volume 2. Oxford.

Davíðsson, G. (1937): Friðun fálkans. *Alþýðublaðið*, 19 July. P. 2.

Dementiew, G. P. (1960): *Der Gerfalke (Falco gyrfalco L. = Falco rusticolus L.)*. A. Ziemsen Verlag, Wittenberg Lutherstadt.

Ditchfield, P. H. (1887): Hawking. *Antiquary* 16: 242–244.

Epstein, H. J. (1943): The Origin and Earliest History of Falconry. *Isis* 34: 497–509.

Ericson, G. P. & Tyrberg, T. (2004): *The Early History of the Swedish Avifauna : A Review of the Subfossil Record and Early Written Sourches.* Stockholm.

Ferguson-Lees, J. & Christie, D. A. (2001): *Raptors of the World.* New York.

Ford, E. (1999): *Gyrfalcon*. London.

Frederick II of Hohenstaufen (1943): *Art of Falconry : Being the De Arte Venandi cum Avibus of Frederick II of Hohenstaufen*. Stanford.

Friðgeirsson, E. (1982): Um fálka: yfirlit um söguna frá 1600 og fram úr. In Guðmundsson, E. (1982): Þjóðsögur og þættir. [Taken from the manuscript *Lbs. 3187, 4to*, first chapter: *Samtíningur um fugla og líf þeirra, m.a. um fálkaveiðar á Íslandi.*] Pp. 247–253.

Friedland, K. (1964): The Hanseatic League and Hanse Towns in the Early Penetration of the North. *Arctic* 37: 539–543.

Gíslason, V. Þ. (1947): *Bessastaðir: þættir úr sögu höfuðbóls*. Akureyri.

Glasier, P. (1980): *Falconry and Hawking*. London.

Grossman, M. L. & Hamlet, J. (1964): *Birds of Prey of the World*. New York.

H., J. (1986): Úr sögu íslenska valsins: Eitt sinn litu Íslendingar fálka og hrafna sömu augum. *Morgunblaðið, B*, 31 August. Pp. 12–13.

Hilmarsson, J. Ó. (2000): Íslenskur fuglavísir. Reykjavík.

Illingworth, F. (1978): *Falcons and Falconry*. Poole.

Jack, T. A. M. (1991): Falconry. In Brooke, M. & Birkhead, T. (eds): *The Cambridge Encyclopedia of Ornithology*. Cambridge. Pp. 301–303.

Jacobsen, P. B. (1848): Bidrag til en Skildring af Falkevæsenet og Falkejagten forhen, navnlig i Danmark. *Nyt Historisk Tidsskrift* 2: 307–414.

Jónsson, Á. I. (1980): Girnast fálkann sem konungar forðum og svífast einskis. *Morgunblaðið*, 2 November. Pp. 14–16.

Karlsson, G. (2000): *A Brief History of Iceland*. Reykjavík.

Krogh, K. J. (1967): *Viking Greenland*. Copenhagen.

Larson, L. M. (ed.) (1917): *The King's mirror: Speculum regale – Konungs skuggsjá*. New York.

Lloyd, G. & Lloyd, D. (1977): *Birds of Prey*. London.

Macdonald, H. (2006): *Falcon*. London.

Magnússon, S. (1935–1936): Lýsing Gullbringu- og Kjósarsýslu. In *Landnám Ingólfs. Safn til sögu þess*, vol. 1. Reykjavík.

Newton, I. (consulting ed.) (1990): *Birds of Prey*. London.

Nielsen, O.K. & Pétursson, G. (1995): Population fluctuations of gyrfalcon and rock ptarmigan: analysis of export figures from Iceland. *Wildlife Biology* 1: 2.

Oggins, R. S. (2004): *The Kings and Their Hawks: Falconry in Medieval England*. New Haven.

Oswald, C. A. (1982): *The History & Practice of Falconry*. Jersey.

Óla, Á. (1965): Úr sögu Reykjavíkur: Fálkahúsið og fálkaverzlun konungs. In *Lesbók Morgunblaðsins*, 19 November, pp. 6–7, 12, and 26 November.

Pollard, J. (1977): *Birds in Greek Life and Myth*. London.

Potapov, E. & Sale, R. (2005): *The Gyrfalcon*. London.

Rowland, B. (1978): *Birds with Human Souls : A Guide to Bird Symbolism*. Knoxville.

Salomonsen, F. (1990): Fugle (Aves). In Muus, B., Salomonsen, F. & Vibe, C. *Grønlands fauna*. København. Pp.

159–361.

Salomonsen, F. (1950–51): *Grønlands fugle*. Volume 1–3. København.

Scott, A.G. O'C. & Kenward, R.E. (1985): Falconry. In Campbell, B. & Lack, E.: *A Dictionary of Birds*. Calton.

Tillisch, C. J. (1949): *Falkejagten og dens historie*. København.

Weidensaul, S. (1996): *Raptors: The Birds of Prey*. New York.

White, C.M. (1994): Gyrfalcon. In del Hoyo, J., Elliott, A. & Sargatal, J. (eds.): *Handbook of Birds of the World*, vol. 2 (New World Vultures to Guineafowl). Barcelona. P. 274.

Þorleifsdóttir, B. (2008): Um listina að klófesta bráð. *Fuglar*, nr. 5 (November): 54–58.

Þorsteinsson, H. (1932-1935): Fálkar. In Þorkelsson, J., et. al. (eds.): *Blanda*. Vol. 5. Reykjavík.

Þórðarson, B. (1955): Réttur konungs til fálkatekju Íslands. In *Afmælisrit helgað Ólafi Lárussyni prófessor dr. juris & philosophiae sjötugum 25. febrúar 1955*. Reykjavík.

Þórðarson, B. (1957): *Íslenzkir fálkar.* Safn til sögu Íslands og íslenzkra bókmennta að fornu og nýju. *Annar flokkur*, I. 5. Hið íslenzka bókmenntafélag, Reykjavík.

Þórðarson, B. (1959): Falkar, Island. In *Kulturhistoriskt lexikon för nordisk meddeltid vol 4. Malmö.*.

Chapter 4

Berdan, F. F., (1992): Economic Dimensions of Precious Metals, Stones, and Feathers: The Aztec State Society. *Estudios de Cultura Nahuatl* 22: 291–323.

Borson, N., F. Berdan, E. Strak, J. States & P.J. Wettstein, (1998): Origins of an Anasazi Scarlet Macaw Feather Artefact. *American Antiquity* 6: 131–142.

Columbus, C., (1893): *The Journal of Christopher Columbus [During His First Voyage, 1492–93] and Documents Relating the Voyages of John Cabot and Gaspar Corte Real.* London.

Cortés, H., (1971): *Letters From Mexico*. New York.

Creel, D. & C. McKusick, (1994): Prehistoric Macaws and Parrots in the Mimbres Area, New Mexico. *American Antiquity* 59: 510–524.

Diaz del Castillo, B., (1956): *The Discovery and Conquest of Mexico 1517–1521*. New York.

Di Peso, C. C., J. B. Rinaldo & G. J. Fenner, (1974): *Casas Grandes: A Fallen Trading Centre of the Gran Chichimeca.* Volume 8, *Bone, Perishables, Commerce, Subsistence and Burials*. Flagstaff.

Duran, F. D., (1964): *The Aztecs*. New York.

Hargrave, L. L., (1970): *Mexican Macaws: Comparative Osteology and Survey of Remains from the Southwest*. Tucson.

Haemig, P. D., (1978): Aztec Emperor Auitzotl and the Great-Tailed Grackle. *Biotropica* 10: 11–17.

del Hoyo, J., A. E. & J. Sargatal, Editors (1997): *Handbook of the Birds of the World*. Volume 4, Sandgrouse to Cuckoos. Barcelona.

Hunn, E. S,. (1977): *Tzeltal Folk Zoology*. New York.

Kisling Jr., V. N. (1998): Colonial menageries and the exchange of exotic faunas. *Archives of Natural History* 25: 349–368.

de La Salle, N., (2003): *The La Salle Expedition on the Mississippi River: A Lost Manuscript of Nicolas de La Salle, 1682*. Austin.

McKusick, C. R., (2001): *Southwest Birds of Sacrifice*. Globe, Arizona: Arizona Archaeologist Number 31, 2001.

Minnis, P. E., M. E. Whalen, J. H. Kelley & J. D. Stewar,t (1993): Prehistoric Macaw Breeding in the North American Southwest. *American Antiquity* 58: 270–276..

Morley, S. G. & G. W. Brainerd, (1983): *The Ancient Maya*. Stanford, 4th edition.

Prescott, W. H., (1936): *History of the Conquest of Mexico and History of the Conquest of Peru*. New York: Modern Library, 1936. [Reprints of History of the Conquest of Mexico 1843 edition and History of the Conquest of Peru 1847 edition]

Reséndez, A., (2007): *A Land So Strange: The Epic Journey of Cabeza de Vaca*. New York.

Roth, W. E., (1924): [Thirty-Eighth] Annual Report of the Bureau of American Ethnology to the Secretary of the Smithsonian Institution, 1916–1917. *An Introductory Study of the Arts, Crafts, and Customs of the Guiana Indians*. Washington.

de Sahagun, B., (1950–82): *General History of the Things of New Spain: Florentine Codex*. Salt Lake City.

Schroeder, A.H., (1968): Birds and Feathers in Documents Relating to Indians of the Southwest. In *Collected Papers in Honour of Lyndon Lane Hargrave,* ed. A. H. Schroeder (= Papers of the Archaeological Society of New Mexico 1):

Santa Fe, Pp. 95–114.

Steward, J.H., General Editor (1946 – 1959): *Handbook of South American Indians.* Volume 1–7. Washington [Bureau of American Ethnology Bulletin 143]

Sturtevant, W.C., ed. (1979): *Handbook of North American Indians.* Volume 9–10, *Southwest* (Alfonso Ortiz, Volume Editor): Washington.

de la Vega, G., (1966): *Royal Commentaries of the Incas.* Austin.

Wauchope, R., ed. (1971): *Handbook of Middle American Indians.* Volume 10–11, *Archaeology of Northern Mesoamerica* (Gordon F. Ekholm and Ignacio Bernal, Volume Editors). Austin.

Chapter 5

Bergman, H-H., (1993): *Der Buchfink. Neues über einen bekannten Sänger.* Wiesbaden.

Broeckhove, M., (1969): *Vinkensport in Vlaanderen.* Gent.

Claeys, J., (1931): De Brugsche vogelmarkt. *Biekorf* 11: 335–339.

De Groote, G. (1947): Eduard de Dene als villonisant. *De Gulden Passer* 25: 313–343.

Defraeye, L., (1954): Vinkenstellingen in West-Vlaanderen en Frans-Vlaanderen anno 1847–1850. *De Vinkenier,* 10 September 1954.

Desmet, J., (1987): *Vogels. Hun levensloop in België, hun wedervaren met de mens.* Brugge.

Gilliodts-Van Severen, L, (1885): *Coutumes des pays et comté de Flandre : coutume du bourg de Bruges* vol. 2. Bruxelles.

Grevendal, L., (1933): *Le Pinson des Pinsonniers.* Verviers.

Knoll, F. (1989): Über die Finkenliebhaberei. *Gefiederte Welt* 113 (5–7).

Matthey, I. (2002): *Vincken moeten vincken lokken. Vijf eeuwen vangst van vogels en kwartels in Holland.* Hilversum.

Neylants, E., (1958): Ieperse vinkeniers ten tijde van Napoleon. *Biekorf* 59: 282

Rembry, E., (1881): *Saint-Gilles. Sa vie, ses reliques, son culte en Belgique et dans le Nord de la France,* vol 2. Bruges.

Rubbrecht, L. (1907): *Geschiedboek der gemeente Rousbrugghe-Haringhe van 1800 tot 1907.* Brugge.

Santens, F. (1995): *Hinke de Vinke. 400 jaar vinkensport in Vlaanderen. 60 jaar Avibo.* Vichte.

Chapter 6

Brehm, A., (1924): *Djurens liv* vol. 6. Stockholm.

Brehm, A., (1925): *Djurens liv* vol. 7. Stockholm.

Brehm, A., (1926): *Djurens liv* vol. 8. Stockholm.

Brusewitz, G., (1982): När brushanen hörde till husdjuren, in the author's *Den nöjsamma nyttigheten: om natur och naturskådande i en svunnen tid.* Stockholm, pp. 218–227.

Brusewitz, G., (2001). De hjälpsamma djuren. In *Människan i naturen: etnobiologi i Sverige* vol. 1. Stockholm, pp. 119–126.

Christoffersson, O., (1936): *Skytts härad: kulturhistorisk beskrivning.* Trelleborg: Allehanda.

Christy, C., (1897): Field-notes on the birds of the island of San Domingo. *Ibis* 3:317–343.

Cameron, L., (1921): *Minor Field Sports.* London.

Falk, J. P., (1786): *Beyträge zur topographischen Kenntnis des Russischen Reichs.* Teil III. St. Peterburg.

Gillmore, R. M., (1950): Fauna and Ethnozoology of South America. *Handbook of South American Indians* 6. Washington D.C.: Smithsonian Institution, pp. 345–463.

Gunda, B., (1969): Die Jagd und Domestikation des Kranichs bei den Ungarn. *Anthropos* 63–64:473–496.

Johansson, L. (1927): Om renlighetsförhållanden i Frostviken (Jämtland). *Svenska Landsmål ock Svenskt Folkliv* 177: 119–138.

Laine, E. (1935): Archive of Suomalaisen Kirjallisuuden Seura KRA. KRK 144:279.

Lehtonen, A. (1930): Archive of Suomalaisen Kirjallisuuden Seura KRA. Paulaharju 18975.

Linden van der, A. (2000): Biological control of caterpillars (Noctuidae) and other pests with insectivorous birds, *Alcippe brunnea* in glasshouses. *Faculteit Landbouwkundige en Toegepaste Biologische Wetenschappen, Universiteit Gent Mededelingen* 65, 2a: 235–243.

Linnaeus, C., (1751): *Skånska Resa.* Stockholm.

Lönnberg, E. (1913): *Linnés föreläsningar öfver djurriket.* Uppsala.

Manzi, M. T. & O. T. Coomes, (2002): Cormorant Fishing in Southwestern China: A Traditional Fishery under Siege, *The Geographical Review* 92: 597–603.

van Nieuwenhuyse, D., J.-C. Génot & D. H. Johnson, (2008): *The Little Owl: Conservation, Ecology and Behavior of* Athene noctua. Cambridge: Cambridge University Press.

Notehelfer, F. G. (1992): *Japan through American Eyes. The Journal of Francis Hall, Kanagawa and Yokohama, 1859–1866.* Princeton N.J.

Orstadius, E. (1909): Vadare som husdjur. *Fauna och flora* 4:44–45.

Osbeck, P., (1996): *Djur och natur i södra Halland under 1700-talet.* Halmstad.

Oesenbry, J., (1575): *Kunst, Weydny oder Vogel Buch.* Zentralbibliothek Zürich: Sign. C 22 (721) fol. 23v.

Pallas, P. S., (1773): *Reise durch verschiedene Theile des Russischen Reichs.* Teil II. St. Petersburg.

Rothof, L. W., (1762): *Hushålls-Magasin. Första delen om Hushålls-Ämnen Til Deras nytta, bruk och skada, Beskrefne Uti Oeconomiska Föreläsningar.* Skara.

Speerschneider, J. (1854): Vergleichende Aufzählung der auf dem S. O. Tühringer Walde und der in der Umgegend von Schlotheim im N. W. Thüringen vorkommenden Vögel. *Naumannia* 4:175–190.

Ståhlberg, S. & I. Svanberg, (2012): Wild animals in Russian, Siberian and Central Asian households according to eighteenth-century travel reports. *Journal de la Société Finno-Ougrienne* 93: 353–383.

Trädgård, I., (1922): *Husflugan och dess syndaregister.* Göteborg.

Tullberg, T., (1919): Familjetraditioner om Linné. *Svenska Linné-Sällskapets Årsskrift* 2:1–20.

Tullgren, A., (1914): *Våra snyltgäster.* Stockholm.

Wetmore, A. (1926): *Observations on the Birds of Argentina, Paraguay, Uruguay and Chile.* Washington D.C.

Wiese, V. (1891): *De nordiske stuefugle, deres røgt og pleje.* Kjøbenhavn.

Zhou, Y. (1990): *A History of Chinese Entomology.* Beijing.

Chapter 7

A new general history of birds, (1745): *A new general history of birds. Including the methods of breeding, managing, and teaching of song birds*, 2 vols., London.

A new way of breeding canary birds, (1742): *A new way of breeding canary birds*, London.

Albin, Eleazar, (1737): *A natural history of English song-birds, and such of the foreign as are usually brought over and esteemed for their singing*, London.

Aldrovandi, Ulysse, (1600): *Ornithologiae. Tomvs alter* [= Ornithologiae. Liber Decimusterius], Bononiae.

Balis, Arnout (1982): Facetten van de Vlaamse dierenschilderkunst. In *Het Aards Paradijs. Dierenvoorstellingen in de Nederlanden van de 16de en 17de eeuw*, exh. cat. Wommelgem. Pp. 37–55.

Beckmann, Johann, (1797): *A history of inventions and discoveries*, [orig. *Beyträge zur Geschichte der Erfindungen*, trans. William Johnston], London.

Bedaux, J.B. & Ekkart, R., (2000): *Kinderen op hun mooist*, exh. cat., Gent – Amsterdam.

Beilby, Ralph, (1797): *History of British birds*, vol. I, Newcastle.

Bergvelt, E. & Kistemaker, R., (1992): *De wereld binnen handbereik. Nederlandse kunst- en rariteitenverzamelingen, 1585–1735*, Zwolle.

Birkhead, Tim (2003): *The Red Canary. A Story of the First Genetically Engineered Animal*, London.

Birkhead, Tim, Schulze-Hagen, Karl & Kinzelbach, Ragnar, (2004): Domestication of the canary, *Serinus canaria* – the change from green to yellow. *Archives of Natural History* 31: 50–56.

Boesmans, Annick, (1984): Huisdieren, *Volkskunde*, 84:2, 129–167.

Buffon, Georges-Louis Leclerc count of (1779): *Histoire naturelle des Oiseaux. Tome Septième*, Paris.

Chomel, Noël, (1743): *Huishoudelyk woordboek. Vervattende vele middelen om zyn goed te vermeerderen, en zyne gezondheid te behouden*, pt. 1, Leiden.

Cox, Nicholas, (1721): *The gentleman's recreation, in four parts. Viz. Hunting, hawking, fowling. Fishing*, London.

Davids, Karel, (1989): *Dieren en Nederlanders. Zeven eeuwen life en leed*, Utrecht.

De Jongh, Eddy, (1967): *Zinne- en minnebeelden in de schilderkunst van de zeventiende eeuw*, Amsterdam.

De Jongh, Eddy (1995): *Kwesties van betekenis. Thema en motief in de Nederlandse schilderkunst van de zeventiende eeuw*, Leiden.

De Roo, Tom (2005): *Dierlijk gezelschap, menselijke reflectie*, Antwerpen.

De Roo, Tom (2007): Vreemde vogels: de papegaai en de

kanarie als uitheemse gezelschapsdieren in Antwerpen van de zestiende tot de achttiende eeuw. In Devisscher, H. (ed.), *Wonderlycke dieren op papier in de tijd van Plantin*, exh. cat. Antwerpen. Pp. 29–47.

De Roo, Tom (2009): Kanarieliefhebberij in de achttiende eeuw – op het kruispunt van wetenschap en vrije tijd? *De Achttiende Eeuw* 41: 218–248.

Duparc, F.J. (ed.), (2004): *Carel Fabritius. 1622–1654*, Den Haag – Schwerin – Zwolle.

Exact and sure directions, [ca. 1710]: *Exact and sure directions how to dress and keep canary-birds, and to preserve them in health, and also, how to use and keep the new fashion bird-cages*, London.

Fairfax, Thomas, (1758): *The compleat sportsman; or country gentleman's recreation*, London.

Gessner, Conrad, (1585): *Historiae Animalium Liber III, qui est de Auium natura*, Francofurdi.

Goldsmith, Oliver, (1779): *An history of the earth, and animated nature*, 8 vols., London.

Grzimek, B. (ed.), (1971): *Het leven der dieren*, part 9, Utrecht – Antwerpen.

Guicciardini, Lodovico, (1612): *Beschryvinghe van alle de Nederlanden*, Amsterdam.

Hervieux, Joseph-Charles Chastanier, (1712): *Traité curieux des Serins de Canarie. Naaukeurige verhandeling van de Kanari Vogels*, trans. A. Moubach, Amsterdam.

Holland, Thomas, (1798): *Exercises for the memory and understanding: consisting of select pieces in prose & verse; together with a series of examinations relative to arts, science, and history*, Manchester.

Moretus-de Neuf, Anna-Maria, (1700–1706): *Household journal 1700–1706*, Museum Plantin-Moretus, Antwerp.

Moretus-de Neuf, Anna-Maria, (1707–1714): *Household journal 1707–1714*, Museum Plantin-Moretus, Antwerp.

Houttuyn, Martinus, (1763): *Natuurlyke historie of uitvoerige beschryving der dieren, planten en mineraalen. Volgens het Samenstel van den Heer Linnaeus. Met naauwkeurige Afbeeldingen*, Amsterdam.

Ionstons, Iohannes, (1660): *Beschrijving vande Natuur der Vogelen neffens haer Beeldenissen in koper gesneden*, [trans. M. Grausius], Amsterdam.

Iunius, A., (1567): *Emblemata. Overgheset in Nederlantsche tale*, Antwerpen.

Kloek, E., (1993): Huwelijk en gezinsleven tijdens het Ancien Régime, 1650–1800. In Zwaan, T. (ed.) e.a., *Familie, huwelijk en gezin in West-Europa. Van Middeleeuwen tot moderne tijd*, Amsterdam – Heerlen, 139–164.

Laroon, Marcellus the Elder, (1990): *The criers and hawkers of London*, edited by Sean Shesgreen. Aldershot.

Laroon, Marcellus the Elder, (1688): *The cryes of the city of London*, London.

Lord, Thomas, (1791–1796): *Lord's entire new system of ornithology. Or oecumenical history, of British birds. ... The writing corrected, & embellish'd, by the Rev.d Dr. Dupree. ...*, London.

Matthey, Ignaz, (2002): *Vincken moeten vincken locken. Vijf eeuwen vangst van zangvogels en kwartels in Holland*, Hilversum.

Muchembled, Robert, (1991): *De uitvinding van de moderne mens. Collectief gedrag, zeden, gewoonten en gevoelswereld van de middeleeuwen tot de Franse Revolutie*, [trans. Siblesz, R. and Visser, T., orig. 1988], Amsterdam.

Nijboer, Harm, (2007): *De fatsoenering van het bestaan. Consumptie in Leeuwarden tijdens de Gouden Eeuw*, Groningen.

Nyland, P., (1702): *Den verstandigen kok*, Amsterdam.

Olina, Giovanni Pietro, (1774): *Les amusemens innocens, contenant le traité des oiseaux de voliere, ou le parfait oiseleur*, [orig. *Uccelliera, overo. Discorso della natura, e proprietà di diversi uccelli*, trans. Pierre Joseph Buc'hoz], Paris.

Percival, Thomas, (1796): *A Father's Instructions; consisting of moral tales, fables, and reflections designed to promote the love of virtue, a taste for knowledge, and an early acquaintaince with the works of nature*, London.

Piercy, John, (1745): *A natural history of Irish Song Birds, and such of the Foreign as are usually brought over and esteemed for their singing*, Dublin.

Roberts, Benjamin, (s.d.): Vaderschap in de achttiende eeuw: de gebroeders Van der Meulen, online at: http://www.richel.org/sbo/vdmuelen.htm (September 2008)

Sabbe, Maurits, (1917): *Dierkennis en diersage bij Vondel*, Antwerpen.

Schockeel, Ignatius, (1758): *Den bloeyenden staet van Antwerpen tydelijk vervallen, maer geestelijk herstelt door [...] de heeren aelmoesseniers, voedster-vaders der armen dezer stad*

Antwerpen, uytgegalmt op de dry-honderd-jaerige jubel-feest naer de eerste instellinge van hunnen over-treffelijcken dienst, plechtiglijk geviert in de cathedrale kerke van O.L. Vrouwe op den 10. december des jaers 1758, Antwerpen

Estienne, Charles & Liebaut, Ian (1582): *De Landtwinninghe Ende Hoeue. Wt de Fransoische spraecke in de Nederduytsche overgheset*, Antwerpen.

Svanberg, Ingvar (2008): Towards a cultural history of the Bengalese Finch (*Lonchura domestica*)". *Der Zoologische Garten* 77: 334–344.

The bird fancier's necessary companion and sure guide, (1760 [1762]): *The bird fancier's necessary companion and sure guide; being an easy way of breeding canary birds, and the best method of chusing and keeping them, both for breeding and song*, London.

The history of singing birds, (1791): *The history of singing birds containing an exact description of their habits & customs, & their manner of constructing their nests their times of incubation with the peculiar excellencies of their several songs. ... translated from the French of the Count de Buffon*, Edinburgh.

The new handmaid, (1790): *The new handmaid to arts, sciences, agriculture, &c.*, London.

Thirsk, Joan, (2000): *Alternative Agriculture. A history from the Black Death to the Present Day*, Oxford.

Thomas, Keith, (1983): *Man and the Natural World. Changing attitudes in England. 1500–1800*. London.

Turner, William, (1544): *Avium Praecipuarum, quarum apud Plinium et Aristotelem mentio est, brevis & succineta historia*, Cologne (integral reprint in Evans, H.A. (ed.), *Turner on Birds. A Short and Succinct History of the Principal Birds Noticed by Pliny and Aristotle*, Cambridge, 1903).

Turner, William, (1562): *The Seconde Partie of Vuilliam Turners Herball*, Cologne (edited reprint in Chapman, G.T.L. e.a., *A New Herball*, Cambridge, 1995).

Valmont de Bomare, Jacques Christophe (1764): *Dictionnaire raisonné universel d'histoire naturelle; contenant l'histoire des animaux, des végétaux et des minéraux, Et celle des Corps célestes, des Météores, & des autres principaux Phénomenes de la Nature*, Paris.

Van de Weijer, Renate, (1992): *Dieren als werknemers. Cultuurhistorische aspecten van mens-dier relaties*, Tilburg.

Van Uytven, Raymond, (2003): *De papegaai van de paus*, Leuven.

Van Wickede, Frans, (1756): *Kanari-uitspanningen of nieuwe verhandeling van de kanari-teelt*, Amsterdam.

Verhandeling van de kanari-teeling, (1759–1784): *Verhandeling van de kanari-teeling ofte den oorsprong, voortteelling, aenkweking, onderscheyden aert, ziektens en gebreken, middelen ter herstelling van de kanarievogels, ende al wat een Liefhebber noodig heeft, 't zy hy de zelve wilt doen broeden ofte niet. Alles nauwkeurig na meer dan XXV–jarige ondervinding beschreven Door eenen Liefhebber der Zelve*, Antwerpen.

Ward, Samuel, (1775): *A modern system of natural history*, vol. VII, London.

Chapter 8

Alexanderson, E., (1955): *Mig som liten är*. Stockholm.

Bernström, J., (1968): Papegojor. In: *Kulturhistoriskt lexikon för nordisk medeltid* 13. Malmö. Pp. 102–104.

Birkhead, T. (2003): *The Red Canary: the Story of the First Genetically Engineered Animal*. London.

Brusewitz, G., (1982): När brushanen hörde till husdjuren. In *Den nöjsamma nyttigheten: om natur och naturskådande i en svunnen tid*. Stockholm. Pp. 218–227.

Olaus Magnus (1555): *Historia de gentibus septentrionalibus*. Roma.

Svanberg, I. (2007): "Kunna lära att hwisla och tala": Linnés och andras papegojor i 1700-talets Sverige. *Uppland: årsbok* 2007: 115–150.

Svanberg, I., (2008): Kammarsångare och rolighetsministrar. In *Signums svenska kulturhistoria* **7**, *Det moderna genombrottet*. Stockholm. Pp 311–341.

Svanberg, I., (2010): Eine Anleitung zur Pflege von Kanarienvögel von 1747: Über des Apothekers Frantz Mikaël von Akens „Hus- und Land-Apotheque". *Blätter aus dem Naumann-Museum* 28: 18–30.

Svanberg, I. (2011): Kanariefågelns svenska historia: ett förbisett 400-årsjubileum. *Svenska Linnésällskapets Årsskrift* 2011: 71–116.

Svanberg, I. (2013): Fågelfängare och sångfåglar i allmogemiljö i 1700- och 1800-talens Sverige. *Svenska Landsmål och Svenskt Folkliv* 135: 97–110.

Svanberg, I. & S. Ægisson (2006): The Black Guillemot (*Cepphus grylle*) in Northern European folk ornithology. *Scripta Islandica* 56: 104–110.

Chapter 9

Adlersparre, A., (1917): Våra rumsfåglars ans och vård. In *Almanack för alla* 22/1918. Stockholm. Pp. 170–173.

Adlersparre, A., (1919): Några iakttagelser till de högre fåglarnas psykologi. *Fauna och flora* 14: 211–217.

Agoramoorthy, G &. M. J. Hsu, (2007): Ritual releasing of wild animals threatens island ecology. *Human Ecology* 35: 251–254.

Albin, E., (1740): *A Supplement to the Natural History of Birds*. London.

Anonymous, (2007): Japanische Mövchen (*Lonchura striata domestica*). *Feder-Kiel*, 2(3): 9–10.

Bechstein, J.M., (1825): *Histoire naturelle des oiseaux de chambre*. Genève.

Birkhead. T.R., (2003): *The Red Canary: the Story of the First Genetically Engineered Animal*. London.

Björkman, L.E. (1902): *Om burfåglar, deras vård och afvel*. Stockholm.

Blakston, W.A., Swaysland, W. , & Wiener, A.F. (1877–80): *The Illustrated Book of Canaries and Cage Birds; British and Foreign*. London.

Brooks, D. (1995): *Love Birds: Unlocking the Secrets of The Emperor's Bird Scroll*. New York.

Brooksbank, A., (1950): *Foreign Birds for Garden Aviaries*. London.

Butler, A.G., (1897): *Foreign Bird Keeping* vol. 1. London.

Butler, A.G., (1899): *Foreign Finches in Captivity*. Hull.

Chaiklin, M. (2005): Exotic-bird collecting in early-modern Japan., In *JAPAnmials: History and Culture in Japan's Animal Life*, eds G. M. Pflugfelder & B L. Walter. Ann Arbor MI. Pp. 125–160.

Dyson, C.E., (1878): *Bird-Keeping. A Practical Guide for the Management of Singing and Cage Birds*. London.

Eguchi, K. & H. E. Amano, (2004): Spread of exotic birds in Japan. *Ornithological Science* 3: 3–11.

Edwards, G., (1743): *A Natural History of Uncommon Birds and of some other rare and undescribed*. London.

Eisner, E., (1957): The Bengalese finch. *Avicultural Magazine* 63: 101–108.

Eisner, E., (1960): The biology of Bengalese finch. *Auk* 77: 281–287.

Ellis, M., (1984): Canary, bengalese and zebrafinch. In *Evolution of Domesticated Animals*, ed. I.L. Mason. London, Pp. 357–360.

Erkes, E., (1942): Vogelzucht im alten China. *T'oung Pao* 37: 15–34.

Finn, F., (1915): *Garden and Aviary Birds of India*. Calcutta.

Gedney, C.W., (1879): *Foreign Cage Birds: Containing full Directions for Successfully Breeding, Rearing, and Managing the Various Beautiful Aviary Birds imported into this Country* vol. 2. London.

Gentry, A., J. Clutton-Brock & C. P. Groves, (2004): The naming of wild animal species and their domestic derivates. *Journal of Archaeological Science* 31: 645–651.

Gray, A.P., 1958: *Bird Hybrids: A Check-List with a Bibliography*. Farnham.

Greene, W.T., (1896): *Feathered Friends Old and New*. London.

Greene, W.T., (1899): *Notes on Cage Birds (Second Series) or Practical Hints on the Management of British and Foreign Cage-Birds, Hybrids and Canaries*. London.

Holden, G.H., (1888): *Canaries and Cage Birds: the Food, Care, Breeding, Diseases And Treatment Of All House Birds For Pleasure And Profit*. New York.

Honda, E. & K. Okanoya, (1999): Acoustical and syntactical comparisions between songs of the White-backed Munia (*Lonchura striata*) and its domesticated strain, the Bengalese Finch (*Lonchura striata* var. *domestica*). *Zoological Science* 16: 319–326.

Immelmann, K., (1965). *Australian Finches in Buch and Aviary*. London.

Isono, N., (2007): Meiji mae dōbutsu torai nenpyō. *Hiyoshi Review of Natural Science, Keio University* 41: 35–66.

Jepson, P. & R. J. Ladle, (2005): Bird-keeping in Indonesia: conservation impacts and the potential for substitution-based conservation responses. *Oryx* 39: 442–448.

Kajishima, T., (2002): *Shiryō Nihon dōbutsushi*. Tokyo.

Kakizawa, R. & M. Watada, (1985): The evolutionary genetics of the Estrildidae. *Journal of the Yamashina Institute for Ornithology* 17: 143–158.

Klören, H., (1983): *Zebrafinken: Fütterung, Unterbringung, Zuchtpraxis, Vererbung, Beschreibung von über 50 Farbschlägen.* Walsrode.

Layton, L., (1991): *Songbirds in Singapore: The Growth of a Pastime.* New York.

Marks, H., (1983): *Huhn-, Struktur- und Warzentauben* (= Neue Brehm-Bücherei 531): Wittenberg-Lutherstadt.

McCarthy, E.M., (2006): *Handbook of Avian Hybrids of the World.* Oxford.

Mizuta, T., H. Yamada, A. F. Lin, Y. Yodogawa & K. Okanoya, (2003): Sexing White-rumped Munias in Taiwan, using morphology, DNA and distance calls. *Ornithological Science* 2: 97–102.

Neunzig, K., (1921): *Fremdländische Stubenvögel.* Magdeburg.

Okanoya, K., (1997): Voco-auditory behavior in the Bengalese finch: A comparison with the zebra finch. *Biomedical Research* 18: 53–70.

Okanoya, K., (2004): The Bengalese finch: a window on the behavioral neurbiology of birdsong syntax. *Annals of the New York Academy of Sciences* 1016: 724–735.

Oppenborn, G., (1992): *Japanische Mövchen. Herkunft, Zucht, Haltung.* Stuttgart.

Oppenborn, G., (2006): Haltung und Zucht von Japanischen Mövchen. *Die Voliere* 29: 274–284.

Pereira, R.A., (1935): The Java Sparrow. *The Hong Kong Naturalist* 6(3–4): 165–166.

Restall, R., (1996): *Munias and Mannikins.* Sussex.

Robbins, L.E., (2002): *Elephant Slaves and Pampered Parrots: Exotic Animals in Eighteenth-Century Paris.* Baltimore.

Robiller, F., (2003): *Das grosse Lexikon der Vogelpflege* vol. 1. Stuttgart.

Robson, C., (2007): The Society Finch: So Many Possibilities. www.centralpets.com

Russ, K., (1872): Neuere und seltene Erscheinungen des deutschen Vogelmarkts. *Die Gefiederte Welt* 1: 60.

Soderberg, P.M., (1956): *Foreign Birds for Cage and Aviaries* 4. London.

Soma, M., (2007): *Maternal effect, sibling competition and the development of physical and sexual traits in a song-bird, the Bengalese finch.* University of Tokyo. Diss.

Sossinka R., (1982): Domestication in birds. In *Avian Biology*, vol. 6, eds. D.S. Farner, J.R. King & K.C. Parkes. New York, Pp. 373–403.

Stölker, C., (1870). Ueber Aufzucht des Reisfinken in Gefangenschaft *(Padda [Oryzornis] oryzovora* Rehb.). *Journal für Ornithologie* 18: 81–84.

Svanberg, I., (2001): Mannen som älskade fåglar: Sten Bergman och Japan. In *Fjärrannära: kontakter mellan Sverige och. Japan genom tiderna*, ed. B. Edström & I. Svanberg. Stockholm. Pp. 119–136.

Svanberg, I. (2008): Towards a cultural history of the Bengalese Finch (*Lonchura domestica*)". *Der Zoologische Garten* 77: 334–344.

Taka-Tsukasa, N., (1922): Aviculture in Japan. *Avicultural Magazine* 13: 15–22.

Takatsukasa, N., (1967): *The Birds of Nippon.* Tokyo.

Thiede, U. & W. Thiede, (1974): Die Dressur von Buntmeisen (*Parus varius*) in Japan. *Ornithologische Mitteilungen* 26: 217–224.

Thunberg, C.P., (1791): *Resa uti Europa, Africa, Asia, Förrättad Åren 1770–1779* vol. 3. Upsala.

Torén, O., (1757): En Ostindisk Resa Til Suratte, China &c, Från 1750 April til 1752 Juni 26. In P. Osbeck, *Dagbok Öfwer en Ostindisk Resa Åren 1750, 1750, 1752.* Stockholm, Pp. 313–376.

Wakasugi, N., (1984): Japanese Quail. In *Evolution of domesticated animals*, ed. I.L. Mason. London, Pp. 319–321.

Webster, N., (1961): *Webster's New International Dictionary* 1. Chicago.

Wetmore, A., (1938): Canaries and other Cage-Bird Friends. *National Geographic Magazine*, 74: 775–806

Wiese, V., (1894–96): *Tropefuglenes Liv i Fangenskab* vol. 1–2. Aarhus.

Wried, A., (2001): *Untersuchungen zur Farbvererbung bei Japanischen Mövchen (*Lonchura striata *f. dom.) unter besonderer Berücksichtigung der Manifestation von Katarakten bei der leuzistischen Farbvariante.* Hannover. Diss.

Chapter 10

Adams, H.G., (1854): *Cage and Singing Birds: How to Catch, Keep, Breed and Rear Them.* London.

Bechstein, J.M., (1881): *The Natural History of Cage Birds.* London.

Burman, F., (1930): *Anteckningar om Jämtland i urval.* Östersund.

Debes, L.J., (1673): *Færoæ et Færoa Researta, det er: Færøernis oc færøiske Indbyggeris beskrifvelse.* Kiøbenhaff.

Dyson, C.E. (1878): *Bird Keeping: A Practical Guide.* London.

Greene, W.T., (1882): *Notes on Cage Birds.* London.

Lazenby, F.D., (1949): Greek and Roman Household Pets. *The Classical Journal* 44: 299–307.

Linnaeus, C., (1746): *Fauna svecica.* Stockholm.

Lönnberg, E., (1913): *Linnés föreläsningar öfver djurriket.* Upsala.

Pliny the Elderly, (1983): *Natural History in Ten Volumes. Libri* VIII–XI. Cambridge, MA.

Verrill, A.H., (1915): *Pets for Pleasure and Profit.* New York.

Wallace, R.L., (1887): *British Cage Birds.* London.

Yapp, W.B., (1982): Birds in captivity in the Middle Ages. *Archives of Natural History* 10: 479–500.

Chapter 11

Adams, M. Drive by history: The Painter Estate. *Cleveland Heights Historical Society – Places.* http://chhistory.org/Places.php?PlacesContent=PainterEstate

Anderson, R. (1996): Early aviculture in the U.S.A. *A.F.A. Watchbird* 23(no.4):46–51.

Audubon, J.J. (1831–1839): *Ornithological Biography.* Edinburgh.

Austin, O.J. (1961): *Birds of the World.* New York: Golden Press

Bahner, E.L. and C. Lynch (2003): *Micronesian Kingfisher Species Survival Plan Complete Analysis and Breeding Plan, 16 December 2003.* American Zoo and Aquarium Association Population Management Center, Chicago, Illinois.

Baratay, E. & E. Hardouin-Fugier (2002): *Zoo: A History of Zoological Gardens in the West.* London: Reaktion Books.

Barbour, T. (1946): *A Naturalist's Scrapbook.* Cambridge, MA: Harvard University Press.

Bartsch, P. (1957): A pet Carolina Parakeet. Pp. 148–151 in *The Birdwatcher's Anthology,* ed. R. T. Peterson, Harcourt, Brace & Co.

Beebe, W. (1918–1922): *A Monograph of the Pheasants.* Witherby.

Beebe, W. (1954): Migration night – 1904. *Animal Kingdom* 57

Beebe, W. & L.S. Crandall (1909): Wild birds bred in captivity in the Eastern United States. [New York] *Zoological Society Bulletin.* No.36:580–583.

Berry, R.J. (1981) Captive Management of Touracos at the Houston Zoological Gardens. *Proceedings – First International Birds in Captivity Symposium – 1978 – Seattle Washington.* International Foundation for the Conservation of Birds: 28–31.

Birdlife International (2000): *Threatened birds of the world.* Lynx Edicions and Birdlife International.

Boehrer, B.T. (2004) *Parrot Culture.* Philadelphia.

Black, A.R. (1930): Breeding the Green-naped Pheasant Pigeon in California. *Avicultural Magazine.* Ser.4, Vol. 8:158–160.

Bridges, W. (1974): *A Gathering of Animals: An unconventional History of the New York Zoological Park.* New York.

Brinkman, G. (1986): Let's blame it on Shakespeare! Rough and tough birds. *A.F.A. Watchbird* 12(No.1):18–19.

Brower, D.R. (1982): *The Condor Question: Captive or Forever Free?* San Francisco.

Browning, W.H. (1909): My aviary and its inmates. [New York] *Zoological Society Bulletin.* No.36:584–586.

Chawner, E.F. (1937): Notes. *Avicultural Magazine* Ser. 5, Vol.2:94.

Clear, V. (1973): The Lady Gould. *American Cage-Bird Magazine.* 45(No.12):32–33.

Cokinos, C. (2000): *Hope is the thing with feathers: a personal chronicle of vanished birds.* New York.

Collar, N.J. & P. Andrew (1988): *Birds to watch.* Washington D.C.

Conant, R. (1957): Arthur Erwin Brown – "Custodian of the Garden" and naturalist of note. *America's First Zoo* 9:1–4.

Conklin, W. A. (1873, 1878, 1879, 1888, 1889): *Report of the Director of the Central Park Menagerie.* New York.

Conway, W.G. (1956) A landscaped pheasantry in St. Louis Zoo. *Avicultural Magazine* 62: 109–111.

Costello, D.P. (1999): Charles Otis Whitman. In *Encyclopedia Americana.* Danbury, CT.

Crandall, L. S. (1911): The Sonneberg aviaries. [New York]

Zoological Society Bulletin 47: 793–795.

Crandall, L.S. (1912): The Carolina Parakeet. *Zoological Society Bulletin* 49: 834–836.

Crandall, L.S. (1913): An American collection of foreign birds. *Zoological Society Bulletin* No.59:1025–1037.

Crandall, L.S. (1915): The breeding of the Emu. *Zoological Society Bulletin* 18: 1261–1263.

Crandall, L.S. (1915): The breeding of the penguin. *Zoological Society Bulletin* 18: 1265–1266.

Crandall, L.S. (1915): Breeding results in the bird department, 1915. *Zoological Society Bulletin* 18: 1270–1271.

Crandall, L.S. (1916): Breeding cranes. *Zoological Society Bulletin* 19: 1409.

Crandall, L.S. (1917a) *Pets: Their history and care.* Henry Holt and company.

Crandall, L.S. (1917b): Breeding birds. *Zoological Society Bulletin* 20: 1442–1443.

Crandall, L. S. (1917c): Wild birds bred in captivity in the eastern United States. *Zoological Society Bulletin* 20:1447–1449.

Crandall, L.S. (1919): Rare birds bred in the Zoological Park. [New York] *Zoological Society Bulletin* 22:16–18.

Crandall, L.S. (1921): Department of birds. *Twenty-sixth Annual Report of New York Zoological Society.* Pp. 71–74.

Crandall, L. S. (1927): Records of birds bred in captivity. *Zoological Society Bulletin.* Pp. 30:33.

Crandall, L.S. (1930): Records of birds bred in captivity. *Aviculture* 2: 112–115.

Crandall, L.S. (1944): Aviculture in America. *Avicultural Magazine* 9: 14–15.

Crandall, L.S. (1964): *The Management of Wild Mammals in Captivity.* Chicago.

Day, D. (1981): *The Doomsday Book of Animals: A Natural History of Vanished Species.* New York.

Delacour, J.T. (1937): American Aviculture (I. The East. & II. The Middle West). *Avicultural Magazine* 2: 109–118.

Delacour, J.T. (1937): American Aviculture (III: California). *Avicultural Magazine* Ser.5, Vol.2:125–139.

Delacour, J.T. (1942): California Aviaries. *Avicultural Magazine* 7: 42–46.

Delacour, J.T. (1950): American Aviculture – 1949 (Part I). *Avicultural Magazine* 56: 62–66.

Delacour, J.T. (1950): American Aviculture – 1949 (Part II). *Avicultural Magazine* 56: 116–121.

Delacour, J.T. (1951): Amalgamation of *Aviculture* and the *Avicultural Magazine. Avicultural Magazine* 57: 3–4.

Delacour, J.T. (1959): *Wild Pigeons and Doves.* Fond du Lac, Wis.

Delacour, J.T. (1966): *The Living Air – Memoirs of an ornithologist.* London.

Delacour, J.T. (1977): *The Pheasants of the World.* Hindhead.

Delacour, J.T. (1996): *American Aviculture. AFA Watchbird* 23: 50–54.

Delacour, J.T. (1996) American Aviculture. *AFA Watchbird* 24: 48– 52.

Dingle, S.. (1996): Shameless Dingle! *AFA Watchbird* 23: 55–56.

Dingle, S. (1998): Prologue. *A.F.A Watchbird* 25:4.

Dolan, J.M. & M. Moran (1970): Historical list of *Aves* exhibited and bred in the San Diego Zoological Gardens. *Der Zoologischer Garten* 39: 41–57.

Ehrlinger, D.B. (1993): *The Cincinnati Zoo and Botanical Garden: From past to present.* Ohio.

Ezra, A. (1931): Rothschild's Grackle (*Leucopsar rotschildi*). *Avicultural Magazine* 9: 93–94.

Ezra, A. (1931): Successful breedinging of *Galeopsar salvadorii* and *Leucopsar rothschildi. Avicultural Magazine* 2: 243–244.

Ezra, A. (1937): Successful rearing of the Western Blue Bird (*Sialia mexicana*). *Avicultural Magazine* 2: 243–244.

Ezra, A. (1937): Breeding particulars of the Yellow-billed Magpie (*Pica nutalli*). *Avicultural Magazine* 2: 257–258.

Ezra, A. (1937): Breeding results at Foxwarren Park in 1937. *Avicultural Magazine* 2: 307–308.

Ezra, A. (1938): Breeding records: the Mountain Blue Bird (*Sialia mexicana*). *Avicultural Magazine* Ser. 5, Vol.3:220–221.

Ezra, A. (1951): Amalgamation of *Aviculture* and the *Avicultural Magazine. Avicultural Magazine* 57: 4–5.

Fiedler, W. C. Wessely (1976): *Schoenbrunn Zoo – History and problems.* Verband der wissenschaftlichen Gesellschaften Österreichs.

Fisher, J. (1967) *Zoos of the World – The story of animals in captivity.* Garden City, N.Y.

Fisher, J., N. Simon, & J. Vincent (1969) *Wildlife in Danger*. New York.

Flint, R.W. (1996): American showmen and European dealers – Commerce in wild animals in Nineteenth Century America. In *New worlds, new animals – From menagerie to zoological park in the Nineteenth Century.* ed. R.J. Hoage & William A. Deiss. Baltimore.

Fonseca, M. (2009): The Jungle Gardens of Avery Island. *Louisiana Garden Archives.* members.cox.net/mryfon/ avery.htm.

Forshaw, J.M. (1973): *Parrots of the world.* Neptune, N.J.

Fuller, E. (1987): *Extinct Birds.* New York.

Gifford, E.W. (1925): The Gray-hooded Quail Dove (*Gallicolumba rubsecens*) of the Marquesas Islands, in captivity. *Auk* 42: 388-396.

Gifford, E.W. (1927): Grayson's Pigeon (*Zenaidura graysoni*) in captivity. *Auk* 44: 512–519.

Gifford, E.W. (1934): Easily sexed doves. *Aviculture* 3: 13–21.

Gifford, E.W. (1941): Taxonomy and habits of pigeons. *Auk* 58: 239–245.

Glukin, N.D. (1977): Pet Birds and Cages of the 18th-Century. *Early American Life* 8:38–59.

Goodwin, D. (1976): *Crows of the World.* Ithaca, N.Y.

Goodwin, D. (1982): *Estrildid Finches of the World.* Ithaca, N.Y.

Gould, S.J. (1991): What the immaculate pigeon teaches the burdened mind. *Natural History* April 1991: 12–21.

Greenwell, G.A. & W.B. Sturgeon (1988): *First Breeding of Wild Waterfowl in North America.* International Wild Waterfowl Association.

Grier, K.C. (2006): *Pets in America: A History.* Chapel Hill.

Hachisuka, M. The Marquis (1940): Aviculture in Japan – Long-lived Micronesian birds. *Avicultural Magazine* 5: 173–176.

Hahn, E. (1967): *Animal Gardens.* Doubleday & co.

Heere, P. (1996): Shame on you, Dingle... *AFA Watchbird* 23: 55.

Hellman, G.T. (1947): *How to Disappear for an Hour.* New York.

Herne, B. (2001): *White Hunters: The Golden Age of African Safaris.* New York.

Hollander, W.F. (1961): Charles O. Whitman's flock. *National Pigeon Association News.* July 1961:24–26.

Hopkinson, E. (1926): *Records of Birds bred in Captivity.* London.

Hopkinson, E. (1932): Senegal Finches. *Avicultural Magazine* 10: 295–296.

Hopkinson, E. (1940): Latham as a bird fancier. *Avicultural Magazine* 5: 128–134.

Hornaday, W.T. (1915): *Popular Official Guide to the New York Zoological Park.* New York.

Hornaday, W.T. (1927): *Hornaday's American Natural History, Eleventh Edition, Revised.* Charles Scribner's Sons.

Howard, H. (2009): *The painter's chair: George Washington and the making of American Art.* New York.

Isenberg, A.H. (1933): Notes from the aviaries. *Aviculture* 5: 56–58.

Isenberg, A.H. (1936): Correspondence. *Aviculture* 6: 59–60.

Jackson, D. (Ed.) (1962): *Letters of the Lewis and Clark Expedition with related documents.* Urbana, IL.

Jaffe, M. (1994): *And no birds sing: The story of an ecological disaster in a tropical paradise.* New York.

Jennings, J. (1987): Model State Law proposed to regulate birds. *AFAA Watchbird* 14 : 44–45.

Job, H. K. (1915): *Propagation of wild birds – A manual of applied ornithology.* Garden City, N.Y.

Kalm, P. (1987): *Travels in North America (The English Version of 1770).* Toronto.

Kastner, J. (1977): *A Species of Eternity.* New York.

Kelly, N. (1992): *Presidential Pets.* New York.

King, W.B. (1979): *Endangered birds of the world – The ICBP bird red data book.* International Union for Conservation of Nature and Natural Resources.

Kinsey, E.C. (1938): Hummingbirds in Japan. *Aviculture* 8: 145

Kinsey, E.C. (1939): California native birdkeeping. *Aviculture* 9: 1–6.

Kinsey, E.C. (1945): Nesting of the California Woodpecker in captivity. *Aviculture* 15: 328–332.

Kinsey, E.C. (1946): Nesting of the Yellow-breasted Chat in Captivity. *Aviculture* 16: 44–54.

Kirchshoffer, R. (1966): *The World of Zoos.* New York.

Kisling, V.N. (2001): Zoological gardens of the United States. in *Zoo and Aquarium History: Ancient Animal Collections to Zoological Gardens*, ed. V.N. Kisling Jr. Boca Raton, FL. Pp. 147–180.

Kolar, K. (1972): Keeping parrots at liberty. Pp. 24–27 in *Encyclopedia of Aviculture* 2, eds. A. Rutgers & K.A. Norris. London.

Kuehler, C. & J. Good (1989) Artificial incubation of bird eggs at the Zoological Society of San Diego. *International Zoo Yearbook* 29:118–136.

Laycock, G. (1966): *The Alien Animals.* Garden City, N.Y.

Lieberman, A. (1988): Chinese pheasant breeding program at the San Diego Zoo. *AAZPA 1988 Annual Conference Proceedings.* Pp. 21–26.

Lindholm, J.H. (1988): Captain Delacour at the Bronx, 1940–1947. *Avicultural Magazine* 94: 31–56.

Lindholm, J.H. (1989): Notes on aviculture at the Cincinnati Zoo. *Avicultural Magazine* 95: 176–191.

Lindhom, J.H. (1993): Kenton C. Lint – March 19, 1912 – December 3, 1992 – Part 2. *AFA Watchbird.* 20 (no.3): 46–49.

Lindholm, J.H. (1996): Reproduction of Asian passerines in American zoos. *Annual Conference Proceedings 1996.* American Federation of Aviculture.

Lindholm, J.H. (1999): An historical review of parrots bred in zoos in the USA. *Avicultural Magazine* 105:145–184.

Lindholm, J.H. (2000): The U.S. zoo pheasant collection in retrospect. *AZA Regional Conference Proceedings 2000*: 72–77.

Lindholm, J.H. (2005): Softbill propagation in U.S. zoos: A thirty year perspective. (Part I – The '70's). *AFA Watchbird* 3: 26–38.

Lindholm, J.H. (2005): Softbill propagation in U.S. zoos: A thirty year perspective. (Part II – The '80's). *AFA Watchbird* 4: 27–41.

Lindholm, J.H. (2007): Cotingas (Cotinginae). In *Encyclopedia of Aviculture,* ed. G. Holland, Hancock House. Pp. 600–611.

Lindholm, J.H. (2008): A Century of Parrots, A book by Rosemary Low. *AFA Watchbird* 1: 42–45.

Lindholm, J.H. (2008): Birds in Germany, Part.1. *AFA Watchbird* 2: 12–22.

Lindholm, J.H. (in litt): The aviculture of hummingbirds – A historical perspective. In *The proceedings of the IV International Symposium on breeding birds in captivity.* ISBBC.

Lint, K.C. (1951): Breeding of the Rose-crested Cockatoo (*Kakatoe moluccensis*). *Avicultural Magazine* 57: 223–224.

Lint, K.C. (1960): Notes on breeding Andean Condors at San Diego Zoo. *International Zoo Yearbook* 3: 82.

Lint, K.C. (1962): White Bali Mynahs seldom leave their island. *ZooNooz* 11: 10–11.

Lint, K.C (1968): A rail of Guam. *ZooNooz* 5: 16–17.

Loisel, G. (1912): *Histoire des ménageries de l'antiquité à nos jours.* II. *(Temps modernes).* Paris.

Low, R. (1980): *Parrots – Their care and breeding.* New York.

Low, R. (1984): *Endangered parrots.* New York.

Low, R. (1996): Aviculture at the crossroads. *AFA Watchbird* 3: 20–21.

Low, R. (2006) *A century of parrots.* Mansfield, Notts, UK.

Lucas, J., N. Duplaix-Hall, & R. Biegler (1972): Species of wild animals bred in captivity during 1970. *International Zoo Yearbook* 12: 311–375.

McClung, R. & G. McClung (1957). Early New Yorkers liked animal shows. *Animal Kingdom* 60: 88–93.

McCowan, T.D. & G.M.Foster (1961): Edward Winslow Gifford, Anthropology: Berkeley. content.cdlib.org.viewdocld=hb2+lnb146&doc.view=frames&chunks.id=div00014&toc.depth=1&toc.id=&brand=calisphere.

McGowan, P.J.K & P.J. Garson (1995): *Status survey and conservation action plan 1995–1999 – Pheasants.* IUCN & World Pheasant Association.

McKinley, D. (1977). The last days of the Carolina Parakeet: Life in the zoos. *Avicultural Magazine* 83: 42–49.

McKinley, D. (1978). Clutch size, laying date, and incubation period in the Carolina Parakeet. *Bird–banding/Journal of Field Ornithology* 49: 223–233.

McNulty, F. (1966) *The Whooping Crane.* E.P. Dutton & Co.

Mann, J. (1848): *The American Bird Keeper's Manual.* Boston.

Mann, W.M. (1930): *Wild animals in and out of the zoo.* New York.

Marshall, A.D. (1994): *Zoo.* New York.

Mearns, B. &.R. Mearns. (1998): *The Bird Collectors.* San Diego, CA.

Mooney, J. (1988): *An Animal Garden in Fairmount Park.* The Zoological Society of Philadelphia.

Morris, D. & C. Jarvis (1960): Species of mammals and birds bred in captivity during 1959. *International Zoo Yearbook* 1: 138–139.

Nowotny, Dr. (1898). The breeding of the Carolina Parakeet in captivity. *Auk* 15: 26–32.

Olney, P.J.S, F.A. Fisken, & L.J. Davolls (1998): Species of wild animals bred in captivity during 1995/1996 and multiple generation births. *International Zoo Yearbook* 36: 363–551.

Olney, P.J.S, F.A. Fisken, & L.J. Davolls (1998): Census of rare animals in captivity. *International Zoo Yearbook* 36: 552–633.

Ossa, H. (1973): *They Saved our Birds: The Battle Won and the War to Win.* New York.

Peterson, R.T. (1957): *The Birdwatcher's Anthology.* Harcourt.

Phillips, J.C. (1922–1926): *A Natural History of the Ducks.* Boston.

Pick, N. (2004): *The Rarest of the Rare: Stories Behind the Treasures at the Harvard Museum of Natural History.* New York.

Plath, K. (1937): The birds at the new Chicago Zoological Park. *Avicultural Magazine* 2: 173–179.

Porter, S. (1956): Obituary – Francis H. Rudkin (1861–1956). *Avicultural Magazine* 62: 119–121.

Pratt, H.D., P.L. Bruner, & D.G. Berrett (1987): *A field guide to the birds of Hawaii and the Tropical Pacific*. Princeton, N.J.

Pratt, J.J. (1996): *The Whooping Crane – North America's symbol of conservation*. Prescott, Ariz.

Prestwich, A.A. (1930): *Who's who in aviculture – First series.* The Avicultural Book Co.

Prestwich, A.A. (1966): Records of breeding the Carolina Parakeet (*Conuropsis carolinensis*) in captivity. *Avicultural Magazine* 72: 20–22.

Reichenbach, H. (1996): A tale of two zoos – The Hamburg Zoological Garden and Carl Hagenbeck's Tierpark, In *New worlds, new animals – From menagerie to zoological park in the Nineteenth Century.* ed. R.J. Hoage & William A. Deiss. Baltimore. Pp. 51–62.

Reuther, R.T. & H.G. Lamm (1959): Successful breeding of the Streamer-tailed Hummingbird (*Trochilus polytmus*). *Avicultural Magazine* 65: 103–106.

Rickman, L.S. (2006): Saving the blues at The Blues Conservatory. *AFA Watchbird* 2: 14–24.

Rickman, L.S. (2009): Saving the blues. *AFA Watchbird* 36(no.4):33–36.

Rickman, L.S. (2009): Model homes. *AFA Watchbird* 4:37–41.

Rickman, L.S. & P. James (2010): Blue-throated Macaws – Parent-rearing: The way forward. *Parrots* 148: 18–24.

Ripley, S.D. (1973): Saving the Wood duck (*Aix sponsa*) through captive breeding. *International Zoo Yearbook* 13 :55–58.

Robbins, L.E. (2002): *Elephant Slaves and Pampered Parrots – Exotic Animals in Eighteenth-Century Paris.* Baltimore.

Roosevelt, T. (2006): *Letters to his children.* Echo Library.

Ross, M., R. Seibels, S. Thompson, & S. Long (2005): *Analysis and breeding/transfer plan – Bali Mynah (Leucopsar rothschildi) Species Survival Plan.* American Zoo and Aquarium Association Population Manage-ment Center.

Sanborn, E.B. (1913): Items of Interest. *Zoological Society Bulletin.*

Sclater, P.L. (1883): *List of the vertebrated animals now or lately living in the gardens of the Zoological Society of London.* London.

Sclater, P.L. (1896): *List of the vertebrated animals now or lately living in the gardens of the Zoological Society of London.* London.

Schorger, A.W. ,(1955): *The Passenger Pigeon: Its natural history and extinction.* Madison.

Sedley, H., (1940): The nest box. *Aviculture* 10: 176–177.

Sellers, C.C., (1979): *Mr. Peale's Museum.* Philadelphia.

Seth-Smith, D., (1925): Two rare lovebirds from Tanganyika Colony. *Avicultural Magazine* 3: 154–155.

Shearing, D., (1996): Opinions. *AFA Watchbird* 6: 26.

Sheppard, C. & D. Bruning, (1999): *AZA Galliformes/Tinamiformes Taxon Advisory Group Regional Collection Plan.* Wildlife Conservation Society/Bronx Zoo.

Sibley, D.A., (2000): *National Audubon Society: The Sibley guide to birds.* New York.

Smith, G.A., (1979): *Lovebirds and related parrots.* London.

Snyder, N.F.R. & H.A. Snyder, (2005): *Introduction to the California Condor*. Berkeley, CA.

Stephan, S.A., (1911): Last living Passenger Pigeon. *Zoological Society Bulletin* 46: 781–782.

Stokes, H.S., (1927): The arrival of Fischer's Loveberds (*Agapornis fischeri*). *Avicultural Magazine* Ser.4, Vol.5:257.

Stoodley, J., (1997): Stoodley responds to Heere's letter. *AFA Watchbird* 24 (no.1):55.

Stoppelmoor, G., (2002): Collecting and breeding the Jamaican Red-billed Streamer-tailed Hummingbird (*Trochilus polytmus*). *AFA Watchbird* 29 (no.2):4–8.

Strasser, P., (1996): Opinions. *AFA Watchbird* 5: 34.

Svanberg, I., (2008): Towards a cultura history of the Bengalese Finch (*Lonchura domestica*). *Zoologischer Garten* 77: 334–344.

Terres, T.K., (1980): *The Audubon Society Encyclopedia of North American birds.* New York.

Thayer, S., (2005): A history of the traveling menagerie. In *American Circus Anthology: Essays of the Early Years*, eds. S. Thayer & W. L. Slout. circushistory.org/Thayer/ Thayer2b. htm

Tongren, S., (1989) *The birds of the National Zoo.* (self-published).

Vogt, W., (1937): *The Birds of America.* New York.

Walters, M., (2003): *A Concise History of Ornithology.* New Haven.

Whitman, C.O., (1919) *Posthumous Works of Charles Otris Whitman.* Washington.

Wilson, A., (1808–14): *American Ornithology, or the Natural History of the Birds of the United States.* Bradford and Inskeep.

Woolfall, S.J., (1990): History of the 13th Earl of Derby's menagerie and aviary at Knowsley Hall, Liverpool (1806–1851) *Archives of natural history.* 17:1–47.

Yamashina, The Marquis, (1940): My experiences with Humming Birds. *Aviculture* Ser.3, Vol10:111–112.

Zoological Society of London, (1960–1998). Birds bred in captivity and multiple generation births. *International Zoo Yearbook.* 1–36.

Chapter 12

Delacour, J. (1924): Colibris in Gefangenschaft. *Der Vogelliebhaber* 1.

Folger, H. (1976): Erfahrungen über die Haltung von Kolibris. *Gefiederte Welt* 100.

Folger, H. (1982): *Kolibris.* Stuttgart.

Hinkelmann, C. (2006): Von Goldenen Fackeln und Grünen Fliegen: aus der Geschichte der Kolibrihaltung. *Gefiederte Welt* 130.

Hüning, W. (1986): Einundzwanzig Jahre Kolibrihaltung in der Tropenvoliere. *Voliere* 9.

Koenig, O., (1959): Die Pflege von Kolibris und Nektarvögeln in Gefangenschaft. *Gefiederte Welt* 83.

Koenig, O. (1977): Kolibrihaltung. *Gefiederte Welt* 101.

Mayr, G. (2004): Old World fossil record of modern-type hummingbirds. *Science* 304.

Mobbs, A.J. (1982): *Hummingbirds.* Hindhead, Surrey.

Neunzig, R. (1932a): Siebzig lebende Kolibris im Vogelhause des Berliner Zoologischen Gartens. *Gefiederte Welt* 61.

Neunzig, R. (1932b): Neues aus dem Vogelhause des Berliner Zoologischen Gartens. *Gefiederte Welt* 61.

Poley, D. (1973): Kolibris in Zahlen. *Gefiederte Welt* 97.

Poley, D. (1976): *Kolibris* (=Die Neue Brehm-Bücherei Bd. 484): Wittenberg.

Poley, D. (1994): *Kolibris* (=Die Neue Brehm-Bücherei Bd. 484, 3. erw. Aufl.) Magdeburg.

Pracht, A. (1932): Meine Kolibri. *Gefiederte Welt* 61.

Ramacher, H.-J., (1986): Erfahrungen bei Haltung und Zucht einiger Kolibri-Arten. *Gefiederte Welt* 110.

Reventlow, A. (1941): Einzelheiten über die Kopenhagener Kolibrihaltung. *Zoologische Garten* (N.F.) 13.

Reventlow, A. (1953): Experience during fifteen years with the feeding and management of hummingbirds in captivity. *Avicultural Magazine* 59.

Robens, L. & W. Robens (1984): Der kleine Einsiedlerkolibri *(Phaethornis ruber ruber)*, seine Haltung und Zucht. *Gefiederte Welt* 108.

Scheithauer, W. (1966a): *Kolibris: fliegende Edelsteine.* München, Basel und Wien.

Scheithauer, W. (1966b): Kolibris, Flugkünstler in Vitrinen, Zimmervolieren und Wohnräumen. *Gefiederte Welt* 90.

Scheithauer, W. (1975): Erfolgreiche Aufzucht eines Kolibris. *Gefiederte Welt* 99.

Scheithauer, W. (1980): 20 Jahre Kolibrihaltung und –zucht – ein Erfolg? *Gefiederte Welt* 104.
Schlippe, H. (1932): Kolibris. *Gefiederte Welt* 61.
Schuchmann, K.-L. (1972a): Schwierige Kolibris in Gefangenschaft *(Phaëtornis-Arten)* . *Gefiederte Welt* 96.
Schuchmann, K.-L. (1972b): Statistische Auswertung der Lebensdauer von Kolibris in Gefangenschaft. *Gefiederte Welt* 96.
Schuchmann, K.-L. (1972c): Bemerkungen zu Fragen und Antworten (Schulze/Pagel) über Kolibris. *AZ-Nachrichten* 19.
Schuchmann, K.-L. (1979): *Kolibris: Haltung und Pflege.* Frankfurt/Main.
Schürer, U. (1983): Zur Geschichte der Kolibrihaltung in europäischen Zoologischen Gärten. *Trochilus* 4.
Stahmer, M. (1978): Kolibris in der Voliere. *Voliere* 1.
Steinbacher, G. (1935a): Kolibri im Berliner Zoologischen Garten. *Gefiederte Welt* 64.
Steinbacher, G. (1935b): Neue Kolibri im Zoologischen Garten Berlin. *Gefiederte Welt* 64.
Steinbacher, G. (1936): Neue Kolibri im Zoologischen Garten Berlin. *Gefiederte Welt* 65.
Steinbacher, G. (1937): Das erste Kolibriei in Gefangenschaft. *Gefiederte Welt* 66.
Strehlow, H. (1985): Zur Frühgeschichte der Kolibrihaltung in Europa. *Trochilus* 6.
Tesdorpf, P.H. (1754): *Versuch einer Beschreibung vom allerschönsten und beynahe allerkleinsten Vogel, der unter dem Namen Colibrit bekannt ist.* Leipzig und Lübeck.
Trier, H., (1936): Aus dem Zoologischen Garten zu Köln. *Gefiederte Welt* 65.
Wiesner, R. (1936): Zur Kolibrihaltung. *Gefiederte Welt* 65.
Wittmann, M. (1980): Werdegang eines Kolibriliebhabers. *Trochilus* 1.
Wittmann, M. (1982): Kolibrihaltung – gestern und heute. *Trochilus* 3.
Zipperlen, A. (1880): Kolibris in der Gefangenschaft. *Gefiederte Welt* 9.

CHAPTER 13

'Angelo', (2003): *Prisoners' Inventions.* Chicago.
Anonymous, (1959): 'What I'm Reading', *Neptune News*, Spring.
Anonymous ,(1965): 16–17 '*Neptune News*', August.
Argus, (1920): Pentridge from Within – Prisoners and their Pets, 14 August 1920.
Berkovits, J. (2000): '"Us Poor Devils:" Prison Life and Culture in Ontario: 1874–1914', PhD Thesis, University of Toronto.
Burgess, P. (1964): Why One Man Wants To Lend a Hand, *The Sun* (Sydney), 28 October.
Davitt, M. (1885): *Leaves From a Prison Diary; or, Lectures to a 'Solitary' Audience*, London.
Dwyer, J. (1949): *Leg-Irons on Wings*. Melbourne.
Gaddis, T. (1955): *The Birdman of Alcatraz: The Story of Robert Stroud*, New York.
Flanagan, J. (2001): Lovebirds transform South Africa's hardest criminals, *Telegraph*, 19 June 2001.
Furst, G. (2007): Without words to get in the way: Symbolic interaction in prison-based animal programs, *Qualitative Sociology Review* 3.
Garvey, J., (ca 1905): *The Indelible Stamp*, Sydney.
Griffiths, A., (1875): *Memorials of Millbank*, London.
Guyon, Madame de la Mothe, (ca 1700): 'Little Bird I Am' in Charles Spears (ed) (1849): *Voices From Prison: a selection of poetry written within the cell, by various prisoners, with autobiographical and critical notes*, Boston. Pp. 40–41.
Hervey, G. (1918): Dyed Sparrows, *The Bulletin* 12.
Hinchy, W. (1931): Comptroller-General of Prisons, to William Bartlett, Officer-in-Charge, Brookfield Afforestation Camp, Mila, 1 September 1931, 'Official Correspondence; Brookfield Afforestation Camp, Mila 1927–1932,' New South Wales State Records SRNSW 3/1765.
Jacobson, L. (2007): Despair takes wing for birdmen of Pollsmoor. *Sydney Morning Herald*, 2 April.
Johnson, R. & N. Chernoff (2002): "Opening a Vein": Inmate Poetry and the Prison Experience. *The Prison Journal* 82(2):141–167.
Kerr, James, (1988): *Out of Sight, Out of Mind: Australia's Places of Confinement*. Sydney.
Lévi-Strauss, C. (1972): *The Savage Mind*, London.
Marshall, V. (1919): *The World of the Living Dead: Prison Poems and Sketches*. Sydney.
Mayhew, H. & J. Binney (1862): *The Criminal Prisons of*

London and Scenes of Prison Life. London.

Mazari, N. & R. Hillman (2008): *The Rugmaker of Mazar-e-Sharif.* Melbourne.

Montgomery, J. (1790): To a Robin Redbreast, The Captive Nightingale and Soliloquy of a Water Wagtail in in Charles Spears (ed) 1849: *Voices From Prison: a selection of poetry written within the cell, by various prisoners, with autobiographical and critical notes.* Boston. Pp. 29–36.

Neitenstein, F. (1897): *Report of the New South Wales Comptroller-General of Prisons for 1896.* Sydney.

New South Wales Prisons Department (1956): *Men Without Faces.*

O'Laughlin, P. (1960): In This New Look Jail Jack's Never A Dull Boy. *Sunday Telegraph,* 4 December.

Pearson, B., (1999): South African prisoners tame and are tamed by exotic birds. *Agence France Presse,* 19 August.

Peltier, L., (1999): *Prison Writing: My life is My Sundance.* New York.

Price, E., (1968): Institutional and Non-Institutional Treatment of Adult Offenders in South Australia, *Seminar on Prisons and the Treatment of Adult Offenders with Special Reference to South Australia.* Adelaide.

Priestley, P. (1985): *Victorian Prison Lives: English Prison Biography, 1830–1914.* London & New York.

Ramsland, J. (1996): *With Just But Relentless Discipline: A Social History of Corrective Services in New South Wales.* Sydney.

Robson, L.C. (1964): Prisons Department, Ex-Prisoners Files, Box 14, Queensland State Archives Item ID 151232.

Rodway, W. (1884): *Songs and Poems by Dr Walter Rodway, composed during his imprisonment.* Sydney.

Rothfels, N. (2002): *Savages and Beasts: The Birth of the Modern Zoo.* London.

Seddon, G., (2005): *The Old Country: Australian Landscapes, Plants and People.* Cambridge

Spear, C. (1849): *Voices From Prison: a selection of poetry written within the cell, by various prisoners, with autobiographical and critical notes.* Boston.

Strimple, E. (2003): A History of Prison Inmate-Animal Interaction Programs. *American Behavioral Scientist* 47.

Strong, G. (1998): Fly Like An Eagle. *The Age* (Melbourne) 14 November.

Stroud, R. (1943): *Stroud's Digest on the Diseases of Birds,* St Paul, Minneapolis.

The Sun (1926): Walled Out of World. 14 March.

Sydney Morning Herald (1949): Dawson May Go On Land: Has Seen The Light. 3 July.

Taylor, B. (2008): Trees of Gold and Men Made Good? Penal Forestry on the New South Wales North Coast, 1913–38. *Environment and History* 14: 545–562.

Taylor, B. (2009): 'It is curious how the convict loves a pet': animal tropes in penal discourse and human-animal relationships in Australian prisons. *Australian Studies* 1.

Wayne, P. (1994): The Landlocked Fleet. In *Architecture of incarceration,* ed. Iona Spens. London.

Wilde, O., (1946): *The Complete Works of Oscar Wilde* vol. II*: De Profundis.* London.

Wise, R., (1990): 'Won Wron', internal Prisons Department fact sheet.

Filmography

Fish, Liz 2007: *Voelvry.*

Frankenheimer, John 1962: *The Birdman of Alcatraz.*

Latham Foundation 1984, *Hi Ya Beautiful.*

Morris, Errol 2008: *Standard Operating Procedure*

Photograph and illustration acknowledgements

The authors and the publishers wish to express their thanks to the following sources of illustrative material and/or permission to reproduce it.

E. Albin, A Natural History of English Songbirds (London 1779), p. 6, 21, 102; E. Albin, Supplement to the Natural History of Birds (London 1740), p. 13, 95; Biblioteca Apostolica Vaticana, p. 46; W.A. Blakston, W. Swayland & A.F. Wiener, The Illustrated Book of Canaries and Cage-Birds (London ca. 1880), p. 18, 19, 129; Bohusläns Museum, p. 130; A.E. Brehm, Tierleben: allgemeine Kunde des Tierreichs (Leipzig 1890), p. 94; British Library, London, p. 48; C. Butler, Foreign Bird Keeping vol. 2 (London 1899) p. 8; Codex Manesse, University Library of Heidelberg, p. 25; Die Gartenlaube (1878), p. 110; G. Edwards, Gleanings of Natural History, vol. 1 (London 1758), p. 93; G. Edwards, Natural History of Birds vol. 3 (London 1750), p. 4; G. Edwards, Natural History of Birds vol. 4 (London 1751), p. 3; Fauna och flora 1908, p. 116; P. H. Gosse, Natural History: Birds (London 1849), p. 136; J. C. Hervieux de Chanteloup, Naaukeurige verhandeling van de kanarievogels (Amstedam 1713), p. 106; G. H. Holden, Canaries and Cage Birds (New York 1888), p. 19, 20, 157, 167, 184, 190; W. Kidd, The Canary, a Cage and Chamber Bird (London 1850), p. 32; © Theo Kleefisch, Germany, p. 203, 206; Library of Congress, Washington D.C., p. 9, 66, 142, 151; C. Linnaeus, Museum Adolpho-Fredericianum (Stockholm 1746), p. 14; Library of Congress, p. 66, 143, 151; Lithografisk Allehanda (Stockholm 1863), p. 113; Lucifer-Ljusbringaren (Stockholm 1893), p. 111; Olaus Magnus, Historia de gentibus septentrionalibus (Roma 1555), p. 23; J. L. Ridter in Før og nu 7 (1921), p. 49: Rijksmuseum, Amsterdam, p. 40, 103: H.R. Robertson, Life on the Upper Thames (London 1875), p. 11, Santangelo Collection, Museo Archeologico di Napoli, p. 37; Filip Santens Collection, p. 71, 72, 74, 82, 83: R. Stroud, Diseases of Canaries (Kansas City 1933), p. 208; © Trustees of the British Museum., London, p. 17, 40, 98.

Thanks also to Sabira at Lecti Book Study and Josef Lindholm, III for help with editing and production.

Notes

Chapter 1. A History of Aviculture

1 See Sparrman 1786: 183.
2 Svanberg & Ægisson 2006:109–110.
3 Malkin 1962: 18; Serpell 1986: 60–72; Rea 2007: 52–54.
4 Berg 1980.
5 Verrill 1915: 218.
6 Simakov 1989; Ståhlberg & Svanberg 2012: 373.
7 Jackson 1997; Naefe 2001; Stoner 1948; Brusewitz 2001
8 Grundel 1993; Rea 2007: 54–56, 120, 128.
9 Verrill 1915: 230; Tschirren & Postma 2010.
10 Kinzelbacher & Hölzinger 2000: 143–148.
11 Hams 2008.
12 Keller 1913: 154–155; Donkin 1991; Boettger 1958: 158–161; Kinzelbach & Hölzinger 2000: 150–153; Roberts 1972:15.
13 Davis 2001: 33–50; Kinzelbach & Hölzinger 2000: 153–156; Boettger 1958: 252.
14 Bender, Sponenberg & Bixby 2000.
15 Boettger 1958: 255–256.
16 Boettger 1958: 139–143.
17 Harper 1972; Keller 1913: 233; Bender, Sponenberg & Bixby 2000.
18 Hesse 1980; Donkin 1989. Kinzelbach & Hölzinger 2000: 97–98.
19 Hornell 1947; Roberts 1972: 21–25; Kinzelbach& Hölzinger 2000:188–194; Green-Armytage 2003; Darwin 1861: 25; Blechman 2007: 164.
20 Boettger 1958: 152–156; Tyrwhitt-Drake 1912: 74.
21 Kinzelbach & Hölzinger 2000: 156–157.
22 Pluis & Stupperich 1986 Kinzelbach & Hölzinger 2000: 198–200.
23 Kinzelbach & Hölzinger 2000: 55–58; White 1975.
24 Ticehurst 1957; Kinzelbach & Hölzinger 2000: 87–88; Bergström 1933, passim.
25 Edwards 1747: 102; Shurtleff & Savage 1996: 177.
26 Laidlay 1914: 205; Raethel & Mancini 2005; Bergström 1933.
27 Kinzelbach & Hölzinger 2000: 149.
28 Albin 1740: 34; Edwards 1747: 68; Robbins 2002: 46; Svanberg 2007b.
29 Barton 1912: 49, 90, 94; Svanberg 2007b.
30 Ives 1951: 17; Verrill 1915: 286; Roberts 1972: 12–15.
31 Gunda 1968: 482, 490; Erkes 1942:18.
32 Gillmore 1950:346: 392–393; Verrill 1915: 306.
33 Svanberg & Ægisson 2005.
34 See edited volume by Dundes 1994 with case studies from all over the world.
35 Lazenby 1949: 250; Fuller et al. 2000:3.
36 Alvez & Noguera 2010: 153.
37 Alvez & Noguera 2010: 153; Jenyns 1930:51
38 Bechstein 1795.
39 Keller 1913: 79–81; Pliny 1983: 345–347; Lazenby 1949; Möller 2011
40 Birkhead 2004; Svanberg & Strand 2012.
41 Chekhov 1922: 235.
42 Ståhlberg & Svanberg 2012; Kinzelbach 2004; Laufer 1939: 38–42; Rahmani 1999.
43 Kawata 2012: 181; Drews 2001; Rea 2007: 42; Belda et al. 2010; Jenyns 1930.
44 Birkhead 2004; Birkhead, Schulze-Hagen & Kinzelbach 2004.
45 Adams 1854: iii; Parsson 1987.
46 Peleman 1920.
47 Holden 1888: 59–60.
48 Finn 1915: 68.
49 Wille & Spormann 2008; Layton 1991: 55–64; Jepson 2010: 320.
50 Yamashina 1961; Sossinka 1982: 378.
51 Ptak 2006: 11–33; Strunden 1984: 7–17; Bigwood 1993, 321–327; Boehrer 2004, 1–22.
52 Bigwood 1993, 321–327; Boehrer 2004, 1–22.
53 Pliny 1983: 367; Toynbee 1973, 247–249; Finsch 1867, 5–6.
54 Boyd 1987: 201
55 Yapp 1982: 484; Killermann 1921: 545–550; Stresemann 1914: 81; Zaunschirm 1984–85.
56 Brooks 1995; Chaiklin 2005.
57 Finsch 1867: 3–26; Robbins 2002: ; Svanberg 2007a: 53–59.
58 Robbins 2002; Sick 1981; Berg 1937.
59 Bergman 1960; Low 1994; Low 1996: 11–72.
60 Pliny 1983: 367; Aristotele 1991: 139.

61 Humboldt 1821: 620
62 Verrill 1915: 218; Grindol & Roudybush 2004.
63 Möller 2015.
64 Blake 1989: 589; Lutwack 1994: 152.
65 Stresemann 1951; Birkhead, Schultze-Hagen & Kinzelbach 2004; cf. Albin 1740; Bechtstein 1795; Robbins 2002; Bezzel & Prinzinger 1990: 515.
66 Brehm 1872: 104–132; Wiese 1896: 199–213; Bezzel & Prinzinger 1990: 515; Erkes 1942; Layton 1991; Birkhead & van Balen 2005; Jepson & Ladle 2005; Dyson 1878.
67 Rahmani 1999; Morris 2001; Jepson & Ladle 2005; Jepson 2010; Alves. & Nogueira 2010.
68 Muse & Muse 1982; Borowski 1998: 154.
69 Wai 2006; Shiu & Stokes 2008.
70 Dundes 1982.
71 Birkhead 2004; Birkhead et al.
72 Svanberg 2008; Verrill 1915: 198–199.
73 Svanberg 2009b; Gould 1865: 82–83.
74 Verrill 1915: 229.
75 Sossinka 1982: 379; Vriends & Heming-Vriends 2002.
76 Strunden 1984: 102–108; Svanberg 2007a, 2009.
77 Svanberg 2012.
78 Holland 2007.

Chapter 2. Aviculture in Ancient Old World Societies

79 Collins 1989; Contenau 1964; Dalley 1993; Finkel 1988; Hawkes 1973; Hughes 1975; Kisling 2001; Loisel 1912; Luckenbill 1926–27; Oppenheim 1964.
80 Breasted 1906–07; Erman 1971; Hawkes 1973; Kees 1961; Kisling 2001; Montet 1962; Naville 1894–1908; Tyldesley 1996.
81 Allchin & Allchin 1982; Auboyer 1965; Fairservis 1971; Hawkes 1973; Kisling 2001; Kosambi 1969; Moynihan 1979; Possehl 1982; Rao 1973; Wheeler 1968.
82 Graham 1938; Huff 1993; Kisling 2001; Menzies 1994; Needham 1954; Polo 1934; Schafer 1968; Snow 1988.
83 Clagett 1955; Glacken 1967; Gowers 1947; Jennison 1937; Kisling 2001; Scullard 1974; Toynbee 1973.
84 Huff 1993; Kisling 2001; Moynihan 1979.
85 Allin 1998; Frederick II 1943; George 1969; Kisling 2001; Laufer 1928; Lloyd 1971; Loisel 1912; Marsh 1965.

Chapter 3. Icelandic Trade with Gyrfalcons from Medieval Times to the Modern Era

86 Jacobsen, 1848: 326–362; Ditchfield, 1887: 242; Bjarnason, 1908: 121–124; Berridge, 1934: 210–215; Thompson, 1936: 116; Epstein, 1943: 497–509; Tillisch, 1949: 73–85, 89–98; Þórðarson, 1957: 9; Grossman & Hamlet,1964: 74–81; Armstrong, 1975: 110; Lloyd & Lloyd, 1977: 144–145; Pollard, 1977: 108–109; Illingworth, 1978: 89; Rowland, 1978: 58; Glasier, 1980: 9–17; Terres, 1980: 276–277; Oswald, 1982: 3; Scott & Kenward, 1985: 203; Newton, 1990: 175; Jack, 1991: 301; Weidensaul, 1991: 106–107; Weidensaul, 1996: 299–300; Ericson & Tyrberg, 2004: 47–49; Oggins, 2004: 37–38; Cocker & Mabey, 2005: 144–145; Potapov & Sale, 2005: 202–203, 207–208, 211–212; Macdonald, 2006: 72–74; Ciesielski, 2007: 151; Þorleifsdóttir, 2008: 54.
87 Þórðarson, 45 :1955; Þórðarson, 31 ,19 :1957; Þórðarson, 152 :1959–153.
88 Þórðarson, 45 :1955; Þórðarson, 153–152 :1959.
89 Þórðarson, 45 :1955; Þórðarson, 19 :1957; Oggins, 40 :2004.
90 Þórðarson, 1955: 45; Þórðarson, 1957: 19, 30.
91 Ericson & Tyrberg, 2004: 47; Oggins, 2004: 10, 12.
92 Þórðarson, 45 :1955; Bernström, 144 :1959–145; Þórðarson, 1959: 153.
93 Óla, 1967b: 6.
94 Þórðarson, 45 :1955; Þórðarson, 19 :1957–30; Þórðarson, 1959: 152–153.
95 Þórðarson, 30 :1957.
96 Þórðarson, 45 :1955–46.
97 Þórðarson, 46 :1955; Þórðarson, 31 :1957–32; Bernström, 1959: 144; Þórðarson, 1959: 153.
98 Friedland, 1964: 539.
99 Friedland, 1964: 540.
100 Ford, 1999: 113.
101 Þórðarson, 46 :1955; Þórðarson, 32 :1957–33; Bernström, 1959: 144; Þórðarson, 1959: 153.
102 Bernström, 1959: 145.
103 Þórðarson, 46 :1955–47; Þórðarson, 1957: 33–34; Bernström, 1959: 145.
104 Þórðarson, 47–46 :1955; Oggins, 63–54 :2004.
105 Þórðarson, 46 :1955–47; Þórðarson, 1955: 34.

106 Þórðarson, 16 :1957.
107 Bernström, 1959: 145.
108 Þórðarson, 34 :1957–35; Þórðarson, 1959: 153; Oggins, 2004: 13.
109 Tillisch, 1949: 131
110 Frederick II of Hohenstaufen, 1943: 111; Oggins, 2004: 13; Þorleifsdóttir, 2008: 56–58.
111 Cf. Jacobsen, 1848: 377, Beebe, 1976: 166, Oggins, 2004: 3, Potapov & Sale, 2005: 212 and Macdonald, 2006: 30.
112 Friedland, 1964: 539–540; Potapov & Sale, 2005: 231
113 Friedland, 1964: 540.
114 Barfod, 1962: 45–51; Krogh, 1967: 7–16, 53–54; Halldórsson, 1978: xv–xxi, 443; Karlsson, 2000: 8; Guðmundsson, 2005: 11–19, 113.
115 Barfod, 1962: 45–51; Krogh, 1967: 126; Halldórsson, 1978: xvii, xix; Karlsson, 2000: 23; Guðmundsson, 2005: 102–112, 121–122.
116 Artzen, 1949: 5; Barfod, 1962: 48; Beebe, 1976: 166; Guðmundsson, 1997: 54, 66; Guðmundsson, 2005: 73–75, 118.
117 Larson, 1917: 144; Halldórsson, 1978: 124; Holm-Olsen, 1983: 30; Oggins, 2004: 13.
118 Oggins, 2004: 13.
119 Birkeland, 1954: 99–100.
120 Cf. Þórðarson, 1957: 104 and Christensen, 1995: 37. Here it is reported that in 1777 ideas had been aired in Denmark of fetching gyrfalcons to Greenland, but after careful consideration all such plans had been abandoned, mainly because of anticipated difficulties in finding adequate provisions to feed the birds during the long voyage.
121 Nørlund, 1972: 94–97; Guðmundsson, 2005: 74–76, 118–119.
122 Halldórsson, 1978: xvii; Guðmundsson, 1997: 55, 66.
123 Cf. Nørlund, 1972: 890, Halldórsson, 1978: 104, 115 and Findarticles.com: Of Monstrous Moles and Unicorn Horns.
124 Shepard, 1982: 254; Bonner, 1989: 146; Reeves, 2002: 322; Findarticles.com: Of Monstrous Moles and Unicorn Horns.
125 Nørlund, 1972: 88, 145; Vibe, 1990: 378–382.
126 Þórðarson, 35 :1957–36.
127 Þórðarson, 31 :1957
128 Halldórsson, 1978: 137; Guðmundsson, 2005: 74.
129 Guðmundsson, 2005: 74.
130 Halldórsson, 1978: 407–408.
131 Salomonsen, 1950–1951: 458.
132 Potapov & Sale, 2005: 222.
133 Nielsen & Pétursson, 1995: 70.
134 Salomonsen, 1950–1951: 455.
135 Salomonsen, 1950–1951: 452, 455; Salomonsen, 1990: 226
136 Salomonsen, 1950–1951: 455; Salomonsen, 1990: 226; Potapov & Sale, 2005: 180.
137 Potapov & Sale, 2005: 10, 180.
138 Jacobsen, 1848: 383; Guðmundsson, 1924: 14; Anonymous, 1932: 26; Oddsson, 1942: 88; Þórðarson, 1957: 105–106; Halldórsson, 1978: 56; 274; Nielsen & Pétursson, 1995: 70; Oggins, 2004: 13.
139 Þórðarson, 106 :1957.
140 Cf. Þórðarson, 1957: 162–164.
141 Nielsen & Pétursson, 1995: 70.
142 Laxness, 1977: 169–170; 238–239; Laxness, 1987: 154–157; Karlsson, 2000: 16–19.
143 Þórðarson, 153 :1959; Friðgeirsson, 253 :1982.
144 Þórðarson, 47 :1955–50; Þórðarson, 1957: 38–46.
145 Þórðarson, 50 :1955; Þórðarson, 46 :1957.
146 Þórðarson, 50 :1955; Þórðarson, 48 :1957.
147 Anonymous, 1932: 2, 11, 18.
148 Christensen, 1995: 120.
149 According to Guttormsson, 2002: 95, Fálka-Jón used an islet (later named after him and called Fálka-Jónshólmi, i.e. the Islet of John the falcon captor) for eyasses.
150 Christensen, 1995: 120.
151 Aðils, 1919: 510–511; Ólason, 1924: 154–155; Þórðarson, 1955: 54; Þorláksson, 2004: 113.
152 Gíslason, 1947: 120; Þórðarson, 1955: 50; Þórðarson, 1957: 47–49; Christensen 1995: 120.
153 Þórðarson, 50 :1955; Þórðarson, 47 :1957–49.
154 Jacobsen, 1848: 397; Ólason, 1924: 153; Tillisch, 1949: 89, 129, Þórðarson, 1957: 85–86.
155 Ólason, 153 :1924; Tillisch, 129 ,89 :1949.
156 Ólason, 153 :1924–155; Kristjánsson, 1999: 23–28; Þorláksson, 2004: 113.
157 Þórðarson, 69 :1957–71
158 Tillisch, 1949: 89, 129–130; Nielsen & Pétursson, 1995: 66; Potapov & Sale, 2005: 232.
159 Tillisch, 1949: 89, 129–130.

160 Ólason, 155 :1924; Anonymous, 25 :1932–26; Þórðarson, 1957: 79–88; Óla, 1967b: 6.
161 Anonymous, 1932: 34; Tillisch, 1949: 96, 134.
162 Horrebow, 1966: 119–122.
163 Horrebow, 1966: 121
164 Horrebow, 1966: 119–122.
165 Tillisch, 1949: 90, 130.
166 Friðgeirsson, 1982: 249.
167 Magnússon, 1935–1936: 80–81
168 Baldursdóttir, 2007: 7.
169 Anonymous, 1932: 26; Magnússon, 1935–1936: 81; Friðgeirsson, 1982: 248.
170 Place indexes (in Icelandic: Örnefnaskrár) of the estates in question (Akurtraðir, Bessastaðir, Búlandsnes, Fremri-Fitjar, Gunnarsholt, Keldunes, Kot, Lönd and Reykjahlíð); Guttormsson, 2002: 95.
171 Gíslason, 1947: 119, 122; Óla, 1965: 13; Óla, 1967a: 11; Óla, 1967b: 6.
172 Gíslason, 1947: 119; Óla, 1965: 13; Óla, 1967b: 6; Óla, 1968: 2; Anonymous, 1969: 10.
173 Helgason, 1936: 3; Óla, 1968: 2.
174 Þorsteinsson, 1932–1935: 80; Óla, 1965: 13; Óla, 1967a: 11; Óla, 1967b: 7, 10.
175 Anonymous, 1932: 34; Óla, 1967b: 10.
176 Óla, 1967b: 10.
177 Óla, 1967b: 10; Anonymous, 1969: 10. This building has now gone, having been demolished in 1868. However, the gables of another house which occupies the same site in Hafnarstræti and was built in the same year, are decorated with carved likenesses of gyrfalcons (by the Icelandic sculptor Ríkharður Jónsson) in memory of the falcon aviary which once stood there (Óla, 1967b: 11; Pétursson, 1988: 9).
178 Tillisch, 1949: 95, 134; Þórðarson, 1957: 102–103, 114; Potapov & Sale, 2005: 7.
179 Óla, 1967b: 7; Christensen, 1995: 37; 96; Þórðarson, 1957: 91–93. Those sources disagree to some extent with regard to numbers, although the difference is not significant.
180 Nielsen & Pétursson, 1995: 67.
181 Tillisch, 1949: 135; Christensen, 1995: 59, 123.
182 Tillisch, 1949: 96,134.
183 Tillisch, 1949: 136.
184 Anonymous, 1932: 40; Þórðarson, 1957: 123–124; Christensen, 1995: 96.
185 Þórðarson, 122 :1957–126.
186 Tillisch, 1949: 130; Þórðarson, 1957: 126, 126.
187 Christensen, 1995: 92.
188 Þórðarson, 128 :1957–129, 137–141; Potapov & Sale, 2005: 230–232.
189 Laxness, 1987: 121–122; Guðjónsson, 1980: 31–43; Skarphéðinsson, 2008: 49–53.
190 Jóhannsson, 2004a: 6–7; Jóhannsson, 2004b: 9.
191 Ólason, 157 :1924; Anonymous, 19 :1932; Þórðarson, 51 :1955–52; Þórðarson, 1957: 55, 129; Laxness, 1977: 144.
192 Þórðarson, 134 :1957–137; Ægisson, 1996: 23; Forseti.is: Ágrip af sögu skjaldarmerkis Íslands.
193 Skarphéðinsson, 2008: 49–53.
194 Þórðarson, 137 :1957–150; Holt & Jóhannesson, 2002: 17–20; Skarphéðinsson, 2008: 49–53.
195 Þórðarson, 134 :1957–137; Ægisson, 1996: 23; Forseti.is: Ágrip af sögu skjaldarmerkis Íslands.
196 Þórðarson, 136 :1957–137; Forseti.is: Ágrip af sögu skjaldarmerkis Íslands.
197 Þórðarson, 137 :1957.
198 Forseti.is: Ágrip af sögu fálkaorðunnar; Forsaetisraduneyti.com: Fálkaorðan.
199 Anonymous, 1937: 5; Davíðsson, 1937: 2; Þórðarson, 1957: 129–137, 141–150; Ægisson, 1996: 23; Arnórsson, 2005: 14–16.
200 Þórðarson, 129 :1957–137, 141–150; Ægisson, 1996: 23.
201 Cf. Anonymous, 1966: 28, Anonymous, 1978: 2, 30, Anonymous, 1982: 2, Anonymous, 1983: 32, Anonymous, 1984a: 26, Anonymous, 1984b: 1, Anonymous, 1984c: 40, Anonymous, 1984d: 32, Anonymous, 1984e: 1, Anonymous, 1984f: 1, [Jónsson], 1984: 4, Þ., 1984: 39, Þ. & H., 1984: 1, Anonymous, 1985a: 4, Anonymous, 1985b: 4, Anonymous, 1985c: 14, H., 1986: 12–13, Pá., 1987: 22, Jónsson, 1980: 14–16, Ó., 1988: 16, H., 1989: 12, Anonymous, 1990: 44, Bruun, 1990: 24, Anonymous, 1992: 4, Anonymous, 1993: 4 and Arnórsson, 2005: 14–16.
202 Petersen, 1998: 127; Hilmarsson, 2000: 106.

Chapter 4. Parrot Aviculture in Ancient New World Societies

203 Del Hoyo & Sargatal 1997.
204 Prescott 1936; Roth 1924; Steward 1946–59; de la Vega 1966.
205 Duran 1964; Morley & Brainerd 1983; Wauchope 1971
206 Berdan 1992; DiPeso, Rinaldo & Fenner 1974; Hunn 1977.
207 Cortés 1971; Haemig 1978.
208 Diaz del Castillo 1956; Haemig 1978.
209 Berdan 1992.
210 Berdan 1992; Haemig 1978; de Sahagun 1950–82.
211 Haemig 1978.
212 Columbus 1893; Kisling 1998; Reséndez 2007.
213 Di Peso, Rinaldo & Fenner 1974.
214 Creel & McKusick 1994; de La Salle 2003.
215 DiPeso, Rinaldo & Fenner 1974; Hargrave 1970; McKusick 2001; Minnis, Whalen, Kelley & Stewart 1993.
216 Borson 1998; Sturtevant 1979.
217 McKusick 2001
218 Schroeder 1968.

Chapter 5. Chaffinch Sport in Flanders

219 Up to the present day only two important books on chaffinch sport in Flanders have been published. This article is mainly baseed on those sources, Santens 1995 and Broeckhove 1969.
220 Based on articles by Albert Deprez, Noël Declercq and Walter Godefroid published in Santens 1995: 613–673. These authors are all prominent fincheneers who have put their lifelong experiences with the chaffinch sitting into words.
221 With the Ancient Régime we mean the period between the Early Middle Ages until the French Revolution (1789).
222 Gilliodts-Van Severen 1885: 402–406.
223 Claeys 1931: 335–339.
224 De Groote 1947: 335.
225 City archive Bruges, Eerste Afdeling, no. 377.
226 Rembry 1881 : 235–238.
227 City Archive Ghent, Ambachten, no. 156/39–12.
228 University Library Ghent, Handschriften, no. 3876.
229 The date of Shrove Tuesday moves from year to year because it is linked to Easter. (The day following Shrove Tuesday is Ash Wednesday.)
230 City Library Kortrijk, Handschriften, no. 296, VI, 343.
231 City Library Ypres, Kroniek van J. Ramaut; Royal library Brussels, Fonds Merghelynck, no. 143; State Archive Kortrijk, Oud-Stadsarchief Kortrijk, at this moment no. 1991, f°260–262; City archive Poperinghe, no. 510.
232 Woordenboek der Nederlandse Taal, XXI, 856.
233 Dictionnaire portatif, Paris-Liège (1770), II, 164.
234 Woordenboek der Nederlandse Taal, XXI, 856.
235 Staatsblad, 11–04–1873.
236 Desmet 1987: 15.
237 Broeckhove 1969: 35–36.
238 Broeckhove 1969: 49–50.
239 Broeckhove 1969: 30.
240 Rubbrecht 1907: 148.
241 De Thouroutnaar, 12–06–1869.
242 De Thouroutnaar, 26–11–1881
243 Broeckhove 1969: 35.
244 S. Wittouck, De lofspraak der Boomgaardvink. In De Bescherming der Kleine Vogelen, Brugge (1907), 37.
245 Desmet 1987: 15.
246 Broeckhove 1969: 36.
247 Broeckhove 1969: 37.
248 Desmet 1987: 16; Het Staatsblad, 11–11–1921
249 Getuigenis van C. Erffelynck in de Vinkenbode, 10–11–1935.
250 Het Staatsblad, 21–09–1924.
251 Het Staatsblad, 22–03–1929.
252 Het Staatsblad, 24–08–1906
253 Het Staatsblad, 31–07–1922.
254 Het Staatsblad, 14–11–1932.
255 Desmet 1987: 17.
256 Het Staatsblad, 24–08–1906.
257 Het Staatsblad, 17–11–1929.
258 The information in this chapter was gathered from the respective magazines of the General Fincheneers Association's De Vinkenbode: De Vinkenbode, 1935, De Vinkenier, 1936–1974 and Het Aviboblad, 1974–.
259 The information of this part was sent to us by Mr Julien Smitz, former chairman of the Sogépin, i.e. the Walloon fincheneers organization. For his text he mainly based himself on the book by Grevendal (1933). Le

Pinson des Pinsonniers of Lambert Grevendal. That work was published in 1933 with G. Nautet – Hans publishers from Verviers.

260 L. Defraeye, Vinkenstellingen in West-Vlaanderen en Frans-Vlaanderen anno 1847–1850 in De Vinkenier, 10–09–1954; Neylants 1958: 282.

261 Oral testimony of Armand Defoor, a Flemish fincheneer who participates in French contests from time to time.

262 The following data was gathered from Mr. Maandag and H.K. Van der Wal, two Dutch birdkeepers. I also gathered some information from the magnificent book by Matthey 2002.

263 The data on the chaffinch sport in Germany was gathered from a book on the chaffinch by Bergman 1993, and from a series of articles by Knoll 1989.

Chapter 6. Keeping Birds as Domestic Pest Control

264 See Brusewitz 2001; Gunda 1969:492; Manzi & Coomes 2002; Ståhlberg & Svanberg 2012

265 Ståhlberg & Svanberg 2012; Pallas 1773:425

266 Brehm 1925:627

267 Zhou 1990:132; Gillmore 1950; Brehm 1925:627

268 Gunda 1969:494; Niewenhuyse et al. 2008 :15–17

269 Johansson 1927:132–133

270 Linnaeus 1751:430–431; Harbe 1956:152; Levander 1947:338–339

271 Osbeck 1996:53

272 Christoffersson 1936:78

273 Oesenbry 1575: fol. 23v; Speerschneider 1854:190; Wiese 1891:35

274 Brehm 1924: 647–648, 650

275 Lönnberg 1913:233, Tullgren 1914:18

276 Orstadius 1909:45

277 Orstadius 1909:45–46

278 Christy 1897; Gillmore 1950: 396, 369; Wetmore 1926: 188

279 Brehm 1926:502; Notehelfer 1992: 590–591

280 Laine 1935

281 Lehtonen 1930. The note was taken in Oulu 1922 from Anni Lehtonen, Vuonninen, Archangel area, northeastern Karelia (Russia)

282 Cameron 1921:83

283 Christopher Gessler in letter

284 Rothof 1762:486

285 Falk 1786:436

286 Falk 1786:393–394

287 Tullberg 1919:6, Lönnberg 1913:234–235

288 Tullgren 1914:218–220; Trädgård 1922

289 van der Linden 2000

Chapter 7. Innovations in Early Modern Aviculture

290 See for a small overview De Roo, 2007, 34–45. The forthcoming publication Few, M. and Totorici, Z. (eds.), Centering Animals: Writing Animals into Latin American History will also contain writings on exotic parrots.

291 Van Uytven, 2003, 156. Parrots are also mentioned anecdotally in writings on exotica, curiosities and collecting, for instance Bergvelt & Kistemaker, 1992.

292 Hue de Boulogne, court-painter of duke of Burgundy Philip III the Good, maintained an aviary in the duke's pleasure garden, see Van Uytven, 2003, 156. Archdukes Albrecht and Isabella of the Habsburgian Netherlands were reported to being truly grieved when their favorite aviary bird passed away, see Balis, 1982, 44.

293 Van de Weijer, 1992, 86.

294 Matthey, 2002, 255–273. In the eighteenth century still, cooking manuals offer recipes that employ larks, starlings and finches, see for instance Nyland, 1702.

295 Ionstons, 1660, 80; Houttuyn, pt. 5, 1763, 534–535; Davids, 1989, 20; Van de Weijer, 1992, 86–87.

296 Houttuyn, pt. 5, 1763, 533.

297 Davids, 1989, 52; Van de Weijer, 1992, 86–87.

298 Thirsk, 2000, 64; Cox, 1721, 264–269 (for the capture of live nightingales).

299 Matthey, 2002, 278–280; Thomas, 1983, 115. See also further in this article.

300 Ionstons, 1660, 80 and 82.

301 This is an application of its natural foraging behavior, see Grzimek (ed.), 1971, 469. See also: Boesmans, 1984, 153; Matthey, 2002, 283. Such a special cage is shown in the work Girl with grapes (1662, Galleria Sabauda, Turin) by Dutch painter Gerrit Dou, see: Duparc (ed.), 2004, 133. Martinus Houttuyn attests how this behavior originated the Dutch vernacular name 'putter', and mentions how this trick can be

taught to other birds as well: Houttuyn, pt. 5, 1763, 539.

302 A book celebrating 300 years of almonership in Antwerp contained an ode to the goldfinch, conferring its industriousness to educating young male orphans as productive members of society. Schockeel, [1758], 32–33.

303 Ionstons, 1660, 79.

304 Valmont de Bomare, pt. 1, 1764, 542.

305 Buffon, 1779, 261

306 De Roo, 2005, 142–143.

307 Grzimek (ed.), 1971, p. 190.

308 Guicciardini, 1612, 24; Ionstons, 1660, 81; Houttuyn, pt. 5, 1763, 461 (citation); Matthey, 2002, 288.

309 Boesmans, 1984, 152.

310 Estienne & Liebaut, 1582, 105–109.

311 Ionstons, 1660, 34 (jackdaw), 37 (crossbil), 97 (tit).

312 Sabbe, 1917, 113; Ionstons, 1660, 77, 82–83 and 88–89; Houttuyn, pt. 5, 1763, 473 (starling).

313 Ionstons, 1660, 100–101; even emblematic literature idiomatically knows how 'a caged nightingale will not sing': Iunius, 1567, 60; Steven & Liebaut, 1582, 107.

314 De Roo, 2005, 115–116. The shipwreck story is mentioned in Olina, 1774, 46–47.

315 Turner, 1544, 108–109 (orig. 88).

316 Turner, 1562, 497 (orig. 193). It is also reported in Birkhead, 2003, 88.

317 Belon du Mans, 1555.

318 Gessner, 1585, 260.

319 Birkhead, Schulze-Hagen & Kinzelbach, 2004, 51–53.

320 Aldrovandi, 1600, 814.

321 Svanberg, 2008, 337.

322 Ionstons, 1660, 80. This book is part of a Dutch translation published in 1660 in Amsterdam of the original Theatrum universale historiae naturalis, dating from 1650–1653.

323 For examples of such non-commercial art, see Birkhead, Schulze-Hagen & Kinzelbach, 2004, 53 ff.

324 The mentioned engraving is 'The canary seller' by Marcellus Laroon the elder: Laroon, 1688, plate 39. See: Laroon, M. and Shesgreen, S. (ed.), 1990, 94–95 (plate 10 in the 1711 edition).

325 Birkhead, Schulze-Hagen & Kinzelbach, 2004, 55.

326 Tyrol is retrospectively mentioned as a commercial center of the canary trade in eighteenth-century literature exclusively treating the canary bird. See, for instance: Buffon, 1779, 70–71, footnote q; Houttuyn, pt. 5, 1763, 548; A new way of breeding canary birds, 1742, 4 ff.; Fairfax, 1728, 227.

327 In the second half of the seventeenth century, 7 out of 81 (or little over 11%) of the probate inventories of the Dutch town of Leeuwarden contained canaries, see: Nijboer, 2007, 54. Antwerp (Southern-Netherlands) probate inventories tell us that about 10% of the inventories at the end of the sixteenth century contained canaries (predominantly with the social elite), whilst this bird was completely absent in the 1630s. It reappeared in the 1680s and the eighteenth century, where it can be found in about 5–10% of the inventories, where it would also begin to appear with the middling classes. See De Roo, 2005, 152 ff.

328 Ward, 1775, 38–39.

329 Buffon, 1779, 1 ff.; Valmont de Bomare, 1764, pt. 5, 227 ff..; Houttuyn, pt. 5, 1763, 545 ff. Martinus Houttuyn should be considered a compiler who popularizes the works of Linneaus, see: Davids, 1989, 37.

330 Bedaux & Ekkart, 2000, 21, 61 and 249–250; De Jongh, 1967, 43–44 and 47–49; De Jongh, 1995, 10 ff.

331 De Roo, 2007, 42.

332 Earlier, he published a treaty on the nightingale, Traité du Rossignol, qui enseigne la manière de les connoître & de les elever, &c, in 1697 (Paris), and a smaller book on canaries and other cage-birds in 1707: Traité du serein de Canarie et autres petits oiseaux de volière, avec la manière de les élever et de guérir leurs maladies [...], (Paris: C. Prudhomme).

333 Valmont de Bomare, pt. 5, 1764, 228; Houttuyn, pt. 5, 1763, 545 ff.; Olina, 1774, 42; and many more.

334 For its musicality see for instance Chomel, 1743, 355. The purchase of canaries is found in two archival sources: Moretus-de Neuf, 1700–1706, fol. 50 (December 1705) and Moretus-de Neuf, 1707–1714, fol. 17 (July 1708).

335 Buffon, 1779, 1–4.

336 Verhandeling van de kanari-teeling, [1759–1784], 12–13, 15–16; Buffon, 1779, 36–37; Hervieux, 1712, 96–108.

337 Simon Schama, for instance, surmises paternal involvement in the well-being of the children, see Roberts, s.d., online. Also see Kloek, 1993, 158–164 and Muchembled, 1991, 273–279.

338 Hervieux, 1712, 272 ff.; Valmont de Bomare, pt. 5, 1764, 228; Buffon, 1779, 43–44; Houttuyn, pt. 5, 1763, 545.

339 Buffon, 1779, 15–68; Hervieux, 1712, 18–232, 175 (citation); Verhandeling van de kanari-teeling, [1759–1784], 21–96; Houttuyn, pt. 5, 1763, 546–547.

340 Fairfax, 1758, 228; Cox, 1721, 269; and many others. Beilby mentions that each year, four of them visit London, carrying with them about 400 birds each – they themselves bought the birds from specialized breeders in their home region for twenty pounds total, and sell them for five shillings a piece – see Beilby, 1797, 170 (note).

341 For instance in the Post Man and the Historical Account, Saturday, February 10, 1700, London.

342 Hervieux, 1712, 2 ff., 235–241; Verhandeling van de kanari-teeling, [1759–1784], 7–8, 11; Houttuyn, pt. 5, 1763, 548; Buffon, 1779, 70–71 and footnote q.

343 Albin, 1737, 84; Fairfax, 1758, 229; The history of singing birds, 1791, 75.

344 Hervieux, 1712, 4–8, 26, 226 and 258–264. Note that he is most probably describing the situation at hand in Paris, an urban heavyweight, therefore this statement cannot be hold to represent the average context.

345 Buffon, 1779, 13–14. The 'tuft'-canary is mentioned by the anonymous author of Verhandeling van de kanari-teeling, [1759–1784], 17, which I presume to be Buffon's 'extra' variant.

346 Hervieux, 1712, 8–10.

347 Hervieux, 1712, 10–12, 18 (citation); Verhandeling van de kanari-teeling, [1759–1784], 30; Buffon, 1779, 51–52, 67–68.

348 Percival, 1796.

349 Beckmann, 1797, 55.

350 On the interplay between commercialization and popularization of scientific knowledge with regard to keeping canaries, see De Roo, 2009.

351 Cages, specific attributes, and food mixes are mentioned in the 'how-to'-literature, see for instance A new way of breeding canary birds, 1742, 16.

352 An example is the anonymous A new way of breeding canary birds, 1742, which is often reprinted throughout the eighteenth century. The author refers to himself as 'a person, who has bred canary birds, several years, with far greater succes [sic] by this New Way, than ever he had before, in the Common Old Troublesome Way of Breeding' – which is quite revealing with regard to the mentality at hand.

353 Beckmann, 1797, 52 ff.; A new way of breeding canary birds, 1742, 5 ff.; Goldsmith, 1779, vol. 5, 339; Lord, [1791–1796], plate XXI. The sources mostly mention their once rarity and exclusiveness hand-in-hand with the lack of attention from natural history towards canaries, as if they presume more than mere correlation.

354 Exact and sure directions, [ca. 1710], 15–16.

355 Piercy, 1745, 1 ff.

356 The aforementioned illustration of Laroon features an ambulant bird-seller. A street crier selling canaries is mentioned in, among others: Holland, 1798, 18.

357 For instance A new general history of birds, 1745; The new handmaid, 1790, 23–35; The history of singing birds, 1791 The bird fancier's necessary companion and sure guide, 1760 [1762], is divided in two parts, one geared towards canaries and the other towards indigenous English birds.

358 Matthey, 2002, 278.

359 For instance in Hervieux, 1712 and Van Wickede, 1756.

Chapter 8. Chamber and Songbirds in the North

360 Olaus Magnus 1555; Bernström 1968.

361 Svanberg 2007.

362 Svanberg 2008, 2013.

363 Svanberg 2007, 2008.

364 Svanberg 2007.

365 Svanberg 2007, 2008.

366 Svanberg 2008.

367 Svanberg 2008.

368 Svanberg 2011.

369 Svanberg 2010

370 Svanberg 2010, 2011.

371 Svanberg 2008.

372 Svanberg 2008, 2013.

373 Brusewitz 1982: 218–227; Svanberg & Ægisson 2006.

374 Svanberg 2008, 2013.

375 Svanberg 2013.

376 Svanberg 2013.

377 Svanberg 2013.

378 Svanberg 2008, 2013.

379 Svanberg 2013.
380 Svanberg 2013.
381 Svanberg 2013.
382 Svanberg 2008.
383 Svanberg 2011.
384 Svanberg 2008.
385 Svanberg 2008.
386 Svanberg 2008.
387 Svanberg 2008.
388 Svanberg 2008.
389 Svanberg 2008.
390 Svanberg 2008, 2011.
391 Svanberg 2011.
392 Svanberg 2008.
393 Svanberg 2008.
394 Svanberg 2008, 2013.
395 Birkhead 2003: 36–40; Svanberg 2011.
396 Svanberg 2011.
397 Reported by Bergman in Avicultural Magazine 1957.
398 Svanberg 2008.
399 Svanberg 2011.
400 Svanberg 2008.
401 The Folklife Archives in Lund, Sweden: LUF 20208.
402 Svanberg 2008.
403 Svanberg 2011.
404 Svanberg 2008, 2011.
405 Svanberg 2008, 2011.
406 Svanberg 2011.
407 Svanberg 2011.
408 Svanberg 2011.
409 Birkhead 2003.
410 Svanberg 2010.
411 Svanberg 2011.
412 Svanberg 2011.
413 Birkhead 2003.
414 Svanberg 2008.
415 Svanberg 2011.
416 Alexanderson 1955: 136–137.

Chapter 9. Two Asian Domesticated Species in Aviculture

417 Brooks 1995; Chaiklin 2005; Thiede & Thiede 1974; Kawata 2012.
418 Sossinka 1982; Klören 1983; Immelmann 1965: 145..
419 Blakston, Swaysland & Wiener 1877– 80: 396; Bechstein 1825: 258.
420 Edwards 1743: 41; Blakston, Swaysland & Wiener 1877– 80: 396; Robbins 2002: 22, 131; Stölker 1870.
421 Neunzig 1921: 386.
422 Butler 1899: 212; Brooksbank 1950: 21
423 Risdon 1965: 67.
424 Wriedt 2001: 9; Oppenborn 1992: 114 f.; Robson 2007; Eisner 1960; Okanoya 1997; Honda & Okanoya 1999; Wriedt 2001; Mizuta et al. 2003; Okanoya 2004; Soma 2007.
425 Blakston, Swaysland & Wiener 1877– 80: 385; Dyson 1878: 209.
426 Gedney 1879: 6–11; Greene 1899; Russ 1872.
427 Dyson 1878: 209; Björkman 1902: 41; Neunzig 1921: 376–377; Wetmore 1938; Soderberg 1956: 61–63; Randall 1996: 83; Anonymous 2007.

428 Ellis 1984: 358; Eisner 1957; Blakston, Swaysland & Wiener 1877–80: 396 f.; Finn 1915: 88; Holden 1888: 235; Webster 1961: 204.
429 Russ 1872; Marks 1983: 60.
430 Wiese 1894: 83; Adlersparre 1917, 1918; Björkman 1902: 41
431 Gray 1958; Oppenborn 1992: 120; McCarthy 2006: 285–286; Russ 1875; Blakston, Swaysland & Wiener 1877– 80; Restall 1996: 81
432 Butler 1899: 212–215; Wetmore 1938: 797; Robiller 2003; Birkhead 2003; Gentry, Clutton-Brock & Groves 2004.
433 Eisner 1960; Oppenborn 1992: 10; Wried 2001: 12; Pereira 1935: 166; Erkes 1942; Wetmore 1938; Layton 1991; Torén 1757, 370.
434 Eguchi & Amano 2004; cf. Sossinka 1982; Taka-Tsuh kasa 1922.
435 Isono 2007; Thunberg 1791: 32; Kajishima 2002; Ta5 ka-Tsukasa 1922.
436 cf. Soma 2007: 10; Blakston, Swaysland & Wiener 1877– 80: 386.
437 Robbins 2002: 331; Holden 1888: 105; Blakston, Swaysland & Wiener 1877–80: 397; Wetmore 1938: 799.
438 Taka-Tsukasa 1922; Chaiklin 2005; Brooks 1995: 252.
439 Holden 1888: 107.
440 Greene 1896: 65; Dyson 1878: 192.
441 Butler 1897: 56.

442 Pereira 1935: 166.
443 Pereira 1935: 166.
444 Taka-Tsukasa 1922; Holden 1888: 107–108; Neumann 1921: 386; Blakston, Swaysland & Wiener 1877– 80: 397.
445 Takatsukasa 1967: 205–208; Wakasugi 1984.
446 Svanberg 2001; Taka-Tsukasa 1922.
447 Kakizawa & Watada 1985; Okanoya 1997; Soma 2077: 10.
448 Honda & Okanaya 1999; Eisner 1957; Restall 1996: 83.
449 cf. Agoramoorthy &. Hsu 2007; Svanberg 2001; Soma 2007: 10.
450 Oppenborn 1992: 121; Oppenborn 2006; Randall 1996: 83.

Chapter 10. Talking Birds

451 Dyson 1878: 30; Verrill 1915: 233.
452 Lazenby 1949: 300; Pliny 1983: 369.
453 Yapp 1982: 482; Dyson 1878: 29.
454 Debes 1673: 121
455 Linnaeus 1746: 24; Lönnberg 1913: 92; Burman 1930: 62.
456 Greene1882: 163.
457 Adams 1854: 125; Verrill 1915: 247.
458 Bechstein 1881: 38.
459 Wallace 1887: 387; Verrill 1915:231

Chapter 11. Aviaries in the Wilderness to Arks in the Metropolis

460 Baratay & Hardouin-Fugier 2003; Loisel 1912, 348–349.
461 Robbins 2002.
462 Ossa 1973.
463 Sellers 197; Pick 2004.
464 Pick 2004.
465 Barbour 1946; Pick 2004; Sellers 1979.
466 Howard 2009; Barbour 1946; Pick 2004; Sellers 1979.
467 Barbour 1946; Pick 2004.
468 Ossa 1983.
469 Kalm 1770/1987, 247.
470 Robbins 2002, 27, 116, 246; Glukin 1977.
471 Hopkinson 1940.
472 Hopkinson 1940.
473 Audubon 1831–39.
474 Hopkinson 1926.
475 Mearns 1998; Kastner 1977; Woolfall 1990.
476 Audubon 1831–39.
477 Hopkinson 1926.
478 Hopkinson 1926; Walters 2003.
479 Hopkinson 1926.
480 Hopkinson 1926; Woolfall 1990; Hopkinson, 1926.
481 Vogt 1937; Audubon 1831–39.
482 Audubon 1831–39.
483 Audubon 1831–39.
484 Kelly 1992.
485 Kelly 1992; Roosevelt 2006; Boehrer 2004.
486 Ossa 1973, Crandall 1944.
487 Crandall 1944.
488 Ossa 1973.
489 Ossa 1973; Jennings 1987.
490 Austin 1961; Sibley 2000.
491 Crandall 1944; Beebe & Crandall 1909; Hopkinson 1926; Crandall 1911; Delacour 1937, 1942; Kinsey 1945; Plath 1937; Kinsey 1938; Yamashina 1940; Chawner 1937; Ezra 1937a, b, c, 1938.
492 Mann 1848.
493 Goodwin 1982.
494 Grier 2006; Kelly 1992.
495 Grier 2006.
496 Mann 1848.
497 Mann 1848.
498 Pick 2004; Kastner 1977; Sellers 1979.
499 Kastner 1977; Sellers 1979; Hopkinson 1932; Wilson 1809.
500 Jackson 1962.
501 Jackson 1962; Kastner 1977.
502 Jackson 1962; Goodwin, 1976.
503 Wilson 1808–14.
504 McClung & McClung 1957; Thayer 2007.
505 Thayer 2007.
506 Thayer 2007.
507 Thayer 2007.
508 Kissling 2001.
509 Conklin 1873, 1877, 1879, 1888, 1889; Schorger 1955, 221.
510 Cokinos, 2000; Ehrlinger 1993; Laycock 1966.
511 Laycock, 1966.
512 Laycock 1966.
513 Laycok 1966; Brinkmann 1986; Terres 1980; Conklin

1889.
514 Terres 1980; Brinkmann 1986.
515 Ehrlinger 1993.
516 Cokinos, 2000; Ehrlinger 1993.
517 Ehrlanger, 1993
518 Ehrlinger 1993: 16; Kirschoffer 1966,106; Crandell 1917.
519 Ehrlanger 1993.
520 Ehrlanger 1993.
521 Kirschoffer 1967.
522 Ehrlanger 1993, 121; Crandall 1917c.
523 Ehrlanger 1993; Flint, 1996; Conant 1957.
524 Mooney 1988; Hahn 1967
525 Ehrlanger 1993.
526 Fiedler & Wessely 1976; Robbins 2002, 27; Mann 1848; Ossa, 1973.
527 McKinley 1977; Reichenbach 1996.
528 McKinley 1977.
529 Hopkinson 1926.
530 McKinley 1978; Prestwich 1966; Nowotny 1898.
531 Mearns & Mearns 1998; Kolar 1972
532 Beebe & Crandall 1909; McKinley, 1977, 1978.
533 McKinley 1978.
534 Prestwich 1966; McKinley 1978.
535 McKinley 1977; Peterson 1957; Bartsch 1957; Cokinos 2000; Low 2006.
536 Crandall 1912; McKinley 1977; Tongren 1989.
537 Lindholm 1989 and 1999; McKinley 1977; Fuller 1987; Ehrlinger 1993.
538 Ossa, 1973; Fuller, 1987; Ehrlinger 1993; Crandall, 1912.
539 Bridges, 1974.
540 Hornaday 1927.
541 Crandall 1912.
542 Ehrlinger 1993; McKinley 1977.
543 Cokinos 2000; Schorger 1955.
544 Sclater 1883; Sclater 1896.
545 Stephan 1911.
546 Ehrlinger 1993; Day 1981
547 Cokinos 2000.
548 Costello 1999.
549 Cosello 1999; Beebe & Crandall 1909.
550 Gould 1991.
551 Whitman 1919; Hollander 1961; Gould 1990.
552 Schorger 1955; Cokinos 2000.
553 Hellman 1947, 155.
554 Crandall 1915.
555 Crandall 1917b.
556 Ossa 1973; Bridges 1974; Beebe 1954.
557 Ossa 1973.
558 Hornaday 1927.
559 Beebe & Crandall, 1909.
560 Cokinos 2000.
561 Hornaday 1927.
562 Crandall 1930.
563 Phillips 1922–1926; Audubon 1831–1839; Hopkinson 1926.
564 Beebe & Crandall 1909; Ripley 1973.
565 Job 1915.
566 Ripley 1973.
567 Beebe & Crandall, 1909; Audubon 1831–1839; Ehrlanger 1993.
568 Bridges 1974.
569 Beebe 1918; Bridges 1974; Marshall 1994.
570 Bridges 1974; Hornaday 1915.
571 Crandall 1917c.
572 Delacour 1977.
573 Beebe & Crandall, 1909.
574 Job 1915; Greenwell & Sturgeon 1988.
575 Browning 1909.
576 Crandall 1917c; Crandall 1911
577 Crandall 1944.
578 Crandall 1930.
579 Beebe & Crandall 1909.
580 Crandall 1930.
581 Crandall 1915a.
582 Crandall 1917b.
583 Crandall 1917a, 1917b, 1915c.
584 Crandall 1915b; Sanborn 1913; Crandall 1919.
585 Fonseca 2009; Crandall 1917.
586 Delacour 1950.
587 Crandall 1916; Crandall 1915c.
588 Ehrlanger 1993; Crandall 1913, 1915.
589 Crandall 1917a.
590 Crandall 1917a.
591 Crandall 1917a.
592 Crandall 1917a.
593 Crandall 1917a.
594 Crandall 1917a.
595 Crandall 1917a.

596 Crandall 1917a.
597 Crandall 1917a.
598 Svanberg 2008
599 Crandall 1917a.
600 Crandall 1917a.
601 Crandall 1917a.
602 Crandall 1927; Crandall 1917a.
603 McCowan & Foster 1961; Delacour 1942; Gifford 1941.
604 Gifford 1934.
605 Gifford 1925.
606 Delacour 1959.
607 Gifford 1927.
608 Greenwell & Sturgeon 1988.
609 Greenwell & Sturgeon 1988.
610 Beebe & Crandall 1909; Crandall 1930.
611 Dolan & Moran 1970; Lindholm 1999; Sedley 1940.
612 Browning 1909.
613 Porter 1956; Seth-Smith 1925.
614 Adams 2006; Herne 2001; Crandall 1921
615 Stokes 1927; Crandall 1930; Lindholm 1999.
616 Isenberg 1933 & 1936; Breese, pers. com.
617 Black 1930; Delacour 1937b & 1959.
618 Tongren 1989; Mann 1930.
619 Crandall 1930.
620 Delacour 1937a, 1937b,1950a, 1950b,1996, 1997; Anderson 1996.
621 Delacour 1951; Ezra 1951.
622 Grier, 2006; Prestwich, 1930; Crandall 1917a.
623 Delacour 1937a.
624 Hellman 1947; Lindholm 1988; Delacour 1950a&b.
625 Delacour 1966.
626 Morris & Jarvis 1960.
627 Lindholm, in litt; Reuther & Lamm 1959; Lindholm 2008b; Stoppelmoor 2002.
628 Lindholm 1993a; Lucas et al.1972.
629 Lucas et al. 1972.
630 Olney et al. 1998a.
631 Frank Todd, pers. com.
632 Berry 1981.
633 Snyder & Snyder 2005; Olney et al. 1998b.
634 Lindholm 1993; Lint 1960; Brower 1982; Snyder & Snyder 2005.
635 McNulty 1966; Pratt 1996.
636 McNulty 1966.
637 Pratt 1996.
638 Ezra 1931a, 1931b; Lindholm 1996, 2005a; Zoological Society of London 1960–1998.
639 Lint 1962.
640 Marvin Jones, pers. com.
641 King 1979; Collar & Andrew 1988; Birdlife International 2000.
642 Lindholm 2005a.
643 Ross et al. 2005.
644 Bahner & Lynch 2006; Lindholm 2005b.
645 Jaffe 1994.
646 Lint 1968; Jaffe 1994.
647 Pratt 1987.
648 Hachisuka 1940.
649 Jaffe 1994.
650 Martin Vince, pers. com.
651 Martin Vince, pers. com.
652 Lindholm 2000.
653 Lindholm 2007.
654 Fisher et al . 1969.
655 Hahn 1967, 30; Fisher 1967, 94; McGowan & Garson 1995.
656 Delacour 1977; Lindholm 2000; Sheppard & Bruning 1999.
657 Delacour 1937a&b; Delacour 1950a&b
658 Conway 1956; Lindholm 2000.
659 Kuehler & Good 1989; Lindholm 2000; Lindholm 2000.
660 Lieberman 1988.
661 Delacour 1977.
662 Lindholm 2005b.
663 Clear 1973.
664 Low 2006.
665 Low 1996 & 2006; Lindholm, 2008a; Heere 1996; Dingle 1996; Shearing 1996; Stoodley 1997; Strasser 1996.
666 Low 2006, 84.
667 Low 2006, 107.
668 Low 2006, 261–262; Low 2006, 266; Lint 1951; International Zoo Yearbook, 85 ; Lindholm 1999.
669 Low 1984, 37.
670 Low 1980; Forshaw 1973; Low 2006; Rickman 2009b.
671 Low 1980; Low 1984.
672 Rickman 2006; Rickman 2009a; Rickman & James 2010.

673 Rickman 2006 & 2009a, 2010).
674 Rickman 2009b.

Chapter 12. Hummingbirds in Europe

675 Delacour 1924.
676 Tesdorpf 1754, Zipperlen 1880, Poley 1976.
677 Poley 1976.
678 Poley 1976.
679 Strehlow 1985.
680 Wittmann 1982; R. Low in Schürer 1983; Delacour 1924, Poley 1976.
681 Schlippe 1932; Trier 1936; Schürer 1983; Neunzig 1932a.
682 Neunzig 1932a.
683 Pracht 1932; Schlippe 1932; Neunzig 1932b, Steinbacher 1935a, 1935b, 1936; Trier 1936.
684 Wiesner 1936; Schürer 1983.
685 Schürer 1983.
686 Koenig 1959.
687 Schürer 1983.
688 Scheithauer 1980; Wittmann 1982.
689 Scheithauer 1966a, 1966b.
690 Wittmann 1980.
691 Schuchmann 1972a, 1972b.
692 Mobbs 1982.
693 Poley 1973; Reventlow 1941, 1953.
694 Schuchmann 1972c.
695 Schürer 1983.
696 Koenig 1977.
697 Scheithauer 1966a, 1966b.
698 Scheithauer 1966a; Poley 1976, 1994; Schuchmann 1979 and Folger 1982; Folger 1976.
699 Scheithauer 1983; Robens & Robens 1984.
700 e.g. Stahmer 1978; Ramacher 1986, Hüning 1986.
701 For an earlier version in German of this essay, see Hinkelmann 2006.

Chapter 13. Gaolbirds

702 For example: Hill 1955, 39.
703 Berkovits 2000, 300.
704 Priestley 1985, 47.
705 Although see Taylor 2009.
706 Rothfels, 2002, 6.
707 This article is based on research carried out on the history of Australian prisons. The story centers therefore on these prisons, although it glances from time to time at British, American and South African experience. According both 'gaolbird' and 'jailbird' appear in the sources. I have used 'gaolbird' throughout as this was the accepted spelling in Australia through most of the twentieth century, although when quoting from American sources, I have followed the original.
708 Mayhew and Binney 1862, 128.
709 Burgess, 1964.
710 'Angelo' 2003, 86.
711 Quoted in Priestly 1989, 164.
712 Griffiths 1875, 223.
713 Davitt 1885, vi.
714 Davitt 1885, vii–viii.
715 The Argus 1920, 21
716 Neitenstein 1897, 67.
717 Hervey 1918, 11.
718 Hervey 1918, 11.
719 Marshall 1919, 17.
720 Marshall 1919, 16.
721 Marshall 1919,17–18.
722 Ramsland 1996, 92–101
723 Garvey nd, 3.
724 O'Laughlin 1960.
725 Anonymous 1965, 16–17 'Neptune News', August.
726 Gaddis, 1955, 119, 125, 129–30.
727 Stroud 1943, 312–13.
728 Anonymous, 1959.
729 Morris, 2008.
730 Ramsland 1996, ch7. See also Taylor 2008.
731 Sydney Morning Herald, 1949.
732 Robson 1964.
733 NSW Department of Prisons, 1956.
734 Price 1968, 42.
735 Hinchy 1931
736 Strimple 2003, 72; See also: Latham Foundation 1984 and Furst 2007.
737 Wise 1990
738 Strong 1998
739 Pearson 1999; Flanagan 2001
740 Jacobson 2007
741 Fish 2007
742 Kerr 1988, 160
743 Mazari 2008, 104
744 Mazari, pers. comm. 14 July 2008

745 Mazari 2008, 206
746 Mazari 2008, 206
747 Seddon, 2005, 2
748 Spears (ed) 1849, 29–36
749 For example: Guyon c.1700; Wayne 1994, 20
750 Gaddis 1955, 314
751 Marshall 1919, 18
752 Marshall 1919, 19; Wilde 1946, 474–75
753 Peltier 1999, 34.
754 Lévi-Strauss, 1972, 204
755 Mazari 2008, 206

INDEX

H

O

P

S

More HANCOCK HOUSE aviculture & zoology titles

Encyclopedia of Aviculture
Glen Holland
978-0-88839-460-6
8½ x 11, hc, 832 pp

Conservation through Aviculture
M. Lamont (ed.)
978-0-88839-731-7
8½ x 11, hc, 482 pp

True Pheasants
John M. Clearwater
978-0-88839-702-7
8½ x 11, hc, 248 pp

Encyclopedia of Lories
Rosemary Low
978-0-88839-413-2
8½ x 11, hc, 432 pp

Pekin Robins and small softbills
Peter Karsten
978-0-88839-606-8
8½ x 11, hc, 252 pp

Cranes: their biology, husbandry and conservation
David Ellis
978-0-88839-385-2
8½ x 11, hc, 308 pp

Estrildid Finches
Matthew Vriends
978-0-88839-493-4
8½ x 11, hc, 256 pp

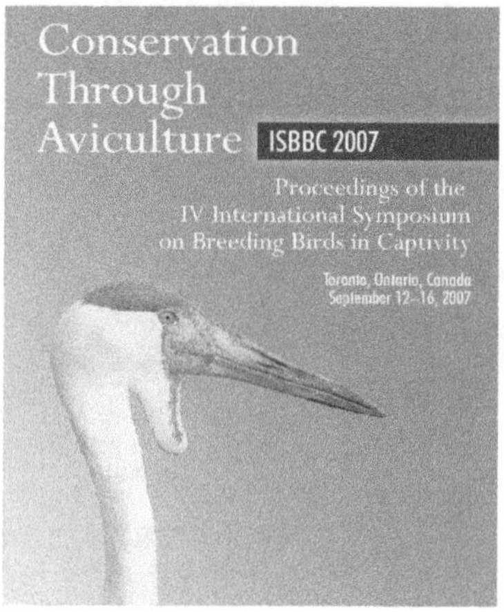

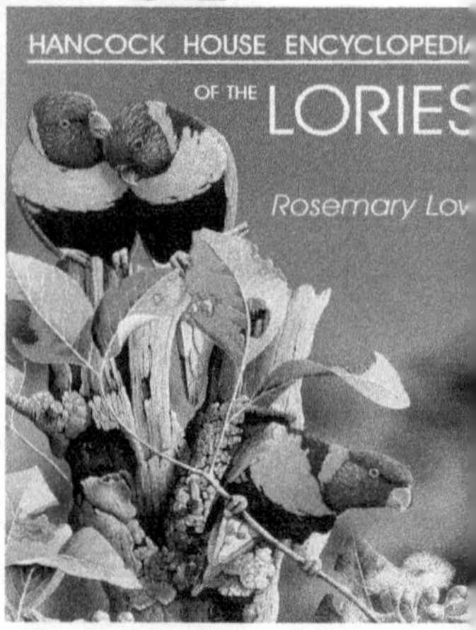

View all HANCOCK HOUSEtwww.hancockhouse.com

www.ingramcontent.com/pod-product-compliance
Lightning Source LLC
LaVergne TN
LVHW062147200926
848484LV00007B/1876